AF509017

Machine Learning and Data Mining Applications in Power Systems

Machine Learning and Data Mining Applications in Power Systems

Editors

Zbigniew Leonowicz
Michał Jasiński

MDPI • Basel • Beijing • Wuhan • Barcelona • Belgrade • Manchester • Tokyo • Cluj • Tianjin

Editors
Zbigniew Leonowicz
Wroclaw University of
Science and Technology
Poland

Michał Jasiński
Wroclaw University of
Science and Technology
Poland

Editorial Office
MDPI
St. Alban-Anlage 66
4052 Basel, Switzerland

This is a reprint of articles from the Special Issue published online in the open access journal *Energies* (ISSN 1996-1073) (available at: https://www.mdpi.com/journal/energies/special_issues/Machine_Learning_Data).

For citation purposes, cite each article independently as indicated on the article page online and as indicated below:

LastName, A.A.; LastName, B.B.; LastName, C.C. Article Title. *Journal Name* **Year**, *Volume Number*, Page Range.

ISBN 978-3-0365-4177-8 (Hbk)
ISBN 978-3-0365-4178-5 (PDF)

Contents

About the Editors

Zbigniew Leonowicz (Professor)

Zbigniew Leonowicz received M.S. and Ph.D. degrees in electrical engineering from Wroclaw University of Science and Technology, in 1997 and 2001, respectively, and a Habilitation degree from the Bialystok University of Technology, in 2012. Since 1997, he has been with the Electrical Engineering Faculty, Wroclaw University of Technology. Since 2019, he has been employed as a Full Professor with the Department of Electrical Engineering, where he is currently the Head of the Chair of Electrical Engineering Fundamentals. He received the title of Professor from the President of Poland and the President of the Czech Republic in 2019.

Michał Jasiński (Doctor)

Michał Jasiński received M.S. and Ph.D. degrees in electrical engineering from the Wroclaw University of Science and Technology, in 2016 and 2019, respectively. Since 2018, he has been with the Electrical Engineering Faculty, Wroclaw University of Technology, where he is currently an Assistant Professor. He is the author and co-author of over 90 scientific publications. His research interests include using big data in power systems, especially regarding points of power quality, as well as optimization in multi-energy systems. He is a Guest Editor of Special Issues in *Energies, Electronics,* and *Sustainability.*

Editorial

Machine Learning and Data Mining Applications in Power Systems

Zbigniew Leonowicz * and Michal Jasinski

Faculty of Electrical Engineering, Wroclaw University of Science and Technology, Wyb. Wyspianskiego 27, 50370 Wroclaw, Poland; michal.jasinski@pwr.edu.pl
* Correspondence: zbigniew.leonowicz@pwr.edu.pl

Citation: Leonowicz, Z.; Jasinski, M. Machine Learning and Data Mining Applications in Power Systems. *Energies* **2022**, *15*, 1676. https://doi.org/10.3390/en15051676

Received: 18 February 2022
Accepted: 21 February 2022
Published: 24 February 2022

Publisher's Note: MDPI stays neutral with regard to jurisdictional claims in published maps and institutional affiliations.

This Special Issue was intended as a forum to advance research and apply machine-learning and data-mining methods in order to facilitate the development of modern electric power systems, grids and devices, smart grids and protection devices, as well as to develop tools for more accurate and efficient power system analysis.

Conventional signal processing is no longer adequate to extract all the relevant information from distorted signals through filtering, estimation, and detection to facilitate decision making and control actions. Machine learning algorithms, optimization techniques and efficient numerical algorithms, distributed signal processing, machine learning, data-mining statistical signal detection, and estimation may help in solving contemporary challenges in modern power systems. The increased use of digital information and control technology can improve the grid's reliability, security, and efficiency; dynamic optimization of grid operations; demand response; incorporation of demand-side resources and integration of energy-efficient resources; distribution automation; and integration of smart appliances and consumer devices. Signal processing offers the tools needed to convert measurement data to information, and to transform information into actionable intelligence.

This Special Issue includes fifteen articles, authored by international research teams from several countries. For a straightforward browsing of the volume content, the articles can be grouped into the following subjects:

- Minority oversampling techniques applied to the detection of injection of false data and commands into communication [1].
- Binary-coded genetic algorithms applied to the intelligent scheduling of smart home appliances [2].
- Adaptive supervised dictionary learning (SDL) for wide-area stability assessment [3].
- Forecasting of consumption in isolated areas using data sequencing, sequential mining, and pattern mining to infer the results into a Hidden Markov Model (MAESHA H2020 project) [4].
- Impact of social distancing, implemented as a result of COVID-19, on residential energy consumption [5].
- Application of the IpDFT spectrum interpolation method to estimate the fundamental frequency of a power waveform [6].
- Application of an adaptive neuro-fuzzy inference system (ANFIS) maximum power point-tracking (MPPT) controller for DFIG-based wind-energy conversion systems (WECS) [7].
- Application of different cluster analysis techniques to evaluate the level of power quality (PQ) parameters of a virtual power plant [8–10].
- Application of Dynamic Differential Annealed Optimization to design of off-grid rural electrification in India using renewable energy resources and battery technologies [11].
- Reviews and studies of power supply quality pollution by voltage and current distortion [12–15].

To conclude, with reference to presented papers, we have seen a broad spectrum of data-mining and modern machine-learning techniques applied to recent problems of operation of power systems.

Data mining is a powerful new technology with great potential to help researchers focus on the most important information in their large databases. Machine learning aims to build computer systems that can learn how to solve complex problems by themselves. Deep learning builds a complex mathematical structure (a neural network) based on vast quantities of data. The two group of methods will eventually merge to provide more powerful tools for the unsupervised analysis of "big data" sets.

Author Contributions: All authors have read and agreed to the published version of the manuscript.

Funding: This research received no external funding.

Acknowledgments: The editors of this Special Issue are grateful to the MDPI Publisher for the invitation to act as guest editors of this Special Issue. All authors are indebted to the editorial staff of *"Energies"* for their kind co-operation, patience and committed engagement.

Conflicts of Interest: The authors declare no conflict of interest.

References

1. Kumar, A.; Saxena, N.; Jung, S.; Choi, B.J. Improving Detection of False Data Injection Attacks Using Machine Learning with Feature Selection and Oversampling. *Energies* **2022**, *15*, 212. [CrossRef]
2. Tutkun, N.; Burgio, A.; Jasinski, M.; Leonowicz, Z.; Jasinska, E. Intelligent Scheduling of Smart Home Appliances Based on Demand Response Considering the Cost and Peak-to-Average Ratio in Residential Homes. *Energies* **2021**, *14*, 8510. [CrossRef]
3. Dabou, R.T.; Kamwa, I.; Tagoudjeu, J.; Mugombozi, F.C. Sparse Signal Reconstruction on Fixed and Adaptive Supervised Dictionary Learning for Transient Stability Assessment. *Energies* **2021**, *14*, 7995. [CrossRef]
4. Guerard, G.; Pousseur, H.; Taleb, I. Isolated Areas Consumption Short-Term Forecasting Method. *Energies* **2021**, *14*, 7914. [CrossRef]
5. Jang, M.; Jeong, H.C.; Kim, T.; Suh, D.H.; Joo, S.-K. Empirical Analysis of the Impact of COVID-19 Social Distancing on Residential Electricity Consumption Based on Demographic Characteristics and Load Shape. *Energies* **2021**, *14*, 7523. [CrossRef]
6. Borkowski, J.; Szmajda, M.; Mroczka, J. The Influence of Power Network Disturbances on Short Delayed Estimation of Fundamental Frequency Based on IpDFT Method with GMSD Windows. *Energies* **2021**, *14*, 6465. [CrossRef]
7. Chhipa, A.A.; Kumar, V.; Joshi, R.R.; Chakrabarti, P.; Jasinski, M.; Burgio, A.; Leonowicz, Z.; Jasinska, E.; Soni, R.; Chakrabarti, T. Adaptive Neuro-Fuzzy Inference System-Based Maximum Power Tracking Controller for Variable Speed WECS. *Energies* **2021**, *14*, 6275. [CrossRef]
8. Aksan, F.; Jasiński, M.; Sikorski, T.; Kaczorowska, D.; Rezmer, J.; Suresh, V.; Leonowicz, Z.; Kostyła, P.; Szymańda, J.; Janik, P. Clustering Methods for Power Quality Measurements in Virtual Power Plant. *Energies* **2021**, *14*, 5902. [CrossRef]
9. Jasiński, M.; Sikorski, T.; Kaczorowska, D.; Rezmer, J.; Suresh, V.; Leonowicz, Z.; Kostyła, P.; Szymańda, J.; Janik, P.; Bieńkowski, J.; et al. A Case Study on Data Mining Application in a Virtual Power Plant: Cluster Analysis of Power Quality Measurements. *Energies* **2021**, *14*, 974. [CrossRef]
10. Jasiński, M.; Sikorski, T.; Kaczorowska, D.; Rezmer, J.; Suresh, V.; Leonowicz, Z.; Kostyła, P.; Szymańda, J.; Janik, P.; Bieńkowski, J.; et al. A Case Study on a Hierarchical Clustering Application in a Virtual Power Plant: Detection of Specific Working Conditions from Power Quality Data. *Energies* **2021**, *14*, 907. [CrossRef]
11. Kumar, P.P.; Suresh, V.; Jasinski, M.; Leonowicz, Z. Off-Grid Rural Electrification in India Using Renewable Energy Resources and Different Battery Technologies with a Dynamic Differential Annealed Optimization. *Energies* **2021**, *14*, 5866. [CrossRef]
12. Michalec, Ł.; Jasiński, M.; Sikorski, T.; Leonowicz, Z.; Jasiński, Ł.; Suresh, V. Impact of Harmonic Currents of Nonlinear Loads on Power Quality of a Low Voltage Network–Review and Case Study. *Energies* **2021**, *14*, 3665. [CrossRef]
13. Wasowski, M.; Sikorski, T.; Wisniewski, G.; Kostyla, P.; Szymanda, J.; Habrych, M.; Gornicki, L.; Sokol, J.; Jurczyk, M. The Impact of Supply Voltage Waveform Distortion on Non-Intentional Emission in the Frequency Range 2–150 kHz: An Experimental Study with Power-Line Communication and Selected End-User Equipment. *Energies* **2021**, *14*, 777. [CrossRef]
14. Kaczorowska, D.; Rezmer, J.; Jasinski, M.; Sikorski, T.; Suresh, V.; Leonowicz, Z.; Kostyla, P.; Szymanda, J.; Janik, P. A Case Study on Battery Energy Storage System in a Virtual Power Plant: Defining Charging and Discharging Characteristics. *Energies* **2020**, *13*, 6670. [CrossRef]
15. Jasiński, M.; Sikorski, T.; Kaczorowska, D.; Rezmer, J.; Suresh, V.; Leonowicz, Z.; Kostyla, P.; Szymańda, J.; Janik, P. A Case Study on Power Quality in a Virtual Power Plant: Long Term Assessment and Global Index Application. *Energies* **2020**, *13*, 6578. [CrossRef]

Article

A Case Study on Power Quality in a Virtual Power Plant: Long Term Assessment and Global Index Application

Michal Jasiński [1,*], Tomasz Sikorski [1], Dominika Kaczorowska [1,*], Jacek Rezmer [1], Vishnu Suresh [1], Zbigniew Leonowicz [1], Paweł Kostyla [1], Jarosław Szymańda [1] and Przemysław Janik [2]

[1] Faculty of Electrical Engineering, Wroclaw University of Science and Technology, 50-370 Wrocław, Poland; tomasz.sikorski@pwr.edu.pl (T.S.); jacek.rezmer@pwr.edu.pl (J.R.); vishnu.suresh@pwr.edu.pl (V.S.); zbigniew.leonowicz@pwr.edu.pl (Z.L.); pawel.kostyla@pwr.edu.pl (P.K.); jaroslaw.szymanda@pwr.edu.pl (J.S.)
[2] TAURON Ekoenergia Ltd., 58-500 Jelenia Góra, Poland; przemyslaw.janik@tauron-ekoenergia.pl
*   Correspondence: michal.jasinski@pwr.edu.pl (M.J.); dominika.kaczorowska@pwr.edu.pl (D.K.); Tel.: +48-7132-02022 (M.J.); +48-7132-02901 (D.K.)

Received: 16 November 2020; Accepted: 11 December 2020; Published: 14 December 2020

Abstract: The concept of virtual power plants (VPP) was introduced over 20 years ago but is still actively researched. The majority of research now focuses on analyzing case studies of such installations. In this article, the investigation is based on a VPP in Poland, which contains hydropower plants (HPP) and energy storage systems (ESS). For specific analysis, the power quality (PQ) issues were selected. The used data contain 26 weeks of multipoint, synchronic measurements of power quality levels in four related points. The investigation is concerned with the application of a global index to a single-point assessment as well as an area-related assessment approach. Moreover, the problem of flagged data is discussed. Finally, the assessment of VPP's impact on PQ level is conducted.

Keywords: virtual power plant (VPP); power quality (PQ); global index; distributed energy resources (DER); energy storage systems (ESS); power systems; long-term assessment

---

1. Introduction

In recent electrical power networks, the number of renewable energy sources (RES) and energy storage systems (EES) have continuously increased. Thus, different approaches to controlling them have appeared, such as microgrids and virtual power plants (VPP) [1,2]. Generally, VPPs are autonomous units equipped with effective power flow control systems. VPPs consist of different elements that are connected to the distribution network. The indicated elements are generators, loads, and energy storage systems [2]. Coordinating the work of the entire VPP is a difficult and demanding task.

The operation of VPPs may be analyzed in different areas. This article is related to power quality (PQ) issues in VPP. Thus, Table 1 presents the current research directions concerning VPPs and PQ. The literature investigation is based only on articles published in the last 5 years.

Table 1. Current research trends concern virtal power pland (VPP) and power quality (PQ).

Current Research Trends Concerning VPP	
optimal active and reactive power scheduling	[3–12]
network voltage control by renewable energy sources integrated	[13–18]
power flow control and analysis	[19–24]
frequency control issues with VPP support	[25–31]
Electric energy storage (EES) sizing, localization and management in VPP	[32–37]
playing a role in energy market	[38–48]
energy management in a VPP	[49–58]
real case study analysis	Denmark [59], Germany [60], Ireland [61], Greece [62], United Kingdom [63], China [64], South Korea [65], India [66], Australia [67].
Current Research Trends Concerning PQ	
development of PQ measurement devices and systems	[68–74]
data mining of PQ measurements	[75–84]
global power quality indices	[85–93]
detection and classification of voltage events	[94–102]
vehicle-to-grid (V2G) impact on PQ	[103–111]

The possibility of a simple assessment of a virtual power plant's impact regarding power quality was the main focus of this paper. The articles that concern PQ issues in VPPs are indicated in this paragraph. The first example is an article [112] which introduces VPP as vehicles to facilitate the cost-efficient integration of distributed energy sources (DER) into the existing power system. The article also presents case studies that demonstrate the application of the concepts on a test system. The result of the system performance includes energy efficiency, power quality, and security. The authors of [113] propose extensions of the IEC 61850 standard to enhance the interaction between the VPP controller and the DER. The article presents the implementation of VPP communication and control architecture in a real case. The investigated case concerns the PQ recorders' issues and demands in accordance with IEC 61850. The authors of [114] attempt to provide a suitable framework for harmonizing the operations of different units of VPP. The authors indicated that decisions and the generation of profit, although complying with the required power quality levels and physical network constraints, are an important element of VPP strategies. The authors of [115] present simulations of a situation with a VPP during an islanded grid with a thermal power plant for baseload. The indicated VPP consists of 200 MW wind power, 100 MW photovoltaic power, and +/− 250 MW pumped storage. The article presents different control strategies of the storage plant to highlight the impact of VPP on power quality. The authors of [116] consider the coordinative operation problem of multi-energy VPP. The bi-objective dispatch model was established for the optimization of the performance of multi-energy VPP in terms of economic cost and PQ. A real case study was performed on Hongfeng Eco-town in Southwestern China. The authors of [117] consider VPP management with priority requirements optimized by the compromised method. The operation optimization model of the virtual power plant was formulated as the fuzzy multiple objective optimization problems. This optimization problem considers the satisfaction of both customers and suppliers, system stability, PQ, and costs with operation limitations. The proposed method was applied in a test system. However, in the literature, there is a lack of research that presents the assessment of PQ in different units of VPP. Thus, this article investigates PQ measurements in long-term performance.

This article presents a case study of analyzing a real VPP that operates in Poland. The investigated VPP consists of a fragment of both low-voltage (LV) and medium-voltage (MV) distribution networks. The VPP consists of a hydropower plant (HPP), a photovoltaic system (PV), and energy storage systems (ESS). In this article, only the part of this VPP is analyzed which is concerned with the 1.25 MW HPP and associated 0.5 MW ESS. The investigation is based on power quality measurements that were obtained synchronically in five measurement points, which are HPP, ESS, associated MV line, and two LV loads. The duration of the measurements was from 1 May 2020 to 28 October 2020.

Therefore, the observable period of time was 26 weeks—182 days. For the indicated time period and measurement points, the assessment of PQ was realized using the global value. The single parameter of a single measurement point for such a long period of time demands analysis of a huge dataset. Thus, this global index approach was used. Moreover, since the measurement points are connected in one network, the proposition of common analysis was proposed. The analysis of how to use the flagging concept was indicated for multipoint measurement. Another element of the analysis was to indicate the different working conditions of the VPP. Fifty of them were indicated concerning HPP and EES working schedule. Then, the comparison of the indicated working conditions was realized using the global index.

To summarize the contributions of this article:

- It contains the analysis of multipoint, synchronic, and long-term power quality data;
- PQ assessment is realized for both single-point and area-related approaches;
- Flagging concept is discussed for both single-point and area-related approaches;
- PQ assessment is realized using the global index approach;
- Different working conditions of the VPP are defined, investigated, and compared in terms of power quality.

To obtain indicated contributions, the article is organized into five sections. In Section 2, the investigated virtual power plant is presented. Section 3 presents results concerning the use of the global index for PQ assessment of each measurement point; different working conditions of the investigated VPP; analyzing the application of the flagging concept for the area-related approach; using the global index for the assessment of different working conditions of the VPP in both single-point and area-related approaches. Section 4 presents a discussion of the obtained results. Section 5 indicates the conclusions.

2. Methodology and Research Object Description

The investigation presented in this article concerns PQ issues in VPP. The proposed approach concerns the application of the global index. Thus, the selection and customization of the global index for VPP issues are presented in Section 2.1. Then, the description of the investigated VPP localized in Poland is presented in Section 2.2. Finally, the specific working conditions of single VPP elements are indicated in Section 2.3.

2.1. Global Power Quality Index

As a current trend in PQ issues, the global index was indicated. Thus, in this article, there is a proposition to use one such index. Aggregated data index (ADI) [86,87] was selected. The used index consists of five classic 10-min PQ parameters, such as:

- frequency—f,
- voltage—U,
- short term flicker severity—P_{st},
- asymmetry factor—k_{u2},
- total harmonic distortion in voltage—THDu,

And additional parameters that are responsible for the enhancement of the sensitivity of the index:

- an envelope of voltage deviation obtained by the difference between the maximum and minimum of 200-millisecond U values identified during the 10-min aggregation interval,
- a maximum of the 200-millisecond value of THDu, similarly identified in the 10-min aggregation interval [86,87].

Indicated parameters refer to the demands of the standard IEC 61000-4-30 [118]. Three-phase values are reduced to one using the mean value of them. All factors that are included in the ADI are based on the differences between the measured 10-min aggregated power quality data and the recommended limits of the selected standard. Therefore, the differences are expressed as a percentage in relation to standard limits. However, to customize the proposed index to VPP issues, the authors decided to exclude frequency. It is known that local changes do not impact frequency. Therefore, to conduct an assessment of VPP, impact on local area frequency was omitted. The calculations of ADI were performed on raw PQ measurement data using Excel software.

2.2. Investigated VPP

The indicated virtual power plant in this article is based on a fragment of the distribution network in Poland [119,120]. It is supplied by two stations of 110/20 kV. The indicated stations are connected to a 110 kV electrical power system. However, in this article, only a certain area of one substation is investigated. The 20 kV network fed from the station is an overhead cable network. The 20 kV network fed from the second station is mainly an urban cable network. The 20 kV network has earth fault current compensation. Connected to the VPP are a 1.25 MW HPP and a 0.5 MW battery ESS, both of which are connected to a medium voltage level.

The simplified scheme of the selected fragment of the VPP area is presented in Figure 1. It consists of a 20 kV distribution network with a hydropower plant (HPP) and an energy storage system (EES) connected with the HV/MV substation by MV line (MV_L). HPP and ESS are connected to the same node of the network. Furthermore, two additional low voltage loads are indicated: LV_L and LV_HE. LV_L is connected with the indicated MV associated line. LV_HE is connected with the node of the HPP and EES. The localization of power quality recorders (denoted as "R") is also indicated in Figure 1. Due to this, the HPP and EES are connected to one node and their PQ recorders use the same voltage transformer. In further research, they are treated as one point, indicated as MV_HPP&EES.

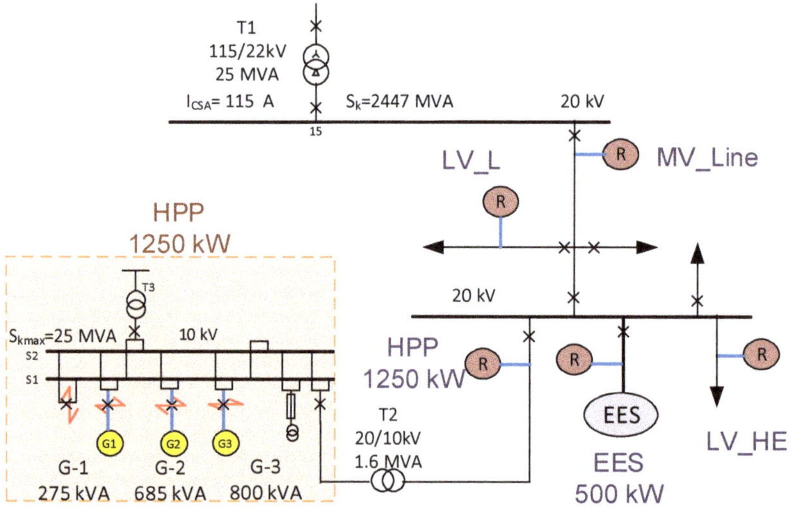

Figure 1. Investigated part of VPP with the placement of PQ recorders.

2.3. Different Working Conditions of Virtual Power Plant (VPP)

To investigate the working conditions of the VPP, different working conditions of the HPP and EES were considered. Regarding the HPP, three different working conditions were considered:

- not working—active power level equal to 0 or equal to own consumption;
- working at a part power—active power higher than own consumption but lower than 1 MW;
- working at a maximal power—active power higher than 1 MW.

Regarding EES, five different working conditions were indicated:

- not working—active power level equal to 0 or equal to own consumption;
- a low power charging—charging power higher than 40 kW and lower than 200 kW;
- a high power charging—charging power higher than 200 kW;
- a low power discharging—discharging power higher than 6 kW and lower than 200 kW;
- a high power discharging—discharging power higher than 200 kW.

The values are based on working circumstances that were discussed with the power distribution company during the VPP settlement.

3. Results

This section presents the results of the investigation of PQ issues in VPP. In Section 3.1, the global index application is presented. The case is realized for EN 50160 [121] demands and realized for each measurement point separately. Section 3.2 analyzes the occurrence of different working conditions of the investigated VPP. Then, in Section 3.3, the flagging concept is applied in accordance with standard IEC 61000-4-30 [114]. Additionally, the cross-analysis between event occurrence in different measurement points is conducted. Finally, Section 3.4 presents the assessment of the indicated working conditions of the VPP using the global index for both single-point and area-related approaches.

3.1. Application of Global Power Quality Index for Single Measurement Points

This section presents the application of ADI to a long-term comparison of power quality level. As indicated in Section 2.1, VPP frequency is omitted so the frequency importance rate was set as 0. Standard EN 50160 [121] was selected to obtain global values. The applied limits based on EN 50160 [121] are presented in Table 2.

Table 2. Limit values of standard EN 50160 [121] used for the global index.

Parameter	Value
U_{limit}	10% of U_d
Pst_{limit}	1.0
$ku2_{limit}$	2%
$THDu_{limit}$	8%

U_d—declared value of voltage.

For such defined ADI, the assessment of PQ level was conducted for all four measurement points (Figures 2–5). Moreover, to enable easier analysis, green-yellow-red colors were included in the figures:

- green for ADI $\leq$ 0.5;
- yellow for 0.5 < ADI $\leq$ 1;
- red for ADI > 1.

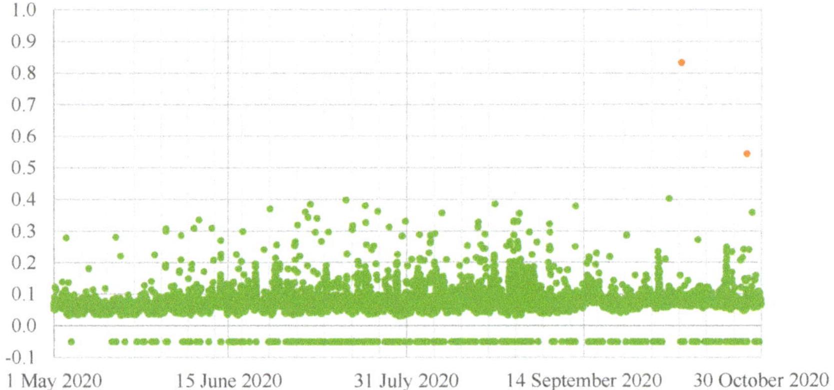

Figure 2. Global power quality index changeability for MV_Line.

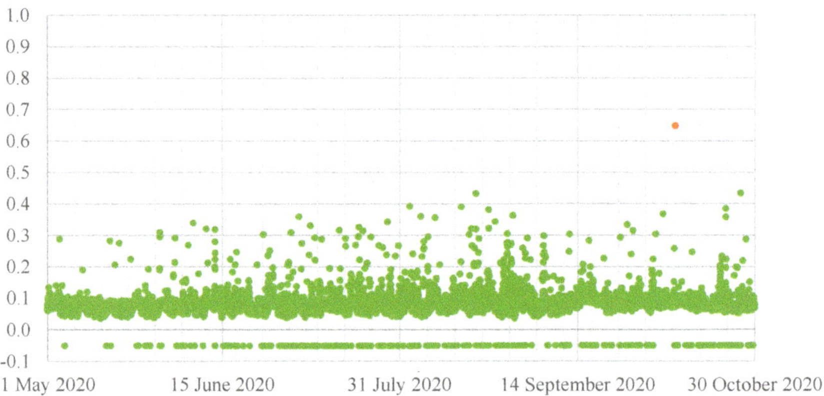

Figure 3. Global power quality index changeability for MV_HPP&EES.

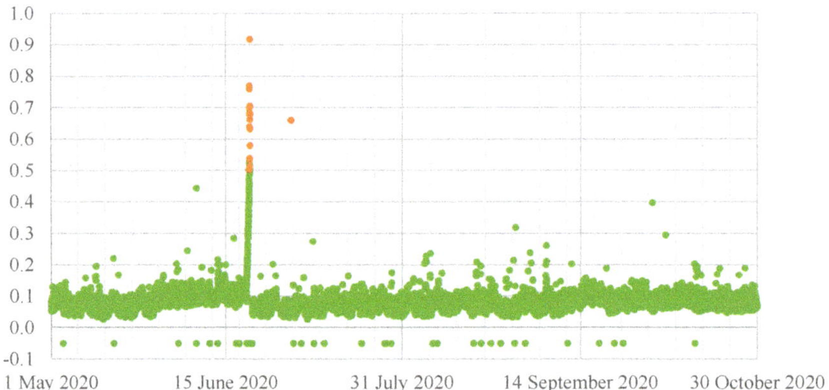

Figure 4. Global power quality index changeability for LV_HE.

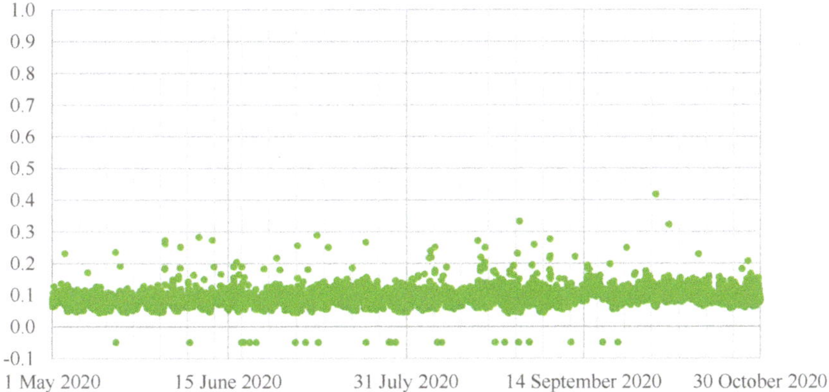

Figure 5. Global power quality index changeability for LV_L.

In the analysis, the time when voltage events occurred was excluded. The event data exclusion was based on the flagging concept of standard IEC 61000-4-30 [118]. Flagged data are represented as "virtual" value of ADI equal to −0.05. A deeper consideration of these voltage events is presented in Section 3.3. To summarize ADI levels for selected points in the selected 26 weeks of measurements, Table 3 was prepared. The results indicated that the highest level of power quality level (the lowest ADI) was in the MV_Line and the lowest level of power quality (the highest ADI) in LV_L.

Table 3. Limit values of standard EN 50160 [121] used for the global index.

Measurement Point	Number of Flagged Data	Mean Value of ADI
MV_Line	327	0.069
MV_HPP&EES	294	0.080
LV_HE	111	0.077
LV_L	47	0.087

During the time domain change in ADI level in LV_HE, a specific period was selected. In long-term assessment, it is challenging to verify the phenomena of this working condition. Therefore, the time shortage to 4 days that were connected with this working condition is presented in Figure 6. It can be observed that the PQ problem is noticed only in LV_HE, so this working condition is not connected with the impact of VPP. However, it is worth noticing that after a continuous (even linear) increase in ADI, a long event time in the LV_HE measurement point was indicated. In this article, in depth analysis of the reason for this working condition is omitted but this is a very interesting topic for future research.

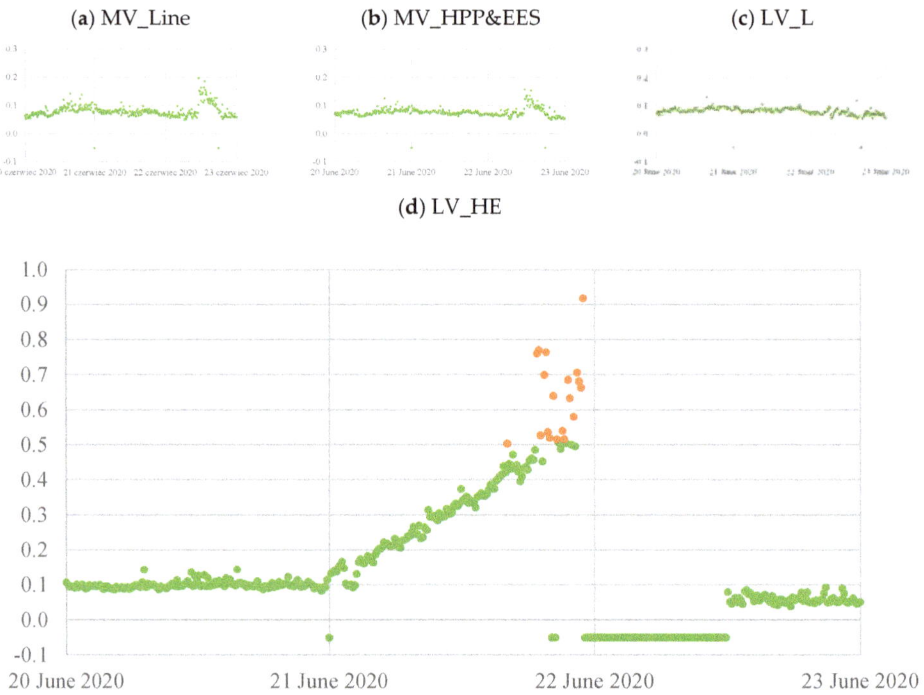

Figure 6. Specific working conditions observed in the level of ADI for measurement point (**a**) MV_Line, (**b**) MV_HPP&EES, (**c**) LV_L, (**d**) LV-HE.

3.2. Indication of the VPP Different Working Conditions

Based on information about HPP and EES possible operation indicated in Section 2.3, the investigated VPP working conditions are as follows:

- "0"—HPP is not working and ESS is not working;
- "1"—HPP is working partially and ESS is not working;
- "2"—HPP is working at maximal power and ESS is not working;
- "3"—HPP is not working and ESS is discharging at low power;
- "4"—HPP is working partially and ESS is discharging at low power;
- "5"—HPP is working at maximal power and ESS is discharging at low power;
- "6"—HPP is not working and ESS is discharging at high power;
- "7"—HPP is working partially and ESS is discharging at high power;
- "8"—HPP is working at maximal power and ESS is discharging at high power;
- "9"—HPP is not working and ESS is charging at low power;
- "10"—HPP is working partially and ESS is charging at low power;
- "11"—HPP is working at maximal power and ESS is charging at low power;
- "12"—HPP is not working and ESS is charging at high power;
- "13"—HPP is working partially and ESS is charging at high power;
- "14"—HPP is working at maximal power and ESS is charging at high power.

Initially, it was determined how long a given state of work occurred. The analyzed period of time was 182 days (26 weeks). Time aggregation of power quality data is 10 min, so the selected period is represented by 26,208 10-min data. However, due to the data coverage based on real measurements, synchronized 10-min data obtained from all points obtained were 25,069. Therefore, data coverage is 97.7%.

Detailed information on the number of 10-min data assigned to the working conditions is included in Table 4. Figure 7 also presents the occurrence of working conditions in the analyzed time period. It is worth noting that the state of operation 9 (HPP is not working and EES is charging at low power) and 12 (HPP is not working and EES is charging at low power) does not occur during measurements. This is because these working conditions were not permitted during the planning of the VPP. It also can be observed that the most common working conditions were "0" (HPP is not working and ESS is not working), "3" (HPP is not working and ESS is discharging at low power), and "2" (HPP is working at maximal power and ESS is not working).

Table 4. Duration of different working conditions.

VPP Working Condition	Number of 10-Min Data	Time of Working Condition Occurring as Percentage of Whole Time
0	9531	38.02
1	962	3.84
2	3382	13.49
3	8194	32.69
4	177	0.71
5	237	0.95
6	72	0.29
7	60	0.24
8	267	1.07
9	0	0.00
10	391	1.56
11	1538	6.14
12	0	0.00
13	127	0.51
14	131	0.52

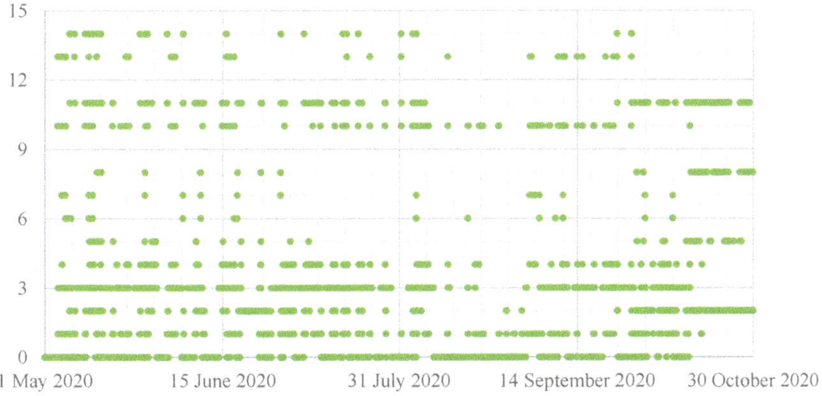

Figure 7. VPP working conditions in analyzed period of time.

3.3. Events Analysis as an Introduction to Area-Related Approach

In this section, the analysis of events occurring is performed. Generally, this is not a strict event analysis such as defining the type of event. The main aim of this subsection is to analyze how events impact the number of flagged data for both single- and multiple-point approaches. The flagging concept is introduced in standard IEC 61000-4-30 [118]. It indicates that if an aggregated value contains the time when an event occurred, these aggregated data must be flagged. This ensures that this problem is not counted twice as an event and as an extension of other parameter values.

For analysis of the number of flagged data, the flagged data matrix is proposed in Table 5. It presents the number of 10-min data that were flagged for single points and together on two measurement points. It can be observed that the highest number of flagged data was for the MV_Line and the lowest for the LV_L. It is also worth noticing that the majority of flagged data were common for MV_Line and MV_HPP&EES measurement points.

Table 5. Matrix of flagged data in analyzed measurement points.

	MV_Line	MV_HPP&EES	LV_HE	LV_L
MV_Line	327			
MV_HPP&EES	277	294		
LV_HE	15	15	111	
LV_L	17	29	10	47

The next step was to check how many of these common flagged data were indicated in more than two points. The results of this investigation are presented in Table 6. The results indicated that the general number of flagged data that are common for all measurement points is equal to 9.

Table 6. Limit values of standard EN 50160 [121] used to the global index.

Measurement Points	Number of Common Flagged Data
MV_Line + MV_HPP&EES + LV_HE	15
MV_Line + MV_HPP&EES + LV_L	17
MV_HPP&EES + LV_HE + LV_L	9
MV_Line + MV_HPP&EES + LV_HE + LV_L	9

The additional investigation included the concept of flagging for all measurement points treated together (area-related approach). This approach is based on set theory. Each measurement point is a single set of flagged data and the area-related value is a union of them all. Generally, the flagging for the whole period for all measurement points together was connected with the flagging of 457 10-min data. Then, the distribution of different working conditions (defined in Section 3.2) was realized. Presented in Section 3.4 is a comparative assessment of VPP working conditions. It is worth noticing that, due to the flagging concept, around 1.7% of data were excluded from the comparative assessment. Detailed results are presented in Table 7.

Table 7. Impact of flagging concept for area-related approach.

VPP Working Condition	Number of Data before Flagging	Number of Data after Flagging	Change in Percentage to the Number of Data before Flagging
0	9531	9373	1.7
1	962	951	1.2
2	3382	3247	4.2
3	8194	8092	1.3
4	177	175	1.1
5	237	235	0.9
6	72	71	1.4
7	60	60	0.0
8	267	264	1.1
9	0	0	-
10	391	389	0.5
11	1538	1501	2.5
12	0	0	-
13	127	127	0.0
14	131	127	3.1

3.4. Assessment of Different Working Conditions of VPP in Area-Related Approach

The next element of this research was the application of ADI to describe the working conditions of VPP indicated in Section 3.2. The analysis of ADI was realized separately for each of the four measurement points. The results are presented in Table 8. The assessment of the impact of individual working conditions of VPP (0–14) was related to the situation when both HPP and the associated EES are not working (working condition 0). Green was used to indicate conditions that are more positive in terms of quality in all measurement points, i.e., lower level of VPP. Additionally, a light green and light red color were introduced to compare the assessment of the ADI of individual measurement points. Light green was used for more favorable conditions and light red was used for worse conditions. The results indicated that working conditions 1, 3, 5, 6, 7, 10, 11, 13, 14 are more positive in terms of power quality than when working condition 0 occurs.

Table 8. Impact of flagging concept for area-related approach.

Working Condition of VPP	MV_Line	MV_HPP&EES	LV_HE	LV_L	Number of Data after Flagging
0	0.069	0.081	0.077	0.088	9373
1	0.067	0.074	0.068	0.084	951
2	0.075	0.077	0.083	0.087	3247
3	0.067	0.080	0.079	0.087	8092
4	0.063	0.077	0.074	0.085	175
5	0.067	0.073	0.082	0.088	235
6	0.065	0.078	0.072	0.085	71
7	0.071	0.080	0.069	0.085	60
8	0.076	0.079	0.074	0.093	264
9	-	-	-	-	0
10	0.065	0.072	0.067	0.083	389
11	0.071	0.075	0.071	0.086	1501
12	-	-	-	-	0
13	0.065	0.073	0.071	0.086	127
14	0.068	0.072	0.065	0.081	127
all	0.069	0.079	0.077	0.087	24,612

Green background—condition is more positive in terms of quality for all measurement points. Light green background—condition is more positive in terms of quality for single measurement point. Light red background—condition is more positive in terms of quality for single measurement point.

However, analysis of the number of data that represent each working condition indicated that some of them occur for a short period of time. Therefore, the next element of the research was to make a connection between them to obtain more numerous working conditions. Thus, the first connection was made for both conditions of HPP working (HPP is working partially and HPP is working at maximal power). Both were connected as a working condition, "HPP working". Moreover, working conditions that do not occur (9 and 12) are omitted. Therefore, the newly proposed working conditions are as follows:

- "20"—HPP is not working and ESS is not working;
- "21"—HPP is working and ESS is not working;
- "22"—HPP is not working and ESS is discharging at low power;
- "23"—HPP is working and ESS is discharging at low power;
- "24"—HPP is not working and ESS is discharging at high power;
- "25"—HPP is working and ESS is discharging at high power;
- "26"—HPP is working and ESS is charging at low power;
- "27"—HPP is working and ESS is charging at high power.

Then, the ADI level for these working conditions is presented in Table 9. Moreover, green/light green/light red colors are used. The results indicated that working conditions 24 and 27 are more positive in terms of power quality than when working condition 0 occurs.

Table 9. Impact of flagging concept for area-related approach.

Working Condition of VPP	MV_Line	MV_HPP&EES	LV_HE	LV_L	Number of Data after Flagging
20	0.069	0.081	0.077	0.088	9373
21	0.073	0.077	0.080	0.086	4198
22	0.067	0.080	0.079	0.087	8092
23	0.066	0.074	0.079	0.087	410
24	0.065	0.078	0.072	0.085	71
25	0.075	0.079	0.073	0.091	324
26	0.070	0.074	0.070	0.085	1890
27	0.066	0.073	0.068	0.083	254

Green background—condition is more positive in terms of quality for all measurement points. Light green background—condition is more positive in terms of quality for single measurement point. Light red background—condition is more positive in terms of quality for single measurement point.

Then, the next connections were made to obtain more numerous working conditions. This time, connection was made for the ESS at low power and a high-power charging/discharging, which is a new working condition that refers only to EES charging/discharging. Thus, the newly proposed working conditions are as follows:

- "30"—HPP is not working and ESS is not working;
- "31"—HPP is working and ESS is not working;
- "32"—HPP is not working and ESS is discharging;
- "33"—HPP is working and ESS is discharging;
- "34"—HPP is not working and ESS is discharging.

Then, the ADI level for these working conditions is presented in Table 10. Moreover, green/light green/light red colors are used. Finally, as a step in the area-related analysis of the investigated part of VPP, one value for all measurement points together was calculated. The value of the global index ADI for the common approach was calculated as a mean value of each measurement point. The results for single points indicated that working condition 30 is more positive in terms of power quality than when working condition 0 occurs. The results of this area-related approach indicate that, generally, there is no negative impact of the different working conditions of HPP and EES in long-term assessment (equal or lower value of ADI).

Table 10. Impact of flagging concept for area-related approach.

Working Condition of VPP	MV_Line	MV_HPP&EES	LV_HE	LV_L	Number of Data after Flagging	ADI for Area-Related Approach
30	0.069	0.081	0.077	0.088	9373	0.079
31	0.073	0.077	0.080	0.086	4198	0.079
32	0.067	0.080	0.079	0.087	8163	0.078
33	0.070	0.076	0.076	0.089	734	0.078
34	0.070	0.074	0.070	0.085	2144	0.075

Green background—condition is more positive in terms of quality for all measurement points. Light Green background—condition is more positive in terms of quality for single measurement point. Light red background—condition is more positive in terms of quality for single measurement point.

Finally, the investigated working conditions were reduced only to two cases:

- "40" when VPP is not working (HPP and EES are not working);
- "41" when VPP is working (HPP or EES is working).

The results are presented in Table 11. Therefore, the final statement is that VPP in long-term assessment has a positive impact on power quality assessment.

Table 11. Impact of flagging concept for area-related approach.

	VPP Working Condition	ADI for Area-Related Approach
40	VPP is not working	0.079
41	VPP is working	0.078

Green background—condition is more positive in terms of quality for all measurement points.

4. Discussion

This article presents a case study analyzing a real VPP that operates in Poland. The investigated VPP consists of a fragment of both low-voltage (LV) and medium-voltage (MV) distribution network. Under investigation are the PQ measurements from five (four after reduction) different PQ recorders. The investigation is concerned with both MV and LV points. The duration of the measurements was 26 weeks (from 1 May 2020 to 28 October 2020).

The first element was an application of the global index ADI [86,87]. This index was successfully applied for the assessment of the multipoint measurement of the mining industry. Therefore, the authors also decided to apply this concept to the VPP. However, during the investigation, the frequency as a part of the global index was omitted because the VPP, even in maximum operational limits, has a negligible impact on frequency. The standard used to apply limits to the global value was EN 50160 [121]. The results indicated that the highest level of power quality (the lowest ADI) was in the associated MV line (MV_Line) and the lowest level of power quality (the highest ADI) in the LV measurement point that is associated with the MV line (LV_L).

During analysis of the changeability of the ADI, specific time periods of LV_HP were denoted. However, an analysis of only the global value enables us to conclude that the reason for such a situation is impossible. Thus, using one global index as a first stage to denote specific working conditions seems interesting. However, after indicating them, a deeper multiparameter assessment is still needed.

As the results concern long-term data, different working conditions occurred. In the first section of the article, the diverse technique was realized in the point of HPP and EES working. Therefore, fifteen different working conditions were indicated. Then, the assessment of PQ for each measurement point using ADI was realized. For the majority of cases, the positive impact of PQ was observed when VPP operated (working conditions 1, 3, 5, 6, 7, 10, 11, 13, 14 in comparison to 0). However, the majority of these working conditions were represented by a short period of time. Therefore, additional agglomeration of working conditions in terms of HPP and EES was realized. The final stage of agglomeration was when only two states occurred, which were when VPP is working and VPP is not working. Additionally, to compare this, the area-related value was calculated. This value was obtained as a mean value of ADI from all measurement points. The conclusion in the long-term approach indicated that the operation of VPP has a generally positive impact on power quality.

The authors are aware that analyzing a single-point assessment for single PQ parameters appears to be more valuable. However, for long-term assessment, this work is time-consuming. Moreover, the main purpose of the VPP is to obtain economic profits, and technical issues are often omitted. Therefore, the proposition of using the PQ global index seems interesting to extend the assessment of VPP in a long-term approach.

5. Conclusions

This article proposes the application of a global index to power quality issues in a virtual power plant. The proposed solution enables simplification of the assessment from many classical power quality parameters to one global value for each measurement point. However, on the other hand, including extremum 200-millisecond values of voltage and harmonics parameters is an extension of the classical approach. Thus, this approach simplifies the assessment and increases the range of used parameters during analysis.

The analysis of the global value changeability for long-term data was indicated as useful to the indication of specific conditions in terms of PQ. However, using this global value enables us to define the reason for this situation. However, it leads to the definition of the period of occurrence. Then, for this period, the classic multiparameter assessment may be realized to analyze the reasons.

This article also indicates the general impact of VPP working conditions on the PQ issue. Both single-point and area-related approaches were realized. The positive and negative results of both approaches were indicated. However, the most important conclusion is that, generally, VPP impact on the long-term performance of PQ issues was positive.

The future research directions are:

- research into using the global index for the identification and then assessment of short/specific working conditions of the selected VPP unit;
- applying data mining techniques to obtain less obvious relations that concern VPP operation and PQ level.

Author Contributions: Conceptualization, M.J. and T.S.; methodology, M.J. and T.S.; software, M.J. and T.S.; validation, D.K., J.R. and V.S.; formal analysis, M.J., T.S., D.K. and J.R.; investigation, M.J. and T.S.; resources, P.K. and P.J.; data curation, V.S. and J.S.; writing—original draft preparation M.J. and T.S.; writing—review and editing, D.K. and J.R.; visualization, D.K. and J.R.; supervision, T.S., J.R. and Z.L.; project administration, T.S. and P.J.; funding acquisition, T.S. and P.J. All authors have read and agreed to the published version of the manuscript.

Funding: This research was funded by the National Center of Research and Development in Poland, the project "Developing a platform for aggregating generation and regulatory potential of dispersed renewable energy sources, power retention devices and selected categories of controllable load" supported by European Union Operational Programme Smart Growth 2014–2020, Priority Axis I: Supporting R&D carried out by enterprises, Measure 1.2: Sectoral R&D Programmes, POIR.01.02.00-00-0221/16, performed by TAURON Ekoenergia Ltd.

Conflicts of Interest: The authors declare no conflict of interest.

References

1. Justo, J.J. Intelligent Energy Management Strategy Considering Power Distribution Networks with Nanogrids, Microgrids, and VPP Concepts. In *Handbook of Distributed Generation*; Springer International Publishing: Cham, Switzerland, 2017; pp. 791–815.
2. Yavuz, L.; Önen, A.; Muyeen, S.M.M.; Kamwa, I. Transformation of microgrid to virtual power plant—A comprehensive review. *IET Gener. Transm. Distrib.* **2019**, *13*, 1994–2005. [CrossRef]
3. Jha, B.K.; Singh, A.; Kumar, A.; Misra, R.K.; Singh, D. Phase unbalance and PAR constrained optimal active and reactive power scheduling of Virtual Power Plants (VPPs). *Int. J. Electr. Power Energy Syst.* **2021**, *125*, 106443. [CrossRef]
4. Sun, H.; Meng, J.; Peng, C. Coordinated Optimization Scheduling of Multi-region Virtual Power Plant With Wind-power/Photovoltaic/Hydropower/Carbon-capture Units. *Dianwang Jishu/Power Syst. Technol.* **2019**, *43*, 4040–4049. [CrossRef]
5. Dong, J.; Zhou, B.; Liu, J.; Gao, H.; Shi, Y.; Liu, J.; Lu, L. Heat and power scheduling of a virtual power plant considering comfort level of customers. *Dianli Jianshe/Electric Power Constr.* **2019**, *40*, 19–26. [CrossRef]
6. Cheng, J.; Xie, S.; Zhang, Y.; Wang, J.; Xiong, S. Optimization model of intelligent power consumption considering Compositive comfort within budget limitation. *Dianli Jianshe/Electric Power Constr.* **2020**, *41*, 88–96. [CrossRef]
7. Qiu, J.; Meng, K.; Zheng, Y.; Dong, Z.Y. Optimal scheduling of distributed energy resources as a virtual power plant in a transactive energy framework. *IET Gener. Transm. Distrib.* **2017**, *11*, 3417–3427. [CrossRef]
8. Jiao, F.; Deng, Y.; Li, D.; Wei, B.; Yue, C.; Cheng, M.; Zhang, Y.; Zhang, J. A self-scheduling strategy of virtual power plant with electric vehicles considering margin indexes. *Arch. Electr. Eng.* **2020**, *69*, 907–920. [CrossRef]
9. Zamani, A.G.; Zakariazadeh, A.; Jadid, S.; Kazemi, A. Stochastic operational scheduling of distributed energy resources in a large scale virtual power plant. *Int. J. Electr. Power Energy Syst.* **2016**, *82*, 608–620. [CrossRef]

10. Ju, L.; Tan, Z.; Yuan, J.; Tan, Q.; Li, H.; Dong, F. A bi-level stochastic scheduling optimization model for a virtual power plant connected to a wind–photovoltaic–energy storage system considering the uncertainty and demand response. *Appl. Energy* **2016**, *171*, 184–199. [CrossRef]

11. Nosratabadi, S.M.; Hooshmand, R.-A.; Gholipour, E. Stochastic profit-based scheduling of industrial virtual power plant using the best demand response strategy. *Appl. Energy* **2016**, *164*, 590–606. [CrossRef]

12. Vahedipour-Dahraie, M.; Rashidizade-Kermani, H.; Shafie-khah, M.; Catalao, J.P.S. Risk-Averse Optimal Energy and Reserve Scheduling for Virtual Power Plants Incorporating Demand Response Programs. *IEEE Trans. Smart Grid* **2020**. [CrossRef]

13. Unger, D.; Spitalny, L.; Myrzik, J.M.A. Voltage control by small hydro power plants integrated into a virtual power plant. In Proceedings of the 2012 IEEE Energytech, Cleveland, OH, USA, 29–31 May 2012; pp. 1–6. [CrossRef]

14. Pavan Kumar, Y.V.; Bhimasingu, R. Improving power quality in microgrids using virtual motor-generator set based control scheme. In Proceedings of the IECON 2016—42nd Annual Conference of the IEEE Industrial Electronics Society, Florence, Italy, 23–26 October 2016; pp. 7173–7178. [CrossRef]

15. Moutis, P.; Georgilakis, P.S.; Hatziargyriou, N.D. Voltage Regulation Support Along a Distribution Line by a Virtual Power Plant Based on a Center of Mass Load Modeling. *IEEE Trans. Smart Grid* **2018**, *9*, 3029–3038. [CrossRef]

16. Dall'Anese, E.; Guggilam, S.S.; Simonetto, A.; Chen, Y.C.; Dhople, S.V. Optimal Regulation of Virtual Power Plants. *IEEE Trans. Power Syst.* **2018**, *33*, 1868–1881. [CrossRef]

17. Paternina, J.L.; Contreras, L.; Trujillo, E.R. Study of voltage stability in a distribution network by integrating distributed energy resources into a virtual power plant. *Contemp. Eng. Sci.* **2017**, *10*, 1441–1455. [CrossRef]

18. Ishihara, H.; Nada, K.; Tanaka, M.; Inoue, S.; Kuwata, A.; Takano, T. A Voltage Control Method for Power Distribution Lines Utilizing Dispersed Customer Resources. In Proceedings of the 2020 22nd European Conference on Power Electronics and Applications (EPE'20 ECCE Europe), Lyon, France, 7–11 September 2020; IEEE: New York, NY, USA, 2020; pp. 1–8.

19. Haque, M.M.; Wolfs, P.; Alahakoon, S. Active Power Flow Control of Three-Port Converter for Virtual Power Plant Applications. In Proceedings of the 2020 IEEE International Conference on Power Electronics, Smart Grid and Renewable Energy (PESGRE2020), Kerala, India, 2–4 January 2020; IEEE: New York, NY, USA, 2020; pp. 1–6.

20. Pudjianto, D.; Djapic, P.; Strbac, G.; Stojkovska, B.; Ahmadi, A.R.; Martinez, I. Integration of distributed reactive power sources through Virtual Power Plant to provide voltage control to transmission network. In Proceedings of the CIRED 2019 Conference, Madrid, Spain, 3–6 June 2019; AIM: Madrid, Spain, 2019.

21. Konara, K.M.S.Y.; Kolhe, M.; Sharma, A. Power flow management controller within a grid connected photovoltaic based active generator as a finite state machine using hierarchical approach with droop characteristics. *Renew. Energy* **2020**, *155*, 1021–1031. [CrossRef]

22. Guerra, G.; Martinez Velasco, J.A. A virtual power plant model for time-driven power flow calculations. *AIMS Energy* **2017**, *5*, 887–911. [CrossRef]

23. Sosnina, E.; Chivenkov, A.; Shalukho, A.; Shumskii, N. Power flow control in a Virtual Power Plant LV network. *Int. J. Renew. Energy Res.* **2018**, *8*, 328–335.

24. Kaczorowska, D.; Rezmer, J.; Sikorski, T.; Janik, P. Application of PSO algorithms for VPP operation optimization. *Renew. Energy Power Qual. J.* **2019**, *17*, 91–96. [CrossRef]

25. Bilbao, J.; Bravo, E.; Rebollar, C.; Varela, C.; Garcia, O. Virtual Power Plants and Virtual Inertia. In *Power Systems*; Springer: Bilbao, Spain, 2020; pp. 87–113. ISBN 16121287.

26. Yang, J.; Huang, Y.; Wang, H.; Ji, Y.; Li, J.; Gao, C. A regulation strategy for virtual power plant. In Proceedings of the 2017 4th International Conference System Informatics (ICSAI 2017), Hangzhou, China, 11–13 November 2017; pp. 375–379. [CrossRef]

27. Ali, J.; Silvestro, F. Conventional Power Plants to TSO Frequency Containment Reserves—A Competitive Analysis for Virtual Power Plant's Role. In Proceedings of the 2019 IEEE 5th International forum on Research and Technology for Society and Industry (RTSI), Firenze, Italy, 9–16 September 2019; IEEE: New York, NY, USA, 2019; pp. 6–11.

28. Dey, P.P.; Das, D.C.; Latif, A.; Hussain, S.M.S.; Ustun, T.S. Active Power Management of Virtual Power Plant under Penetration of Central Receiver Solar Thermal-Wind Using Butterfly Optimization Technique. *Sustainability* **2020**, *12*, 6979. [CrossRef]

29. Kim, J.; Muljadi, E.; Gevorgian, V.; Mohanpurkar, M.; Luo, Y.; Hovsapian, R.; Koritarov, V. Capability-coordinated frequency control scheme of a virtual power plant with renewable energy sources. *IET Gener. Transm. Distrib.* **2019**, *13*, 3642–3648. [CrossRef]

30. Zhong, W.; Murad, M.A.A.; Liu, M.; Milano, F. Impact of Virtual Power Plants on Power System Short-Term Transient Response. *Electr. Power Syst. Res.* **2020**, *189*, 106609. [CrossRef]

31. Alhelou, H.H.; Siano, P.; Tipaldi, M.; Iervolino, R.; Mahfoud, F. Primary Frequency Response Improvement in Interconnected Power Systems Using Electric Vehicle Virtual Power Plants. *World Electr. Veh. J.* **2020**, *11*, 40. [CrossRef]

32. Han, N.; Wang, X.; Chen, S.; Cheng, L.; Liu, H.; Liu, Z.; Mao, Y. Optimal Configuration of Energy Storage Systems in Virtual Power. *IOP Conf. Ser. Earth Environ. Sci.* **2019**, *295*, 042072. [CrossRef]

33. Michiorri, A.; Lugaro, J.; Siebert, N.; Girard, R.; Kariniotakis, G. Storage sizing for grid connected hybrid wind and storage power plants taking into account forecast errors autocorrelation. *Renew. Energy* **2018**, *117*, 380–392. [CrossRef]

34. Stuhlenmiller, T.; Koenigsdorff, R. Optimum thermal storage sizing in building services engineering as a contribution to virtual power plants. *J. Build. Perform. Simul.* **2010**, *3*, 17–31. [CrossRef]

35. Sadeghian, O.; Oshnoei, A.; Khezri, R.; Muyeen, S. Risk-constrained stochastic optimal allocation of energy storage system in virtual power plants. *J. Energy Storage* **2020**, *31*, 101732. [CrossRef]

36. Kim, S.; Kwon, W.-H.; Kim, H.-J.; Jung, K.; Kim, G.S.; Shim, T.; Lee, D. Offer Curve Generation for the Energy Storage System as a Member of the Virtual Power Plant in the Day-Ahead Market. *J. Electr. Eng. Technol.* **2019**, *14*, 2277–2287. [CrossRef]

37. Sun, J.; Li, X.; Ma, H. Study on Optimal Capacity of Multi-type Energy Storage System for Optimized Operation of Virtual Power Plants. In Proceedings of the 2018 China International Conference on Electricity Distribution (CICED), Tianjin, China, 17–19 September 2018; IEEE: New York, NY, USA, 2018; pp. 2989–2993.

38. Heimgaertner, F.; Ziegler, U.; Thomas, B.; Menth, M. A Distributed Control Architecture for a Loosely Coupled Virtual Power Plant. In Proceedings of the 2018 IEEE International Conference on Engineering, Technology and Innovation (ICE/ITMC), Stuttgart, Germany, 17–20 June 2018; IEEE: New York, NY, USA, 2018; pp. 1–9.

39. Candra, D.; Hartmann, K.; Nelles, M. Economic Optimal Implementation of Virtual Power Plants in the German Power Market. *Energies* **2018**, *11*, 2365. [CrossRef]

40. Moreno, B.; Díaz, G. The impact of virtual power plant technology composition on wholesale electricity prices: A comparative study of some European Union electricity markets. *Renew. Sustain. Energy Rev.* **2019**, *99*, 100–108. [CrossRef]

41. Sikorski, T.; Jasiński, M.; Ropuszyńska-Surma, E.; Węglarz, M.; Kaczorowska, D.; Kostyła, P.; Leonowicz, Z.; Lis, R.; Rezmer, J.; Rojewski, W.; et al. A Case Study on Distributed Energy Resources and Energy-Storage Systems in a Virtual Power Plant Concept: Economic Aspects. *Energies* **2019**, *12*, 4447. [CrossRef]

42. Loßner, M.; Böttger, D.; Bruckner, T. Economic assessment of virtual power plants in the German energy market—A scenario-based and model-supported analysis. *Energy Econ.* **2017**, *62*, 125–138. [CrossRef]

43. Khorasany, M.; Raoofat, M. Bidding strategy for participation of virtual power plant in energy market considering uncertainty of generation and market price. In Proceedings of the 2017 Smart Grid Conference (SGC), Tehran, Iran, 20–21 December 2017; IEEE: New York, NY, USA, 2017; pp. 1–6.

44. Mohy-ud-din, G.; Muttaqi, K.M.; Sutanto, D. Transactive energy-based planning framework for VPPs in a co-optimised day-ahead and real-time energy market with ancillary services. *IET Gener. Transm. Distrib.* **2019**, *13*, 2024–2035. [CrossRef]

45. Shafiekhani, M.; Badri, A.; Shafie-khah, M.; Catalão, J.P.S. Strategic bidding of virtual power plant in energy markets: A bi-level multi-objective approach. *Int. J. Electr. Power Energy Syst.* **2019**, *113*, 208–219. [CrossRef]

46. Foroughi, M.; Pasban, A.; Moeini-Aghtaie, M.; Fayaz-Heidari, A. A bi-level model for optimal bidding of a multi-carrier technical virtual power plant in energy markets. *Int. J. Electr. Power Energy Syst.* **2021**, *125*, 106397. [CrossRef]

47. Zhou, Y.; Wei, Z.; Sun, G.; Cheung, K.W.; Zang, H.; Chen, S. Four-level robust model for a virtual power plant in energy and reserve markets. *IET Gener. Transm. Distrib.* **2019**, *13*, 2036–2043. [CrossRef]

48. Al-Awami, A.T.; Amleh, N.A.; Muqbel, A.M. Optimal Demand Response Bidding and Pricing Mechanism with Fuzzy Optimization: Application for a Virtual Power Plant. *IEEE Trans. Ind. Appl.* **2017**, *53*, 5051–5061. [CrossRef]

49. Yin, S.; Ai, Q.; Li, Z.; Zhang, Y.; Lu, T. Energy management for aggregate prosumers in a virtual power plant: A robust Stackelberg game approach. *Int. J. Electr. Power Energy Syst.* **2020**, *117*, 105605. [CrossRef]

50. Ciupageanu, D.-A.; Barelli, L.; Ottaviano, A.; Pelosi, D.; Lazaroiu, G. Innovative power management of hybrid energy storage systems coupled to RES plants: The Simultaneous Perturbation Stochastic Approximation approach. In Proceedings of the 2019 IEEE PES Innovative Smart Grid Technologies Europe (ISGT-Europe), Bucharest, Romania, 29 September–2 October 2019; IEEE: New York, NY, USA, 2019; pp. 1–5.

51. Rahimiyan, M.; Baringo, L. Real-time energy management of a smart virtual power plant. *IET Gener. Transm. Distrib.* **2019**, *13*, 2015–2023. [CrossRef]

52. Othman, M.M.; Hegazy, Y.G.; Abdelaziz, A.Y. Electrical energy management in unbalanced distribution networks using virtual power plant concept. *Electr. Power Syst. Res.* **2017**, *145*, 157–165. [CrossRef]

53. Mears, A.; Martin, J. Fully Flexible Loads in Distributed Energy Management: PV, Batteries, Loads, and Value Stacking in Virtual Power Plants. *Engineering* **2020**, *6*, 736–738. [CrossRef]

54. Liu, T.; Zhang, J.; Li, S.; Yue, L.; Zhou, X. Home energy management method for realizing demand response based on virtual power plant platform. *IOP Conf. Ser. Mater. Sci. Eng.* **2020**, *768*, 052114. [CrossRef]

55. Sheidaei, F.; Ahmarinejad, A. Multi-stage stochastic framework for energy management of virtual power plants considering electric vehicles and demand response programs. *Int. J. Electr. Power Energy Syst.* **2020**, *120*, 106047. [CrossRef]

56. Maanavi, M.; Najafi, A.; Godina, R.; Mahmoudian, M.; Rodrigues, E.M.G. Energy Management of Virtual Power Plant Considering Distributed Generation Sizing and Pricing. *Appl. Sci.* **2019**, *9*, 2817. [CrossRef]

57. Nosratabadi, S.M.; Hooshmand, R.-A. Stochastic electrical energy management of industrial Virtual Power Plant considering time-based and incentive-based Demand Response programs option in contingency condition. *Int. J. Emerg. Electr. Power Syst.* **2020**, *21*. [CrossRef]

58. Shayegan Rad, A.; Badri, A.; Zangeneh, A.; Kaltschmitt, M. Risk-based optimal energy management of virtual power plant with uncertainties considering responsive loads. *Int. J. Energy Res.* **2019**, *43*, 2135–2150. [CrossRef]

59. Gabderakhmanova, T.; Engelhardt, J.; Zepter, J.M.; Meier Sorensen, T.; Boesgaard, K.; Ipsen, H.H.; Marinelli, M. Demonstrations of DC Microgrid and Virtual Power Plant Technologies on the Danish Island of Bornholm. In Proceedings of the 2020 55th International Universities Power Engineering Conference (UPEC), Torino, Italy, 1–4 September 2020; IEEE: New York, NY, USA, 2020; pp. 1–6.

60. Heimgaertner, F.; Schur, E.; Truckenmueller, F.; Menth, M. A Virtual Power Plant Demonstration Platform for Multiple Optimization and Control Systems. In Proceedings of the International ETG Congress 2017, Bonn, Germany, 28–29 November 2017; pp. 1–6.

61. Van Summeren, L.F.M.; Wieczorek, A.J.; Bombaerts, G.J.T.; Verbong, G.P.J. Community energy meets smart grids: Reviewing goals, structure, and roles in Virtual Power Plants in Ireland, Belgium and the Netherlands. *Energy Res. Soc. Sci.* **2020**, *63*, 101415. [CrossRef]

62. Nikolaou, T.; Stavrakakis, G.S.; Tsamoudalis, K. Modeling and Optimal Dimensioning of a Pumped Hydro Energy Storage System for the Exploitation of the Rejected Wind Energy in the Non-Interconnected Electrical Power System of the Crete Island, Greece. *Energies* **2020**, *13*, 2705. [CrossRef]

63. Jenkins, A.M.; Patsios, C.; Taylor, P.; Khayrullina, A.; Chirkin, V. Optimising Virtual Power Plant Response to Grid Service Requests at Newcastle Science Central by Coordinating Multiple Flexible Assets. In Proceedings of the CIRED Workshop 2016, Helsinki, Finland, 14–15 June 2016; Institution of Engineering and Technology: London, UK, 2016; p. 212.

64. Zhao, H.; Wang, B.; Pan, Z.; Sun, H.; Guo, Q.; Xue, Y. Aggregating Additional Flexibility from Quick-Start Devices for Multi-Energy Virtual Power Plants. *IEEE Trans. Sustain. Energy* **2020**. [CrossRef]

65. Jeon, W.; Cho, S.; Lee, S. Estimating the Impact of Electric Vehicle Demand Response Programs in a Grid with Varying Levels of Renewable Energy Sources: Time-of-Use Tariff versus Smart Charging. *Energies* **2020**, *13*, 4365. [CrossRef]

66. Sharma, H.; Mishra, S. Techno-economic analysis of solar grid-based virtual power plant in Indian power sector: A case study. *Int. Trans. Electr. Energy Syst.* **2020**, *30*. [CrossRef]

67. Behi, B.; Baniasadi, A.; Arefi, A.; Gorjy, A.; Jennings, P.; Pivrikas, A. Cost–Benefit Analysis of a Virtual Power Plant Including Solar PV, Flow Battery, Heat Pump, and Demand Management: A Western Australian Case Study. *Energies* **2020**, *13*, 2614. [CrossRef]

68. Bolshev, V.; Vinogradov, A.; Jasinski, M.; Sikorski, T.; Leonowicz, Z.; Gono, R. Monitoring the Number and Duration of Power Outages and Voltage Deviations at Both Sides of Switching Devices. *IEEE Access* **2020**, *8*, 137174–137184. [CrossRef]

69. Matthee, A.; Moonen, N.; Leferink, F. Versatile High-Sample Frequency Power Quality Measurement Device. In Proceedings of the 2020 IEEE International Symposium on Electromagnetic Compatibility & Signal/Power Integrity (EMCSI), Reno, NV, USA, 27–31 June 2020; IEEE: New York, NY, USA, 2020; pp. 213–215.

70. Saravanakumar, T.; Sainarayanan, G.; Porkumaran, K. Design of a Cost Effective Optimized Power Factor Measurement Device for Nonlinear Single Phase Home Appliances. *Res. J. Appl. Sci. Eng. Technol.* **2015**, *10*, 454–463. [CrossRef]

71. Vinogradov, A.; Bolshev, V.; Vinogradova, A.; Kudinova, T.; Borodin, M.; Selesneva, A.; Sorokin, N. A System for Monitoring the Number and Duration of Power Outages and Power Quality in 0.38 kV Electrical Networks BT—Intelligent Computing & Optimization. In Proceedings of the Intelligent Computing & Optimization, Pattaya, Thailand, 4–5 October 2018; Vasant, P., Zelinka, I., Weber, G.-W., Eds.; Springer International Publishing: Cham, Switzerland, 2019; pp. 1–10.

72. Chintakindi, R.; Mitra, A. Execution of Real-time Wide Area Monitoring System with Big Data Functions and Practices. In Proceedings of the 2020 IEEE 9th Power India International Conference (PIICON), Sonepat, India, 28 February–1 March 2020; IEEE: New York, NY, USA, 2020; pp. 1–6.

73. Kitzig, J.-P.; Bumiller, G. Evaluation of Power Quality Measurement System Concept using an experimental setup. In Proceedings of the 2019 IEEE International Instrumentation and Measurement Technology Conference (I2MTC), Auckland, New Zealand, 20–23 May 2019; IEEE: New York, NY, USA, 2019; pp. 1–6.

74. Xie, Z.; Chen, Y.; Wu, W.; Luo, A.; Zhou, L.; Zhou, X.; Yang, L.; Tan, W.; Wang, Y. UPQC-Based High Precision Impedance Measurement Device and its Switching Control Method. In Proceedings of the IECON 2018-44th Annual Conference of the IEEE Industrial Electronics Society, Washington DC, USA, 21–23 October 2018; IEEE: New York, NY, USA, 2018; pp. 1556–1561.

75. Deng, P.; Xu, X.; Wang, F.Y.; Chen, Z. Data mining and data driving of harmonic in AC arc furnaces based on functional analysis. *IOP Conf. Ser. Earth Environ. Sci.* **2019**, *354*, 012122. [CrossRef]

76. Jasiński, M.; Sikorski, T.; Borkowski, K. Clustering as a tool to support the assessment of power quality in electrical power networks with distributed generation in the mining industry. *Electr. Power Syst. Res.* **2019**, *166*, 52–60. [CrossRef]

77. Jasiński, M.; Sikorski, T.; Leonowicz, Z.; Borkowski, K.; Jasińska, E. The Application of Hierarchical Clustering to Power Quality Measurements in an Electrical Power Network with Distributed Generation. *Energies* **2020**, *13*, 2407. [CrossRef]

78. Zhang, Y. Method for Extracting Typical Characteristics of Regional Grid with Scale Distributed PV Based on Time Series. In Proceedings of the 2018 2nd IEEE Conference on Energy Internet and Energy System Integration (EI2), Beijing, China, 20–22 October 2018; IEEE: New York, NY, USA, 2018; pp. 1–4.

79. Jasiński, M.; Borkowski, K.; Sikorski, T.; Kostyla, P. Cluster Analysis for Long-Term Power Quality Data in Mining Electrical Power Network. In Proceedings of the 2018 Progress in Applied Electrical Engineering (PAEE), Kościelisko, Poland, 18–22 June 2018; IEEE: New York, NY, USA, 2018; pp. 1–5. [CrossRef]

80. Gorjani, O.M.; Bilik, P.; Vanus, J. Application of Optimized Deterministic Methods in Long-term Power Quality. In Proceedings of the 2019 20th International Scientific Conference on Electric Power Engineering (EPE), Jeseniky Mountains, Czech Republic, 15–17 May 2019; IEEE: New York, NY, USA, 2019; pp. 1–5.

81. Zhong, Q.; Yao, W.; Lin, L.; Wang, G.; Xu, Z. Data Analysis and Applications of the Power Quality Monitoring. In Proceedings of the 2018 International Conference on Power System Technology (POWERCON), Guangzhou, China, 24–26 October 2018; IEEE: New York, NY, USA, 2018; pp. 4035–4039.

82. Jasinski, M.; Sikorski, T.; Leonowicz, Z.; Kaczorowska, D.; Suresh, V.; Szymanda, J.; Jasinska, E. Different working conditions identification of a PV power plant using hierarchical clustering. In Proceedings of the 2020 12th International Conference on Electronics, Computers and Artificial Intelligence (ECAI), Bucharest, Romania, 25–27 June 2020; IEEE: New York, NY, USA, 2020; pp. 1–8.

83. Borges, F.A.S.; Fernandes, R.A.S.; Silva, I.N.; Silva, C.B.S. Feature Extraction and Power Quality Disturbances Classification Using Smart Meters Signals. *IEEE Trans. Ind. Inform.* **2016**. [CrossRef]

84. Asha Kiranmai, S.; Jaya Laxmi, A. Data mining for classification of power quality problems using WEKA and the effect of attributes on classification accuracy. *Prot. Control Mod. Power Syst.* **2018**, *3*, 29. [CrossRef]

85. Ignatova, V.; Villard, D.; Hypolite, J.M. Simple indicators for an effective Power Quality monitoring and analysis. In Proceedings of the 2015 IEEE 15th International Conference on Environment and Electrical Engineering, EEEIC 2015-Conference Proceedings, Rome, Italy, 10–13 June 2015; pp. 1104–1108.
86. Jasiński, M.; Sikorski, T.; Kostyła, P.; Leonowicz, Z.; Borkowski, K. Combined Cluster Analysis and Global Power Quality Indices for the Qualitative Assessment of the Time-Varying Condition of Power Quality in an Electrical Power Network with Distributed Generation. *Energies* **2020**, *13*, 2050. [CrossRef]
87. Jasinski, M.; Sikorski, T.; Kostyla, P.; Borkowski, K. Global power quality indices for assessment of multipoint Power quality measurements. In Proceedings of the 2018 10th International Conference on Electronics, Computers and Artificial Intelligence (ECAI), Iasi, Romania, 28–30 June 2018; IEEE: New York, NY, USA, 2018; pp. 1–6.
88. Raptis, T.E.; Vokas, G.A.; Langouranis, P.A.; Kaminaris, S.D. Total Power Quality Index for Electrical Networks Using Neural Networks. *Energy Procedia* **2015**, *74*, 1499–1507. [CrossRef]
89. Ge, B.; Pan, T.; Li, Z. Synthetic assessment of power quality using relative entropy theory. *J. Comput. Inf. Syst.* **2015**, *11*, 1323–1331. [CrossRef]
90. Khramshin, V.R.; Khramshin, R.R.; Karandaev, A.S.; Medvedev, V.N. Methodic of calculation of the non-sinusoidal voltage index within electrical networks with high-voltage frequency convertors. In Proceedings of the 2015 International Siberian Conference on Control and Communications (SIBCON), Omsk, Russia, 21–23 May 2015; IEEE: New York, NY, USA, 2015; pp. 1–6.
91. Lee, B.; Sohn, D.; Kim, K.M. *Development of Power Quality Index Using Ideal Analytic Hierarchy Process*; Springer: Singapore, 2016; pp. 783–793.
92. Langouranis, P.A.; Kaminaris, S.D.; Vokas, G.A.; Raptis, T.E.; Ioannidis, G.C.; General, A. *Fuzzy Total Power Quality Index for Electric Networks*; Institution of Engineering and Technology: London, UK, 2014; pp. 1–6.
93. Kaushal, J.; Basak, P. A Novel Approach for Determination of Power Quality Monitoring Index of an AC Microgrid Using Fuzzy Inference System. *Iran. J. Sci. Technol. Trans. Electr. Eng.* **2018**, *42*, 429–450. [CrossRef]
94. Strack, J.L.; Carugati, I.; Orallo, C.M.; Maestri, S.O.; Donato, P.G.; Funes, M.A. Three-phase voltage events classification algorithm based on an adaptive threshold. *Electr. Power Syst. Res.* **2019**, *172*, 167–176. [CrossRef]
95. Kapoor, R.; Gupta, R.; Son, L.H.; Jha, S.; Kumar, R. Detection of Power Quality Event using Histogram of Oriented Gradients and Support Vector Machine. *Meas. J. Int. Meas. Confed.* **2018**, *120*, 52–75. [CrossRef]
96. Yildirim, Ö.; Erişti, B.; Erişti, H.; Ünal, S.; Erol, Y.; Demir, Y. An online electric power quality disturbance detection system. In Proceedings of the 2016 51st International University Power Engineering Conference UPEC, Coimbra, Portugal, 6–9 September 2017. [CrossRef]
97. Ucar, F.; Alcin, O.F.; Dandil, B.; Ata, F. Machine learning based power quality event classification using wavelet-Entropy and basic statistical features. In Proceedings of the International Conference on Methods and Models in Automation and Robotics, MMAR 2016, Dzyzdroje, Poland, 29 August–1 September 2016. [CrossRef]
98. Rajeshbabu, S.; Manikandan, B.V. Detection and classification of power quality events by expert system using analytic hierarchy method. *Cogn. Syst. Res.* **2018**. [CrossRef]
99. Katic, V.A.; Stanisavljevic, A.M. Smart Detection of Voltage Dips Using Voltage Harmonics Footprint. *IEEE Trans. Ind. Appl.* **2018**. [CrossRef]
100. Balouji, E.; Salor, O. Classification of power quality events using deep learning on event images. In Proceedings of the 3rd International Conference on Pattern Recognition and Image Analysis, Shahrekord, Iran, 19–20 April 2017. [CrossRef]
101. Ucar, F.; Alcin, O.F.; Dandil, B.; Ata, F. Power quality event detection using a fast extreme learning machine. *Energies* **2018**, *11*, 145. [CrossRef]
102. Biswal, B.; Biswal, M.; Mishra, S.; Jalaja, R. Automatic classification of power quality events using balanced neural tree. *IEEE Trans. Ind. Electron.* **2014**, *61*. [CrossRef]
103. Casaleiro, Â.; Amaro e Silva, R.; Teixeira, B.; Serra, J.M. Experimental assessment and model validation of power quality parameters for vehicle-to-grid systems. *Electr. Power Syst. Res.* **2021**, *191*, 106891. [CrossRef]
104. Krishna, B.; Anusha, D.; Karthikeyan, V. Ultra-Fast DC Charger with Improved Power Quality and Optimal Algorithmic Approach to Enable V2G and G2V. *J. Circuits Syst. Comput.* **2020**, *29*, 2050197. [CrossRef]
105. Luo, H.; Nduka, O.S. Power Quality Ancillary Services Support from Customer-owned Electric Vehicles in Low Voltage Distribution Networks. In Proceedings of the 2020 12th IEEE PES Asia-Pacific Power and Energy Engineering Conference (APPEEC), Xi'an, China, 17–19 April 2020; IEEE: New York, NY, USA, 2020; pp. 1–5.

106. Baraniak, J.; Starzyński, J. Modeling the Impact of Electric Vehicle Charging Systems on Electric Power Quality. *Energies* **2020**, *13*, 3951. [CrossRef]

107. Watson, N.; Watson, R.; Paterson, T.; Russell, G.; Ellerington, M.; Langella, R. Power Quality of a bidirectional Electric Vehicle charger. In Proceedings of the 2020 19th International Conference on Harmonics and Quality of Power (ICHQP), Dubai, UAE, 22–25 March 2020; IEEE: New York, NY, USA, 2020; pp. 1–5.

108. Gupta, J.; Kushwaha, R.; Singh, B. Improved Power Quality Charger Based on Bridgeless Canonical Switching Cell Converter for a Light Electric Vehicle. In Proceedings of the 2020 IEEE 9th Power India International Conference (PIICON), Sonepat, India, 28 February–2 March 2020; IEEE: New York, NY, USA, 2020; pp. 1–6.

109. Iqbal, S.; Xin, A.; Jan, M.U.; ur Rehman, H.; Masood, A.; Salman, S.; Abbas Rizvi, S.A.; Aurangzeb, M. Role of Power Electronics in Primary Frequency Control and Power Quality in an Industrial Micro-grid Considering V2G Technology. In Proceedings of the 2019 IEEE 3rd Conference on Energy Internet and Energy System Integration (EI2), Changsha, China, 8–10 November 2019; IEEE: Kraków, Poland, 2019; pp. 1188–1193.

110. Cardoso, K.R.; Lima, R.S.; Borba, B.S.M.C.; Larrea, J.A.S.; Fortes, M.Z. Electric Vehicles Insertion in Power Grids and Impacts on Power Quality: A Conventional and Inductive Charging Comparative Study. In Proceedings of the 2019 IEEE PES Innovative Smart Grid Technologies Conference-Latin America (ISGT Latin America), Gramado City, Brazil, 15–18 September 2019; IEEE: New York, NY, USA, 2019; pp. 1–6.

111. Zhang, L.; Zhang, X.; Li, D.; Tan, H. Research on Power Quality Control Method of V2G System of Electric Vehicle Based on APF. In Proceedings of the 2019 International Conference on Advanced Mechatronic Systems (ICAMechS), Shiga, Japan, 26–29 August 2019; IEEE: New York, NY, USA, 2019; pp. 186–189.

112. Pudjianto, D.; Ramsay, C.; Strbac, G. Microgrids and virtual power plants: Concepts to support the integration of distributed energy resources. *Proc. Inst. Mech. Eng. Part A J. Power Energy* **2008**, *222*, 731–741. [CrossRef]

113. Etherden, N.; Vyatkin, V.; Bollen, M.H.J.J. Virtual Power Plant for Grid Services Using IEC 61850. *IEEE Trans. Ind. Inform.* **2016**, *12*, 437–447. [CrossRef]

114. Caldon, R.; Patria, A.; Turri, R. Optimal Control of a Distribution System with a Virtual Power Plant. In Proceedings of the Bulk Power System Dynamics and Control-VI, Cortina d'Ampezzo, Italy, 22–27 August 2004; pp. 278–284.

115. Beguin, A.; Nicolet, C.; Kawkabani, B.; Avellan, F. Virtual power plant with pumped storage power plant for renewable energy integration. In Proceedings of the 2014 International Conference Electricity Machines ICEM, Berlin, Germany, 2–5 September 2014; pp. 1736–1742. [CrossRef]

116. Zhang, J.; Xu, Z.; Xu, W.; Zhu, F.; Lyu, X.; Fu, M. Bi-Objective Dispatch of Multi-Energy Virtual Power Plant: Deep-Learning-Based Prediction and Particle Swarm Optimization. *Appl. Sci.* **2019**, *9*, 292. [CrossRef]

117. Gong, J.; Xie, D.; Jiang, C.; Zhang, Y. Multiple Objective Compromised Method for Power Management in Virtual Power Plants. *Energies* **2011**, *4*, 700–716. [CrossRef]

118. *IEC 61000 4-30 Electromagnetic Compatibility (EMC)—Part 4-30: Testing and Measurement Techniques—Power Quality Measurement Methods*; International Electrotechnical Commission: Geneva, Switzerland, 2015.

119. Jasinski, M.; Sikorski, T.; Kaczorowska, D.; Kostyla, P.; Leonowicz, Z.; Rezmer, J.; Janik, P.; Bejmert, D. Global Power Quality Index application in Virtual Power Plant. In Proceedings of the 2020 12th International Conference and Exhibition on Electrical Power Quality and Utilisation-(EPQU), Krakow, Poland, 14–15 September 2020; IEEE: New York, NY, USA, 2020; pp. 1–6.

120. Sikorski, T.; Jasiński, M.; Ropuszyńska-Surma, E.; Węglarz, M.; Kaczorowska, D.; Kostyla, P.; Leonowicz, Z.; Lis, R.; Rezmer, J.; Rojewski, W.; et al. A Case Study on Distributed Energy Resources and Energy-Storage Systems in a Virtual Power Plant Concept: Technical Aspects. *Energies* **2020**, *13*, 3086. [CrossRef]

121. *EN 50160: Voltage Characteristics of Electricity Supplied by Public Distribution Network*; British Standards: London, UK, 2010.

Publisher's Note: MDPI stays neutral with regard to jurisdictional claims in published maps and institutional affiliations.

Article

A Case Study on Battery Energy Storage System in a Virtual Power Plant: Defining Charging and Discharging Characteristics

Dominika Kaczorowska [1,*], **Jacek Rezmer** [1], **Michal Jasinski** [1,*], **Tomasz Sikorski** [1], **Vishnu Suresh** [1], **Zbigniew Leonowicz** [1], **Pawel Kostyla** [1], **Jaroslaw Szymanda** [1] and **Przemyslaw Janik** [2]

[1] Faculty of Electrical Engineering, Wroclaw University of Science and Technology, 50-370 Wroclaw, Poland; jacek.rezmer@pwr.edu.pl (J.R.); tomasz.sikorski@pwr.edu.pl (T.S.); vishnu.suresh@pwr.edu.pl (V.S.); zbigniew.leonowicz@pwr.edu.pl (Z.L.); pawel.kostyla@pwr.edu.pl (P.K.); jaroslaw.szymanda@pwr.edu.pl (J.S.)
[2] TAURON Ekoenergia Ltd., 58-500 Jelenia Gora, Poland; przemyslaw.janik@tauron-ekoenergia.pl
* Correspondence: dominika.kaczorowska@pwr.edu.pl (D.K.); jasinski@ieee.org (M.J.); Tel.: +48-713-202-901 (D.K.); +48-713-202-022 (M.J.)

Received: 17 November 2020; Accepted: 14 December 2020; Published: 17 December 2020

Abstract: A virtual power plant (VPP) can be defined as the integration of decentralized units into one centralized control system. A VPP consists of generation sources and energy storage units. In this article, based on real measurements, the charging and discharging characteristics of the battery energy storage system (BESS) were determined, which represents a key element of the experimental virtual power plant operating in the power system in Poland. The characteristics were determined using synchronous measurements of the power of charge and discharge of the storage and the state of charge (SoC). The analyzed private network also includes a hydroelectric power plant (HPP) and loads. The article also examines the impact of charging and discharging characteristics of the BESS on its operation, analyzing the behavior of the storage unit for the given operation plans. The last element of the analysis is to control the power flow in the private network. The operation of the VPP for the given scenario of power flow control was examined. The aim of the scenario is to adjust the load of the private network to the level set by the function. The tests of power flow are carried out on the day on which the maximum power demand occurred. The analysis was performed for four cases: a constant value limitation when the HPP is in operation and when it is not, and two limits set by function during normal operation of the HPP. Thus, the article deals not only with the issue of determining the actual characteristics of charging and discharging the storage unit, but also their impact on the operation of the entire VPP.

Keywords: virtual power plant (VPP); distributed energy resources (DER); battery energy storage systems (BESS); power systems; smart grids

---

1. Introduction

VPP is defined as an integration of generators and energy storage systems (ESS). The characteristic features of a VPP include [1]:

- All the components included in the VPP participate in the energy market as one entity;
- The basic power generators in the VPP are distributed energy resources (DER);
- Electricity consumers in a VPP with DER are called prosumers;
- The prosumers can act as both loads and generators.

The main advantages of a VPP are environmental protection, high efficiency, synergy, interactivity, and balance [2].

In the literature, the most frequently discussed topics related to VPP are:

- Topics associated with reducing the peak loads;
- Topics associated with power flow management;
- Topics associated with power quality (frequency, voltage);
- Economy-related topics (costs and profits);
- Energy storage related topics.

1.1. Peak Loads Reduction, Power Flow Management

Due to the constantly growing demand for electricity, the importance of the issue of reducing peak loads is increasing. On the one hand, there is the possibility of building a conventional power plant that will work to reduce peak loads, but this is an expensive solution, causing additional carbon dioxide emissions. On the other hand, there is a continuous increase of renewable energy sources in the power system. Intelligent management of such sources in cooperation with energy storage is an effective way to reduce peak loads. The solution for such an integration is a virtual power plant presented in [3]. The application of the VPP to reduce load peaks was also presented in [4]. The authors of the paper achieved a reduction of load peaks by finding an optimal operating plan for the ESS. The optimization method was heuristic and the obtained solution was then corrected using genetic algorithm.

One of the difficulties that can be encountered in the case of central management of DER facilities, which are located in different parts of the system, is the power flow management. The problem is described in detail in [5]. The authors of the study proposed an algorithm to optimize the power flow. In general, the topic of power flow optimization is the subject of many interesting studies. In [6], optimal power flow is achieved by optimizing the installed PV capacity and the ESS parameters. Optimization is based on the power flow scenario and was performed using particle swarm optimization algorithm. The results of [7] show optimal power flow for VPP with two levels of DER integration. The first level includes wind power plants and HPPs. The power generated is dispatched directly to the grid. The second level is PV and the generated power is consumed in the VPP. The proposed model is then applied to an actual VPP in Aragon (Spain).

1.2. Power Quality, Electric Vehicles and Economy

The integration of DER within a VPP not only serves to reduce load peaks or to optimize power flow. In [8], an interesting issue was raised, in which electric vehicles (EV) are connected within a VPP. Using the energy stored in EVs, the aim is to improve the network frequency response. A similar problem is presented in [9], where the authors of the paper examined the possibility of using a VPP for frequency control in the power system, in which the solar thermal system, wind turbine generator, and electric vehicles have a large share.

The use of EVs in a VPP has been very popular recently. In [10], the use of electric vehicle demand response to reduce operating costs and improve the reliability of the power system with a large share of variable renewable sources and electric vehicles is shown. The study was based on the example of Jeju Island, located in South Korea. Furthermore, reducing operating costs and maximizing profits in a VPP is also the subject of many interesting studies. For example, the authors of [11] focus on the problem of maximizing the operating profit of a virtual power plant. Part of this VPP is based on pumped storage HPP, whose work plan consists of the objective function discussed in the article. Another example is presented in a study in Western Australia, in which the authors present an analysis of a VPP profitability [12]. This VPP consists of, among others, a solar power plant, anadium redox flow battery, and a heat pump for hot water systems. The model analyzed in the article also allows to control some devices, e.g., washing machines, dishwashers, and dryers, located in apartments, which are also part of the VPP. Another example where operational costs were minimized is [13].

The optimization was performed for IEEE 30 and the IEEE 57 bus test cases. Power flow was optimized using mixed integer distributed ant colony optimization.

A VPP may contain more than one type of power plant using renewable energy sources (RES). It is possible to use one power plant to support the operation of another, as presented in [14], which details a case study of a VPP located on the island of Crete, Greece. Additional, unused energy produced by wind power plants is used to pump water in HPPs.

1.3. Storage of Electric Energy

In the era of the rapid development of renewable energy, a major technological challenge is the efficient storage of electric energy [15]. Depending on the way the energy is stored, the energy storage facilities can be classified into [16]:

- Electrochemical (batteries, flow batteries, hydrogen);
- Electromagnetic (electric, magnetic);
- Thermodynamic (pressure, heat);
- Mechanical (gravity, kinetic).

Among the energy storage devices, chemical batteries are increasingly used in the professional power industry [17]. The desired characteristics of batteries are high energy density, high charging and discharging power, and a long life cycle. Other aspects are also important, for example, methods to determine the state of charge and the possibility of recycling [18].

Currently, research on BESS facilities is focused on their optimal distribution, system location, e.g., to minimize power loss [19], or to improve frequency [20]. A large part of the research is also devoted to the use of BESS devices to improve energy quality parameters [21,22]. In [23], the authors analyze an economically interesting case of using old BESS from EVs as a stationary energy storage system. Studies on the performance of BESS are mostly focused on the diagnosis of the condition of the storage unit. In [24], the health condition of the BESS is analyzed based on parameters related to the capacity, internal resistance, and voltage in the open circuit of the battery. This system is located in a laboratory in Choco, Colombia. The study in [25] focuses on the performance of the power conversion system performance among four topologies and on estimating the SoC of a lithium-ion cell by combining two methods: Coulomb and open circuit voltage measurement. The study in [26] focuses on testing the reliability of BESS, based on Markov models. The BESS models focus on the electrochemical model, the analytical model, and the substitute circuit model [27,28].

To our best knowledge, there is a shortage of publications analyzing the actual characteristics of charging and discharging of BESS facilities, as well as models based on these characteristics, especially in the power systems industry. The publications written so far focus on the study of single battery cells, like [29], where the discharge voltages are tested depending on aging levels. The analysis is carried out for lithium nickel–cobalt–manganese-oxide battery cell. The charging and discharging voltages are also tested in [30]. The study presents charging and discharging voltages for different charging rates. This test was also carried out for lithium nickel-cobalt-manganese-oxide battery. However, the tests are also not carried out on BESS but on a single battery. Research close to applications in the power system industry was conducted in [31]. Although the study was not performed for BESS located in the power grid, it was performed for EV, which are operating in the distribution grid.

The main contributions of the article include:

1. Analyzing the operation of the actual BESS. The BESS is located in a VPP in Poland;
2. Determining the actual charging and discharging characteristics of the BESS;
3. Using the determined characteristics to create the BESS model;
4. Investigating the possibility of controlling the BESS.

The rest of the article has the following structure. Section 2 presents and discusses the most frequently used technologies and parameters of BESSs used in VPPs. Attention was paid to the issues of appropriate control of energy storage units and the importance of the charging and discharging characteristics of such facilities. Section 3 presents the results. The structure of the examined VPP. The methodology of determining the actual charging and discharging characteristics of the BESS located in the analyzed VPP. The actual charging and discharging characteristics of the BESS were determined. It was also checked how the determined characteristics performed when the storage facility followed a given operational plan. The analysis was completed by examining the operation of the VPP for a given scenario of power flow control on the day with the highest demand. The presented scenario realizes the adjustment of the VPP operation to a given curve. The study was carried out for four cases: adjustment of the power consumed from the distribution system to a given level for normal power plant operation; adjustment of the same to a given level in case of lack of power plant operation; adjustment of the power consumed from the grid to two curves set as a function for normal power plant operation. Section 4 contains a discussion of the results obtained. Section 5 finishes with the conclusions.

2. Battery Energy Storage System

Lithium-ion batteries are most commonly used in BESSs [18,32]. This type of battery is recommended for long-term use due to its high energy density of 100–250 Wh/kg, high power density of 230–340 W/kg, and long lifetime of 500–6000 cycles [33]. The disadvantages of a lithium-ion battery include its high cost and the need to use a ventilation system to ensure the cooling of the batteries. The battery requires a built-in controller to ensure that each cell operates correctly within the specified voltage, temperature, and current range during charging and discharging. To optimize the operation of the battery, the cell controller communicates with the battery management system (BMS) [33,34].

The battery management system provides many functions. In power system applications, this is very important because the battery usually operates under difficult conditions. The BMS must react to external conditions that can cause damage. An important function of the BMS is to determine the SoC of the battery. The BMS monitors and calculates the SoC of each battery cell. SoC is also used to determine the end of charging and discharging cycles [35]. Excessive charging and discharging is often the cause of a battery failure; therefore, the BMS is designed to keep the battery in its operating range to deliver the required power, and to receive energy without overcharging the cells [33,34].

Especially the charging and discharging speed of the BESS is influenced by:

- Construction;
- State of charge (SoC);
- Temperature.

These factors are described in detail in [34]. For the selected BESS type, it is possible to determine the dependence of charging (discharging) power based on the SoC [36]. In this case, the charging (discharging) speed is determined by power expressed in [W] or by a relative value in relation to the rated power of the storage unit. The SoC can be defined as energy in [Wh] or as a relative value in relation to the storage capacity.

It should be noted that these characteristics have a double meaning. Firstly, they determine the operating limitations introduced by the manufacturer, depending on the design, and secondly, they can be a tool to control the operation of the storage unit by the system operator [32]. In the latter case, the characteristic parameters must not exceed the limits specified by the manufacturer.

Characteristics of the dependence of the charging and discharging power of the storage facility on the state of charge are the basic dependencies required to properly control the BESS [37]. An example of the actual characteristics of a BESS equipped with lithium-ion batteries and supervised by the BMS is shown in Figure 1. The shape of the characteristics depends on the technical limitations of the batteries, the manufacturing technology, operating temperature, degree of usage, and the way the BESS is controlled and managed by the BMS [38,39].

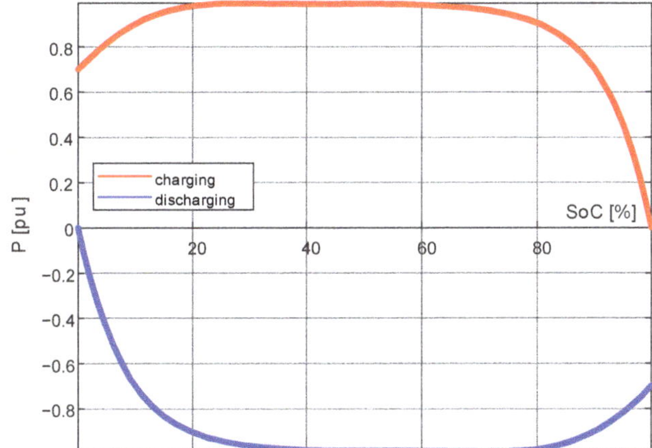

Figure 1. Dependence of the storage unit power on the state of charge (SoC).

3. Results

The chapter is divided into six subchapters. The first part is a description of the VPP in which the analyzed energy storage is located. Then, the methodology for determining the charging and discharging characteristics of the energy storage is discussed. The next part is the determination of the characteristics based on actual measurements. Then, the analysis of the influence of the characteristics of the BESS on VPP operation. The chapter ends with the analysis of the operation of the private network, in which the VPP operates for a given power flow control scenario for four cases. The analysis is performed for the day of the year with the highest peak demand.

3.1. Virtual Power Plant

In a modern power system, virtual power plants are solutions equipped with efficient and safe power flow control systems. VPPs are equipped with power generators and storage facilities. A VPP in connection with local electricity consumers creates a private network [40]. These devices are equipped with controllers, usually, they are power-electronic converters allowing power control. Most often such networks are connected to a public distribution network. An appropriate strategy of energy storage in the BESS allows for optimal use of the electrical energy generated locally [41].

The virtual power plant, designed to simulate the power flow, was designed on the basis of a simplified scheme of the power system in which the VPP operates in Poland. The private network consists of the hydroelectric power plant, load (including the power plant's own needs), and the BESS. Figure 2 presents a diagram of the dispersed control network connected to the distribution grid.

The regulations governing the operation of the electricity market must take into account the characteristics of the operation of the power system, such as the need to ensure a continuous balance of supply and demand and reliable system operation [42,43]. On the other hand, until recently, it was assumed that there was no possibility of storing energy immediately, which resulted in low energy price flexibility and limited generation flexibility. The development of DER, microgrid systems in which large BESS and VPP structures operate is currently of decisive importance in the transformations taking place in the modern power system.

Therefore, experimental research of the VPP for different power flow control approaches allows to verify the assumptions, scenarios, and algorithms.

The BESS, which is a part of the experimental VPP, is of key importance for the effective operation of the system in this case. Therefore, it seems necessary to point out the potential possibilities of using the storage in experimental research.

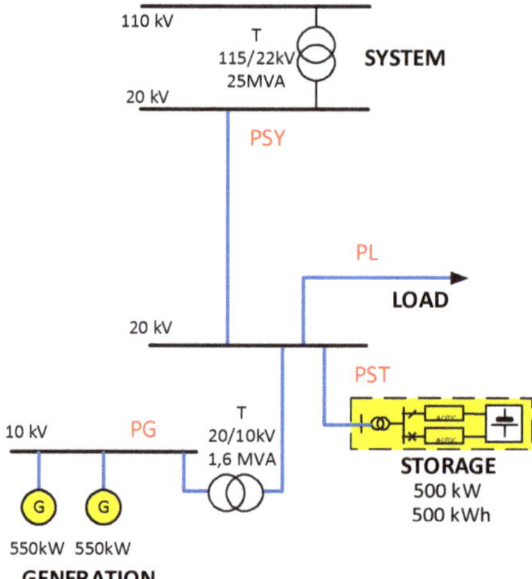

Figure 2. Diagram of the analyzed active grid. PG—power generation, PL—load, PSY—power distribution system, PST—BESS.

3.2. Experimental Determination of BESS Characteristics

One of the practical ways of utilizing BESS in the private network and as an important resource of a VPP may be to experimentally determine the characteristics of storage operation. Knowledge of the basic technical limitations of the installation related to the power of charging and discharging the storage, which is also associated with the speed of response to changes in covering the power demand and response to changes in local generation, is the basis for proper and effective control within the VPP. The adopted and proven method of measuring the characteristics of the energy storage operation will be able to be used in other installations of this type. The obtained results will allow for correct and optimal use of the VPP resources.

An effective method of determining the BESS characteristics may be the use of measurements during the storage facility operation. The prerequisite is access to the synchronized recording of the storage unit's charging and discharging power, the SoC, and, if possible, the operating temperature of the batteries. As shown in [31], with a large number of measurements, it is possible to try to determine the actual operating characteristics of BESS. The boundary areas of such determined characteristics at many points of the storage facility operation can then be approximated by the polynomial function.

The proposed way of determining points of the storage characteristics on the basis of current registrations is schematically presented in Figure 3. With the synchronous measurement of the storage power P, (P+ for charging, P− for discharging) and SoC, it is possible to determine points of BESS characteristics for selected points of time.

Based on the technical documentation of the BESS control and parameter monitoring systems, assumptions have been made about performing measurements of the examined storage.

- The tested energy storage is treated as a uniform unit. This means that the internal design, the number, and layout of the battery unit connections, as well as the supervisory and security systems and algorithms are omitted. All technological limitations and those resulting from BMS procedures are taken into account by the characteristics of the storage unit operation.
- The maximum power of charging and discharging BESS is 500 kW.
- The storage can be charged to a maximum of 100% SoC (500 kWh) and discharged to a minimum of 20% SoC (100 kWh). The limitation of the discharge results from the warranty provisions.
- Charging the storage unit with full power is only possible during the operation of the HPP, so the test time with charging sequences can be a maximum of 3 h.

- Due to the parameters of the control systems implemented in the VPP, the changes in the storage unit's operating schedule (charging and discharging power changes) can be performed with a minimum step of 5 min.
- Measurement of the electric parameters of the storage unit, such as: charging and discharging power, SoC, voltage, and current can be performed with a step of minimum 1 min (data recorded in the BMS system).
- The average temperature of the batteries and the ambient temperature is recorded in a 15 min step. Initially, it is assumed that the storage unit, thanks to the temperature monitoring of the batteries by the BMS, operates within the permissible range, thus limiting the influence of the temperature on the charging and discharging characteristics.

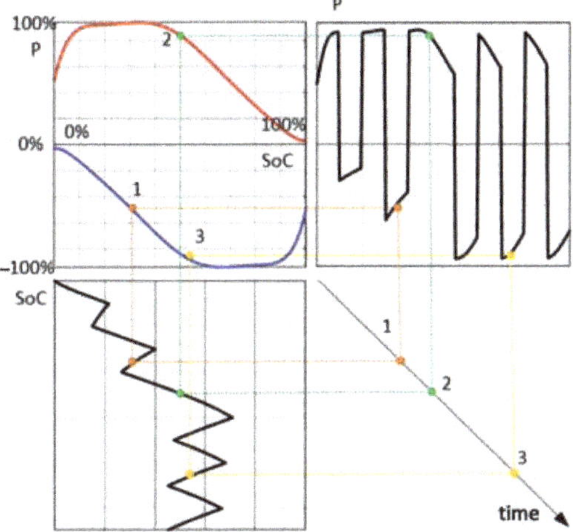

Figure 3. The scheme of determining points of BESS characteristics on the basis of current monitoring of storage operation. P—the power of charging and discharging the storage, SoC—state of charge.

3.3. Results of Short Term Tests for the Identification of the Actual Charging and Discharging Characteristics of BESS

The characteristics were determined using samples measured at 1 min intervals, obtaining a set of points describing the characteristics. The measurements are shown in Figure 4.

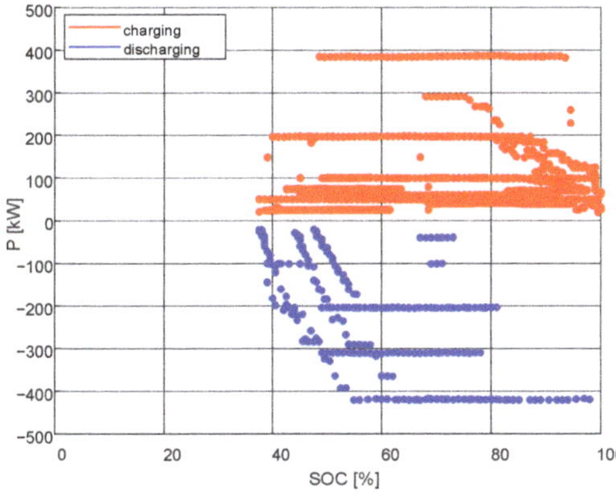

Figure 4. A set of points that forms the charging and discharging characteristics of the BESS for 1 min samples.

Then the obtained results, for the extreme values, were approximated with the polynomial function, obtaining the characteristics presented in Figure 5.

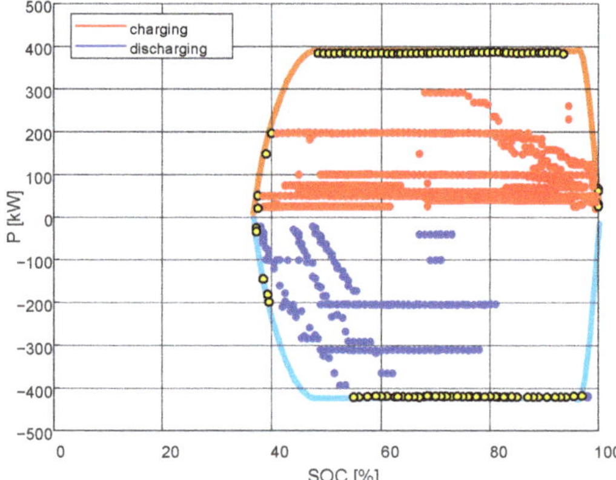

Figure 5. The result of the approximation of the results with the 1 min data polynomial function.

The final characteristics determined on the basis of 1 min data are as in Figure 6.

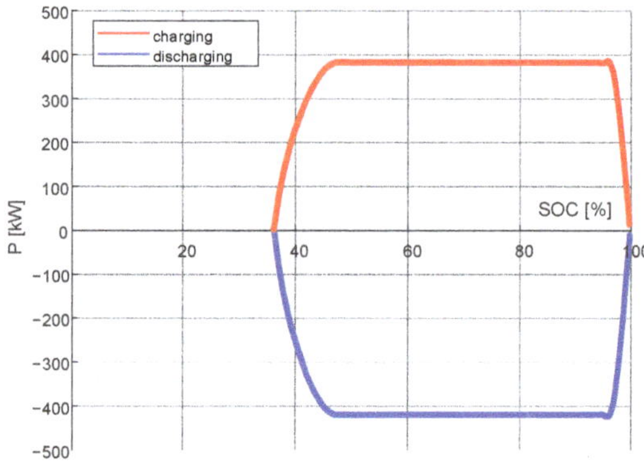

Figure 6. Determined charging and discharging characteristics of the BESS based on 1 min data.

The obtained shape of the characteristics approximately corresponds to the shape of typical characteristics of a storage unit based on chemical batteries, as shown in Figure 1. The difference is the used range of the storage unit's state of charge, which indicates that the unit has not been discharged below 30% SoC. Additionally, the charge and discharge power of the accumulator does not exceed 80% of the rated power.

Using the 1 min data, a map of "attendance" of the storage unit's operating states, shown in Figure 7, was also drawn up. The map allows to check what is the most frequently used power range and what is the SoC at that time.

The map closely correlates with the charging and discharging characteristics of the storage unit. The range of work of the storage depicted on the map is within the range of work defined by the characteristics. In addition, when analyzing the map, it can be seen that the BESS usually works with SoC in the range 90–100%. However, it rarely works at a low SoC level, within 40%.

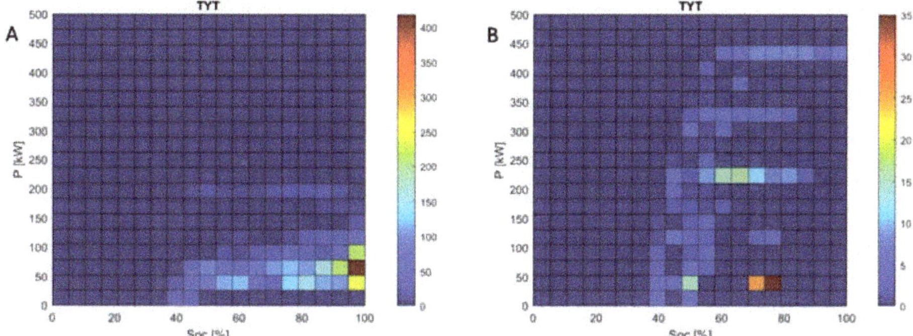

Figure 7. Map of "attendance" of the storage unit operation states during charging (**A**) and discharging (**B**) for 1 min data; the color scale represents a value proportional to the operation time in the P and SoC range.

Using 1 min data, the dependence of the storage power on the operation plan was also checked (Figure 8). The points where the BESS worked according to the plan are on a straight line, inclined at 45°. Deviations from this line show the difference between the setpoint and actual storage power. Measurements also show that the system only applied the storage power at levels 50 kW apart.

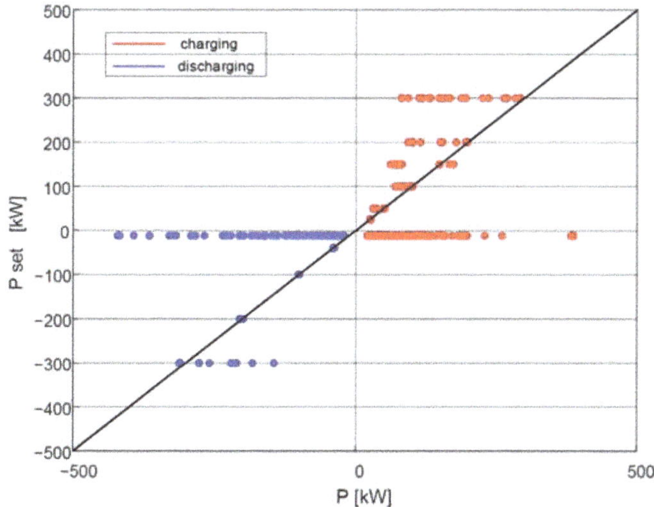

Figure 8. Dependence of storage power on the operation plan.

3.4. Control of the Charging and Discharging Power of the BESS for the Assumed Operation Plan

This chapter presents the results for two operation plans. The first operation plan is shown in Figure 9 and the simulation results are shown in Figure 10. The second operation plan is shown in Figure 11, while the results are presented in Figure 12. The capacity of the 500 kWh storage device and its rated power of 500 kW corresponds to the parameters of the BESS storage device under test.

The conducted simulations allow to determine the changes in the power of the storage unit and the SoC over time for the determined real characteristics.

The analysis was performed for two storage operation plans, presented in Figures 9 and 11.

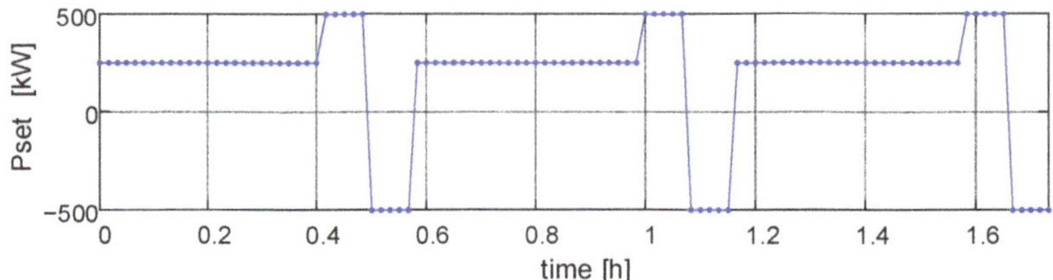

Figure 9. Set operation plan of the BESS.

For the presented plan (Figure 9), the operation of the storage device, resulting from its charging and discharging characteristics, is presented in Figure 10.

It can be seen that, initially, the storage facility works according to the plan. The difference occurs after approx. 1.3 h, despite the assumed constant performance, the storage unit gradually reduces the charging power to 0 kW. Then, after about 1.7 h, the storage facility does not increase the discharge power to 500 kW, but to about 250 kW.

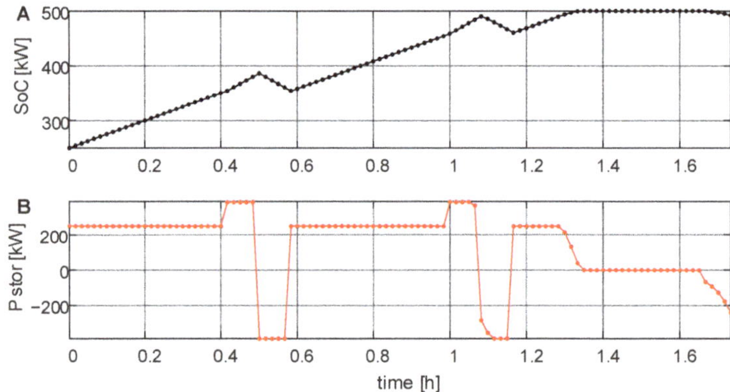

Figure 10. Operation resulting from the charging and discharging characteristics of the BESS with the real characteristics; change of energy accumulated in the BESS over time (**A**); change of BESS charging and discharging power over time (**B**).

A similar analysis was performed for the work plan presented in Figure 11.

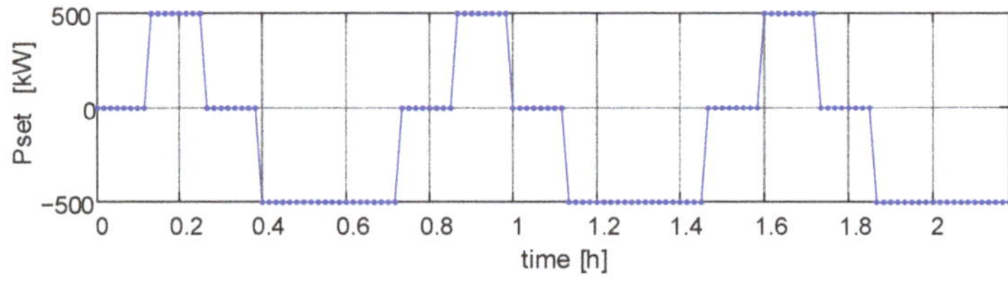

Figure 11. Set operation plan of the BESS.

For the presented plan (Figure 11), the operation of the storage device, resulting from its charging and discharging characteristics (Figure 12), is presented.

As before, the storage unit initially works according to the operation plan, the difference occurs after approx. 0.6 h. The storage starts to reduce the discharge power faster than the plan would have resulted. In about 0.9 h, the facility reaches a lower maximum charging power than planned. After approx. 1.3 h, the unit's output is around 0 kW.

The difference between the operation plan and the actual operation of the storage device is most visible in the case of a low SoC (about 200 kWh).

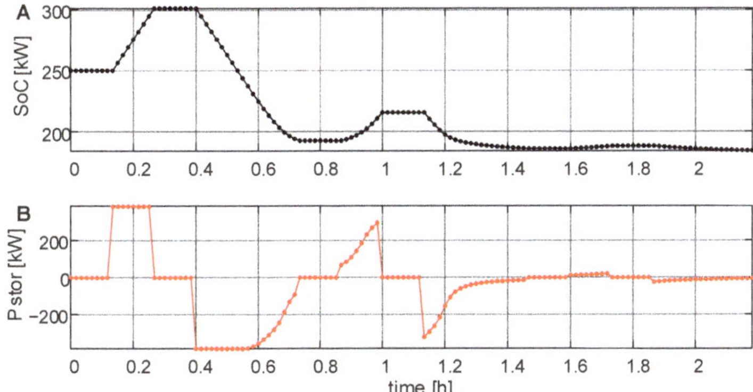

Figure 12. Operation resulting from the charging and discharging characteristics of the BESS with the real characteristics; change of energy accumulated in the BESS over time (**A**); change of BESS charging and discharging power over time (**B**).

3.5. Controlling the BESS in the Private Network

The main objectives of the VPP systems concern the economic aspects related to electricity trading. The VPP control algorithms provide for charging the storage facility at low energy prices from the distribution system and discharging and selling energy at peak demand at high prices.

However, technical issues cannot be overlooked when planning such solutions. For example, the voltage levels (slow and fast voltage changes) at all points of the distribution network should be within the range allowed by the standards. The same applies to current values in lines and transformer windings. Issues concerning the operation of the BESS itself are also important.

The impact of the VPP on the power system can be examined by analyzing the power flows. The daily analysis should give an answer to the question whether the changes of voltages and currents in the network caused by the generation and storage control are within acceptable limits.

Simulations of the operation of the 0.5 MWh BESS with the permissible voltage changes at the PCC site were carried out. The actual measurements of power demand by local loads were used, as well as measurements of power generation in the hydroelectric power plant. The diagram of the analyzed private network is presented in Figure 2. For the analysis, the characteristic day with the maximum power demand in the 110 kV/SN station was selected. The power flow simulation was conducted using the DC power flow method, which only takes into account only active power. This approach is often used in developmental analyses, with high data uncertainty [44–47].

The storage control scenario (Figure 13) is based on the technical aspect of reducing the power taken from the distribution system to a given level. The demand of local electricity consumers for power exceeding the preset margin should be covered from the local generation and from the BESS. This is one of the possible scenarios for controlling the storage unit. The comprehensive scenario should make optimal use of the energy of the BESS both for technical and economic reasons related to the costs of operating the equipment and current prices on the energy market.

The algorithm works with a given step, using measurement or forecast of power generation, load, and current SoC. The simulations assume average values on a ten minutes step basis. The basis for operation of the algorithm is to check in the first step the sum sign of power generation and load. As a result, it is possible to determine whether the generator itself can cover the demand locally. In the next steps, the algorithm chooses such power of discharging or charging the energy storage so that the energy exchange between the private network and the system remains at the set level.

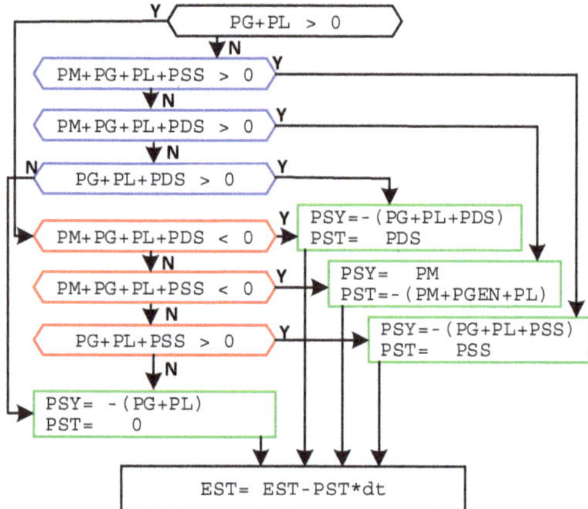

Figure 13. Storage control algorithm, PG—power generation (positive), PL—load (negative), PSS—storage charging power (negative), PDS—storage discharging power (positive), EST—storage energy, PM—set level of power exchange between the private network and the distribution system.

The values of power and energy required to control the BESS are forecast (PG—power generation, PL—load) and measured (EST—storage energy, i.e., SoC). The storage charging or discharging power (PST) is determined as the amount set to the energy storage controller. The algorithm takes into account the limitations resulting from the characteristics of the storage facility operation (PSS—permissible power to charge the storage facility, PDS—permissible power to discharge the storage facility). The characteristics introduce limits to the storage power depending on the SoC. The discharge power decreases as the energy in the storage unit decreases to a zero value with a set minimum stored energy. Changing the shape of the characteristics enables regulation of the BESS power depending on the parameters of the active network. System power (PSY) is the excess or deficit of power while performing power balancing in the private network. As an objective function for the scenario, the power exchange between the private network and the distribution grid (PM) is set. System power (PSY) tries to match this level in subsequent steps.

It is not necessary to directly indicate the intervals for the variables used by the algorithm, because the limitations result directly from the adopted characteristics of the storage unit operation (e.g., at maximum SoC, the charging power reaches zero). Therefore, it is important to correctly define the energy storage performance.

For the analysis of the impact of the VPP on the conditions of power grid operation, the day on which the maximum power demand in the 110 kV/SN station occurs was selected, that is 19 January 2017.

3.6. Analysis of Power Flow for the Day of Maximum Demand in 110 kV/SN Station (19 January 2017)

The presented analysis covers 19 January 2017—the hydroelectric power plant operates with a power of 1 MW between 14:00 and 19:00 and the load peak is between 06:00 and 11:00. In the private network, there is a BESS with a nominal power of 500 kW and a capacity of 500 kWh. The profile of generation and load on that day is shown in Figure 14. Four scenarios of the private network operation have been considered:

- Power exchange margin with the external grid at the level of 400 kW, during normal operation of the HPP (Figures 15 and 16);
- Power exchange margin with the external grid at 400 kW and no HPP operation (Figures 17 and 18);
- Power exchange margin with the external grid set according to the curve, variable from 300 kW to 600 kW, in normal operation of the HPP (Figures 19 and 20);

- Power exchange margin with the external grid set according to the step curve, between −400 kW and 400 kW, in normal operation of the HPP (Figures 21 and 22).

The operation of the private network with a storage facility with the actual charging and discharging characteristics was examined.

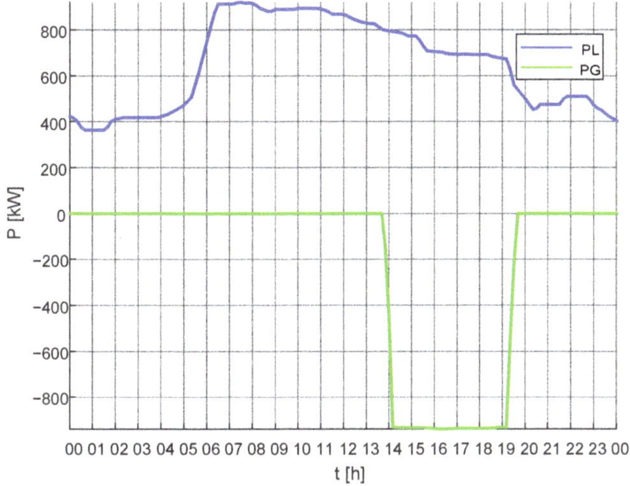

Figure 14. Power generation—PG and load PL in the private network for 19 January 2017.

In the first step, the power flow in the private network for power generation and load was checked (Figure 14), using a scenario (Figure 13) that reduces the exchange power between the private and the distribution grid to a set level of 400 kW for a grid with a storage unit with real characteristics (Figures 15 and 16).

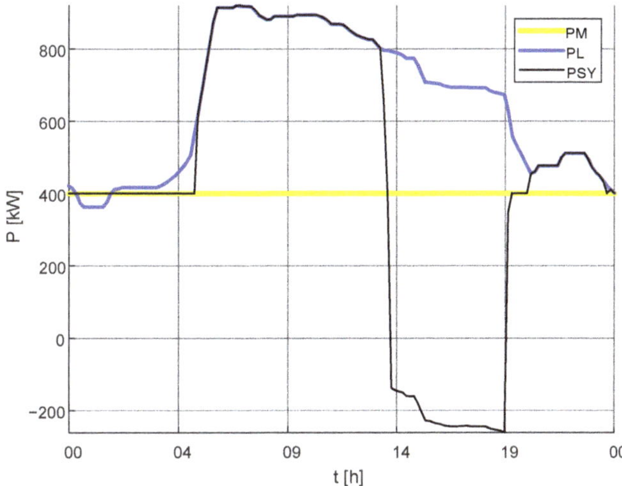

Figure 15. Results of the simulation of the operation of the private network with the BESS with real characteristics for the set exchange power PM 400 kW; PSY—exchange power with the distribution grid and PL—load.

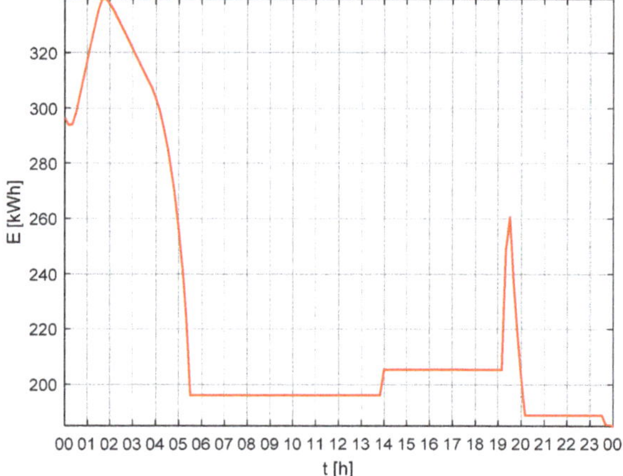

Figure 16. Results of the simulation of the operation of the private network with the BESS with real characteristics for a set exchange power of PM 400 kW; change of energy accumulated in the BESS over time.

Analyzing Figure 15, it can be seen that the power taken from the distribution grid is at a set margin from 00:00 to about 05:00 and then for an hour, about 19:00 to 20:00. From 05:00 to about 13:00 and from 20:00 to 00:00, the power demand is covered from the distribution grid. Between 13:00 and 19:00, there is a decrease in the power taken from the network due to the operation of the HPP. The operation of the private network is reflected in the diagram of the change of energy accumulated in the storage device in time (Figure 16), it can be seen that the storage device operates mainly between 00:00 and 05:00 and between 19:00 and 20:00, i.e., when the network takes the power from the distribution system by a given margin.

In the next step, it was checked how the grid will be handled in the same scenario if the HPP is not in operation. The results are shown in Figures 17 and 18.

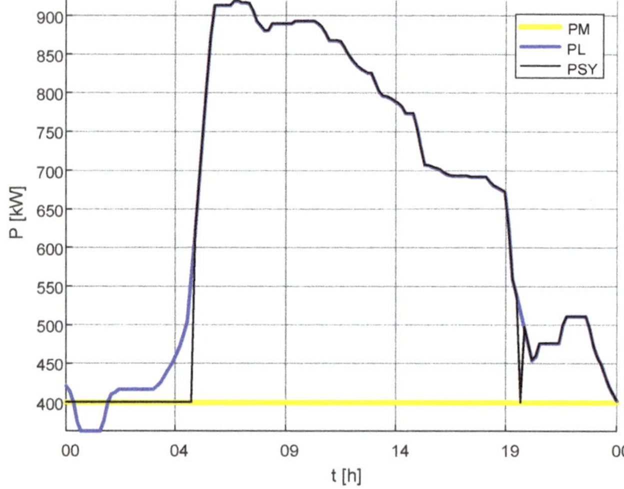

Figure 17. Results of the simulation of the operation of the private network with the BESS with real characteristics for the set PM 400 kW exchange power in the absence of generation; PSY—exchange power with the distribution grid and PL—load.

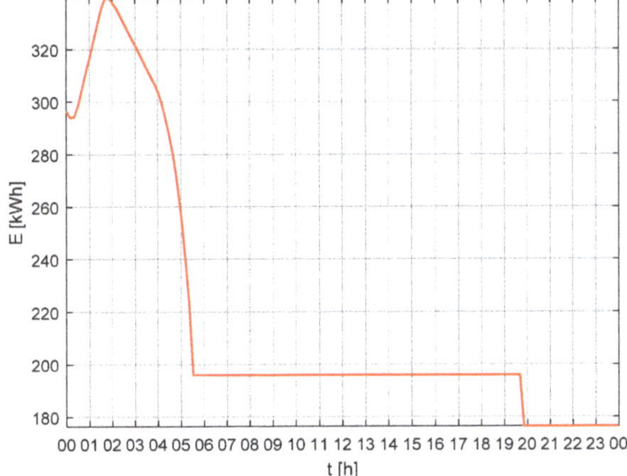

Figure 18. Results of the simulation of the operation of the private network with the BESS with real characteristics for the set PM 400 kW exchange power in the absence of generation; change of energy accumulated in the BESS over time.

As in the previous analysis, from 00:00 to 05:00, the private network works with a margin setpoint power (Figure 17). For the rest of the simulation time, the power of the demand is covered by the distribution system, except for a few minutes around 20:00. The profile of the VPP operation is also visible on the graph of energy stored in the storage unit in time (Figure 18). The storage device operates between 00:00 and 05:00. Between 05:00 and 20:00, the SoC does not change. Around 20:00, the storage unit discharges to 180 kWh.

In the next step, the power flow for the private network was checked if the power margin is set as a function between 300–600 kW (Figures 19 and 20).

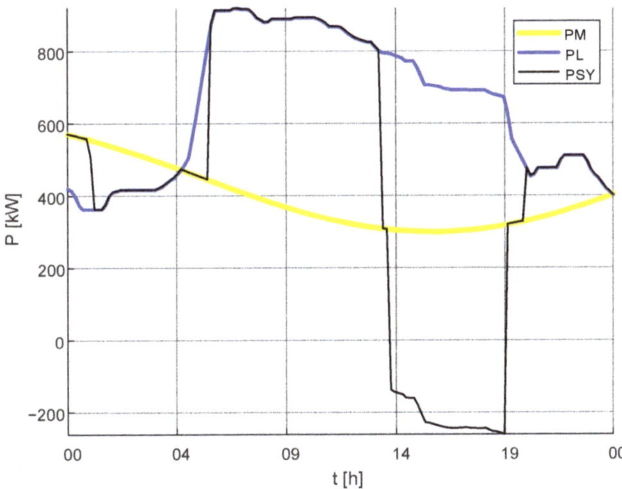

Figure 19. Results of the simulation of the operation of the private network with the BESS with real characteristics for the set exchange power according to the curve, variable in the range PM 300–600 kW; PSY—exchange power with the distribution grid and PL—load.

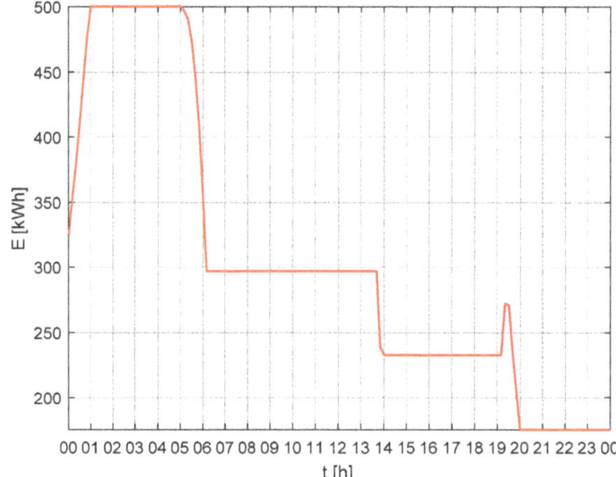

Figure 20. Results of the simulation of the operation of the private network with the BESS with real characteristics for the set exchange power according to the curve, variable in the range PM 300–600 kW; change of energy accumulated in the BESS over time.

In the case under consideration, the private network works with a power value of the setpoint by a margin between 00:00 and 01:00, between 04:00 and 05:00, between 13:00 and 14:00 and between 19:00 and 20:00 (Figure 19). The nature of the operation of the private network can be seen on the diagram of the change of energy accumulated in the storage unit in time (Figure 20). In the analyzed case, while the network is working on the given margin, the storage unit is charged or discharged. The exception is the hours from 13:00 to 19:00, when the profile of the private network operation is dictated by the operation of the HPP.

The last scenario considered was to check the power flow for the private network if the power margin is set as a step function between −400 and 400 kW (Figures 21 and 22).

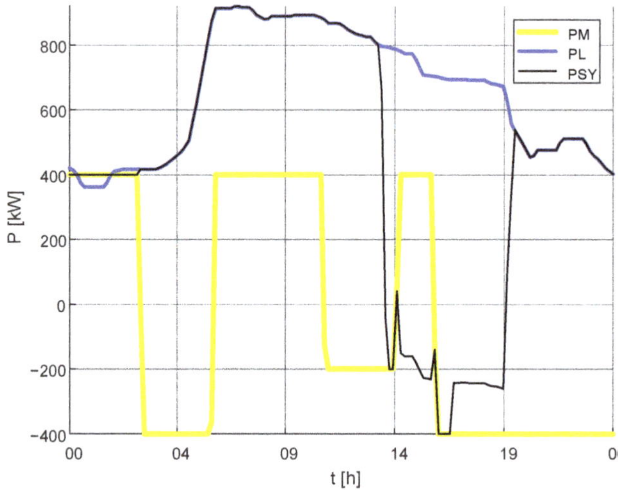

Figure 21. Results of the simulation of the operation of the private network with the BESS with real characteristics for the power margin PM set as a step function between −400 and 400 kW; PSY—exchange power with the distribution grid and PL—load.

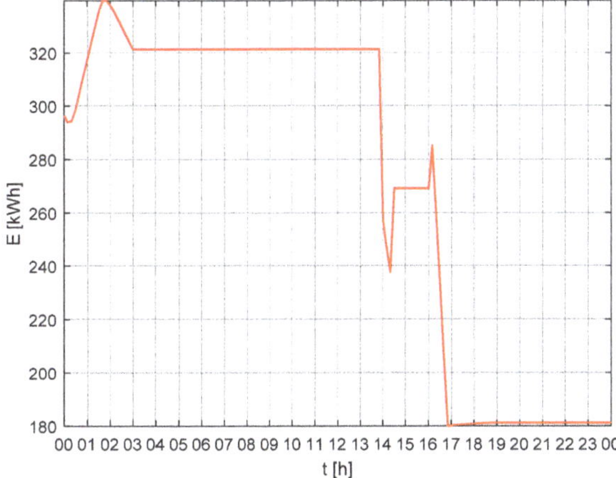

Figure 22. Results of the simulation of the operation of the private network with the BESS with real characteristics for the power margin PM set as a step function between −400 and 400 kW; change of energy accumulated in the BESS over time.

If the margin is set as a step function, the network runs at the set level from 00:00 to about 03:00, then up to about 13:00 the demand is covered by the distribution system (Figure 21). Between 13:00 and 20:00, the power taken from the distribution network is reduced due to the operation of the HPP. From 20:00 to 0:00, the demand is again covered by the distribution system. In the diagram of the change of energy accumulated in the BESS over time (Figure 22), it is possible to see the charging and discharging of the BESS between hours 00:00 and 03:00—the operation of the network at the set level and the discharging and charging of the BESS between 14:00 and 17:00—the influence of the operation of the HPP.

Table 1 shows a summary of the operation of the private network, for different scenarios with a BESS with real characteristics (Figures 15–22). The table shows the time for which the private network works exactly on a given margin. The data ARE shown in percentage of when the power exchange with the grid maintains the set margin to the total simulation time.

Table 1. The time for which the private network works on a given margin.

	PM = Const. PM = 400 kW	PM = Const. PM = 400 kW PG = 0 kW	PM ≠ Const. PM ∈ [300, 600] kW	PM ≠ Const. PM ∈ [−400, 400] kW
operating time at a given level	28.97%	24.14%	13.79%	18.62%

For the longest percentage of the time, the private network operates on a preset margin, when it is set to a straight line with a constant value of 400 kW (28.97% of simulation time). The shortest one, when it is set with the function between 300–600 kW (13.79% of the simulation time). For the same margin, 400 kW, the difference between a scenario where the HPP operated from 14:00 to 19:00 and a scenario where the HPP did not operate is about 5% of simulation time. For the margin set by the step curve, it was maintained for 18.62% of the simulation time.

The advantage of the applied scenario may be the value of energy transmitted to the distribution system, which in this case is much smaller. Taking into account the significant length of the line connecting the storage and the HPP with the distribution system, lower power transferred to the distribution system guarantees lower power losses.

4. Discussion

In the fast-growing field of electricity storage, tests are necessary to ensure their efficient operation and minimize the risk of failure. The development of new standards and guidelines for the energy storage devices to be delivered to the customer is a continuous process, and documents are constantly growing. As standard, energy storage facilities are tested according to [48]:

- IEC 61427, IEC 62133, IEC 62619;
- UL 1642, UL 1973, UL 9540A;
- IEEE 1547, IEEE P2030.3.

In particular, IEC 61427 [49] describes the exact procedure of the test, introducing the "standard cyclic test" and defines on this basis the performance characteristics of the storage. The IEC 61427 standard [49] defines a special method for identifying the parameters of a battery that works with a RES. The standard test protocol assumes an elevated temperature (40 °C) and a series of cycles that mimic the photovoltaic system. The energy storage test procedure consists of shallow cycles at low SoC and shallow cycles at high SoC. The test carries out 50 cycles with 30% depth of discharge (between 5% and 35% SoC), followed by 100 cycles between 75% and 100% SoC. After 150 cycles, the parameters are measured. The test lasts about 50 days and the total BESS test period requires 3 to 10 tests, which take 5 to 16 months to complete, despite increasing the operating temperature of the batteries to 40 °C, in order to speed up degradation and thus reduce the test time. The article proposes an alternative way to identify the BESS characteristics by using synchronous recording of power, BESS charge level, and temperature of the batteries during normal use of the storage. This is of particular importance at a time when there are limitations in performing tests on industrial facilities.

The article presents an algorithm of control in the active network using active power balances to determine the value of the power of the energy storage at a given value of power exchanged with the distribution network. The algorithm assumes that the exchange power should be constant and its value should depend on the average power of local consumers and generation equipment. Such an assumption, for technical reasons, has a number of advantages related to generation efficiency, power supply reliability, and energy quality. In the case of weather-dependent uncontrollable sources (photovoltaic sources and wind turbines) or with limited controllability (water turbines), it may be difficult to maintain exchange capacity between the active and the distribution grid. Therefore, it is possible to set the exchange capacity value at a level determined by a function. Such a function may depend on the seasons or weather forecasts. Another objective function of the VPP control scenario may be economic aspects. Then, the level of exchange power will be variable and will depend, for example, on exchange prices or balancing market prices resulting from the economic forecasting model.

Charging and discharging characteristics were introduced into the BESS control algorithm. The characteristics introduce limits to the power of the BESS depending on the SoC. The shape of the characteristics results from the technology of storage of energy in the device and system limitations introduced by the manufacturer. Typical charging characteristics of a chemical storage facility based on lithium-ion technology are presented in Figure 1.

The course of charging (discharging) characteristics is conditioned by physical and chemical changes inside the battery cells. Such charging characteristics are designed to ensure the safety of the battery and maintain maximum capacity of the cell even after many cycles of operation.

Storage facilities that use other energy storage technology, e.g., supercapacitors, fuel cells, kinetic storage units, pumped-storage power plants are characterized by different power dependence on the degree of charge.

In this work, the simulation of power flow control in the active network for the selected scenario was carried out using the direct-current distribution method. Such an approach is often used in the first step for design analyses. The obtained results allow to generally determine the way of operation of the separate system for the proposed control algorithm. Target analyses should also include reactive power distribution. Therefore, it is justified to test the AC model. The possibility of generating reactive power in HPP systems and the BESS allows to regulate voltage levels in network nodes. Additionally, the reactive power of local consumers should be compensated in order to obtain the required power factor at the point of connection of the active network to the parent network.

In the private (local) network, the power flow control scenarios should provide for increased generation and discharge of the energy storage in the morning peaks, while in relation to global demand, generation and discharge of the storage should be increased in the evening peaks.

The task of the VPP operator is therefore to select and optimize the method. In the VPP control scenarios, the best economic result is always preferred; however, this is not always related only to energy market activities. Conducted experiments in the actual network may answer the questions to what extent one should engage in the energy market and ancillary services.

5. Conclusions

In the article, the method of experimental determination of charging and discharging characteristics of the BESS located in a real VPP in Poland is presented. However, the presented methodology can also be applied to other BESS, including ESS based on other technologies. Based on the determined charging and discharging characteristics, the BESS model was established. Such a methodology for determining the BESS model can also be applied to the BESS located in the microgrids or in the distribution system in general.

The article checks how, for the given scenario of power flow control, the private network will be preserved when a storage device with real characteristics is installed in it. The analysis was performed for the day when the maximum demand occurred (19th January).

The usefulness of the scenario on the technical market can be twofold. The first concerns the adoption of an appropriate control strategy and selection of devices already at the stage of designing the private network (microgrid) and the VPP, the second is related to the application of the VPP concept to the existing infrastructure. In the first case, greater savings are possible at the investment stage, in the second case, based on intelligent control algorithms, power flow (current) can be controlled and reduced, which in consequence may have a positive impact on network equipment, such as electrical equipment operating in the distribution network which when subject to current exposure, results in aging, wear and tear and damage. The overloading of current circuits and equipment contacts leads to excessive heating, which accelerates corrosion processes, reduces their mechanical strength, and weakens the electrical strength of insulation. The processes of equipment insulation degradation due to the flow of large currents are associated with changes in the physical properties of the insulation material and chemical changes in the long term.

The considered private network consists of VPP, BESS and a generator. Using the correct power flow scenario, it is possible to ensure optimal operation of a private network connected to the distribution grid.

Author Contributions: Conceptualization, D.K. and J.R.; methodology, D.K. and J.R.; software, D.K. and J.R.; validation, M.J., T.S. and V.S.; formal analysis, D.K., J.R., M.J. and T.S.; investigation, D.K. and J.R.; resources, P.K. and P.J.; data curation, J.S. and V.S.; writing—original draft preparation, D.K. and J.R.; writing—review and editing, M.J. and T.S.; visualization, M.J. and T.S.; supervision, Z.L., T.S. and J.R.; project administration, T.S. and P.J.; funding acquisition, T.S. and P.J. All authors have read and agreed to the published version of the manuscript.

Funding: This research was funded by the National Center of Research and Development in Poland, the project "Developing a platform for aggregating generation and regulatory potential of dispersed renewable energy sources, power retention devices and selected categories of controllable load" supported by European Union Operational Programme Smart Growth 2014–2020, Priority Axis I: Supporting R&D carried out by enterprises, Measure 1.2: Sectoral R&D Programmes, POIR.01.02.00-00-0221/16, performed by TAURON Ekoenergia Ltd.

Conflicts of Interest: The authors declare no conflict of interest.

References

1. Liu, C.; Yang, R.J.; Yu, X.; Sun, C.; Wong, P.S.P.; Zhao, H. Virtual power plants for a sustainable urban future. *Sustain. Cities Soc.* **2020**, 102640. [CrossRef]
2. Yu, S.; Fang, F.; Liu, Y.; Liu, J. Uncertainties of virtual power plant: Problems and countermeasures. *Appl. Energy* **2019**, *239*, 454–470. [CrossRef]
3. Luo, Z.; Hong, S.; Ding, Y. A data mining-driven incentive-based demand response scheme for a virtual power plant. *Appl. Energy* **2019**, *239*, 549–559. [CrossRef]
4. Agamah, S.U.; Ekonomou, L. Energy storage system scheduling for peak demand reduction using evolutionary combinatorial optimisation. *Sustain. Energy Technol. Assess.* **2017**, *23*, 73–82. [CrossRef]
5. Yang, H.; Yi, D.; Zhao, J.; Dong, Z. Distributed Optimal Dispatch of Virtual Power Plant via Limited Communication. *IEEE Trans. Power Syst.* **2013**, *28*, 3511–3512. [CrossRef]
6. Gubanski, A.; Kaczorowska, D. Power flow optimization between microgrid and distribution system. In Proceedings of the 2018 Innovative Materials and Technologies in Electrical Engineering (i-MITEL), Sulecin, Poland, 6–8 May 2018; pp. 1–4.
7. Naval, N.; Sánchez, R.; Yusta, J.M. A virtual power plant optimal dispatch model with large and small-scale distributed renewable generation. *Renew. Energy* **2020**, *151*, 57–69. [CrossRef]
8. Alhelou, H.H.; Siano, P.; Tipaldi, M.; Iervolino, R.; Mahfoud, F. Primary Frequency Response Improvement in Interconnected Power Systems Using Electric Vehicle Virtual Power Plants. *World Electr. Veh. J.* **2020**, *11*, 40. [CrossRef]
9. Dey, P.P.; Das, D.C.; Latif, A.; Hussain, S.M.S.; Ustun, T.S. Active Power Management of Virtual Power Plant under Penetration of Central Receiver Solar Thermal-Wind Using Butterfly Optimization Technique. *Sustainability* **2020**, *12*, 6979. [CrossRef]
10. Jeon, W.; Cho, S.; Lee, S. Estimating the Impact of Electric Vehicle Demand Response Programs in a Grid with Varying Levels of Renewable Energy Sources: Time-of-Use Tariff versus Smart Charging. *Energies* **2020**, *13*, 4365. [CrossRef]
11. Naval, N.; Yusta, J.M. Water-Energy Management for Demand Charges and Energy Cost Optimization of a Pumping Stations System under a Renewable Virtual Power Plant Model. *Energies* **2020**, *13*, 2900. [CrossRef]
12. Behi, B.; Baniasadi, A.; Arefi, A.; Gorjy, A.; Jennings, P.; Pivrikas, A. Cost–Benefit Analysis of a Virtual Power Plant Including Solar PV, Flow Battery, Heat Pump, and Demand Management: A Western Australian Case Study. *Energies* **2020**, *13*, 2614. [CrossRef]
13. Suresh, V.; Janik, P.; Jasinski, M. Metaheuristic approach to optimal power flow using mixed integer distributed ant colony optimization. *Arch. Electr. Eng.* **2020**, *69*, 335–348. [CrossRef]
14. Nikolaou, T.; Stavrakakis, G.S.; Tsamoudalis, K. Modeling and Optimal Dimensioning of a Pumped Hydro Energy Storage System for the Exploitation of the Rejected Wind Energy in the Non-Interconnected Electrical Power System of the Crete Island, Greece. *Energies* **2020**, *13*, 2705. [CrossRef]
15. Energy storage in high renewable penetration power systems: Technologies, applications, supporting policies and suggestions. *CSEE J. Power Energy Syst.* **2020**. [CrossRef]
16. Al Shaqsi, A.Z.; Sopian, K.; Al-Hinai, A. Review of energy storage services, applications, limitations, and benefits. *Energy Rep.* **2020**. [CrossRef]
17. Tiba, S.; van Rijnsoever, F.J.; Hekkert, M.P. Firms with benefits: A systematic review of responsible entrepreneurship and corporate social responsibility literature. *Corp. Soc. Responsib. Environ. Manag.* **2019**, *26*, 265–284. [CrossRef]
18. Liu, C.; Xu, D.; Weng, J.; Zhou, S.; Li, W.; Wan, Y.; Jiang, S.; Zhou, D.; Wang, J.; Huang, Q. Phase Change Materials Application in Battery Thermal Management System: A Review. *Materials* **2020**, *13*, 4622. [CrossRef]
19. Yuan, Z.; Wang, W.; Wang, H.; Yildizbasi, A. A new methodology for optimal location and sizing of battery energy storage system in distribution networks for loss reduction. *J. Energy Storage* **2020**, *29*, 101368. [CrossRef]
20. Yuan, Z.; Wang, W.; Wang, H.; Yıldızbaşı, A. Allocation and sizing of battery energy storage system for primary frequency control based on bio-inspired methods: A case study. *Int. J. Hydrogen Energy* **2020**, *45*, 19455–19464. [CrossRef]

21. Mansor, M.A.; Hasan, K.; Othman, M.M.; Noor, S.Z.B.M.; Musirin, I. Construction and Performance Investigation of Three-Phase Solar PV and Battery Energy Storage System Integrated UPQC. *IEEE Access* **2020**, *8*, 103511–103538. [CrossRef]
22. Joshua, A.M.; Vittal, K.P. Transient behavioural modelling of Battery Energy Storage System supporting Microgrid. In Proceedings of the 2020 IEEE International Conference on Power Electronics, Smart Grid and Renewable Energy (PESGRE2020), Cochin, India, 2–4 January 2020; pp. 1–6.
23. Rallo, H.; Canals Casals, L.; De La Torre, D.; Reinhardt, R.; Marchante, C.; Amante, B. Lithium-ion battery 2nd life used as a stationary energy storage system: Ageing and economic analysis in two real cases. *J. Clean. Prod.* **2020**, *272*, 122584. [CrossRef]
24. Banguero, E.; Correcher, A.; Pérez-Navarro, Á.; García, E.; Aristizabal, A. Diagnosis of a battery energy storage system based on principal component analysis. *Renew. Energy* **2020**, *146*, 2438–2449. [CrossRef]
25. Leow, Y.Y.; Ooi, C.A.; Hamidi, M.N. Performance evaluation of grid-connected power conversion systems integrated with real-time battery monitoring in a battery energy storage system. *Electr. Eng.* **2020**, *102*, 245–258. [CrossRef]
26. Pham, T.T.; Kuo, T.C.; Bui, D.M.; Cao, M.T.; Nguyen, T.P.; Nguyen, T.D. Reliability Evaluation of an Aggregate Battery Energy Storage System Under Dynamic Operation. In Proceedings of the 2020 IEEE International Conference on Power Electronics, Smart Grid and Renewable Energy (PESGRE2020), Cochin, India, 2–4 January 2020; pp. 1–8.
27. Bai, B. Estimate the Parameter and Modelling of a Battery Energy Storage System. In Proceedings of the 2020 Chinese Control And Decision Conference (CCDC), Hefei, China, 22–24 August 2020; pp. 5444–5448.
28. Shi, T.; Yang, H.; Zhang, N.; Hua, G. Research on Verification Method of Electromechanical Transient Simulation Model of Battery Energy Storage System. In Proceedings of the 2020 Asia Energy and Electrical Engineering Symposium (AEEES), Chengdu, China, 29–30 May 2020; pp. 767–770.
29. Kumar, J.; Parthasarathy, C.; Västi, M.; Laaksonen, H.; Shafie-Khah, M.; Kauhaniemi, K. Sizing and Allocation of Battery Energy Storage Systems in Åland Islands for Large-Scale Integration of Renewables and Electric Ferry Charging Stations. *Energies* **2020**, *13*, 317. [CrossRef]
30. Kwon, S.-J.; Lee, S.-E.; Lim, J.-H.; Choi, J.; Kim, J. Performance and Life Degradation Characteristics Analysis of NCM LIB for BESS. *Electronics* **2018**, *7*, 406. [CrossRef]
31. Collin, R.; Miao, Y.; Yokochi, A.; Enjeti, P.; von Jouanne, A. Advanced Electric Vehicle Fast-Charging Technologies. *Energies* **2019**, *12*, 1839. [CrossRef]
32. Khodadoost Arani, A.A.; Gharehpetian, G.B.; Abedi, M. Review on Energy Storage Systems Control Methods in Microgrids. *Int. J. Electr. Power Energy Syst.* **2019**, *107*, 745–757. [CrossRef]
33. Ghiji, M.; Novozhilov, V.; Moinuddin, K.; Joseph, P.; Burch, I.; Suendermann, B.; Gamble, G. A Review of Lithium-Ion Battery Fire Suppression. *Energies* **2020**, *13*, 5117. [CrossRef]
34. Sikorski, T.; Jasiński, M.; Ropuszyńska-Surma, E.; Węglarz, M.; Kaczorowska, D.; Kostyla, P.; Leonowicz, Z.; Lis, R.; Rezmer, J.; Rojewski, W.; et al. A Case Study on Distributed Energy Resources and Energy-Storage Systems in a Virtual Power Plant Concept: Technical Aspects. *Energies* **2020**, *13*, 3086. [CrossRef]
35. Silva, V.A.; Aoki, A.R.; Lambert-Torres, G. Optimal Day-Ahead Scheduling of Microgrids with Battery Energy Storage System. *Energies* **2020**, *13*, 5188. [CrossRef]
36. Waclawek, Z.; Rezmer, J.; Janik, P.; Naneworter, X. Sizing of Photovoltaic Power and Storage System for Optimized Hosting Capacity. In Proceedings of the 2016 IEEE 16th International Conference on Environment and Electrical Engineering (EEEIC), Florence, Italy, 7–10 June 2016; pp. 1–5.
37. Kaczorowska, D.; Rezmer, J.; Jasinski, M.; Kostyla, P.; Leonowicz, Z.; Sikorski, T.; Janik, P.; Bejmert, D. Identification of Technological Limitations of a Battery Energy Storage System. In Proceedings of the 2020 12th International Conference and Exhibition on Electrical Power Quality and Utilisation- (EPQU), Cracow, Poland, 14–15 September 2020; pp. 1–6.
38. Chen, S.; Zhao, Z.; Gu, X. The Research on Characteristics of Li-NiMnCo Lithium-Ion Batteries in Electric Vehicles. *J. Energy* **2020**, *2020*, 3721047. [CrossRef]
39. Zhang, S.S.; Xu, K.; Jow, T.R. Charge and discharge characteristics of a commercial LiCoO2-based 18650 Li-ion battery. *J. Power Sources* **2006**, *160*, 1403–1409. [CrossRef]
40. Dulau, L.I.; Abrudean, M.; Bica, D. Distributed generation and virtual power plants. In Proceedings of the 2014 49th International Universities Power Engineering Conference (UPEC), Cluj-Napoca, Romania, 2–5 September 2014; pp. 1–5.

41. Alahyari, A.; Ehsan, M.; Mousavizadeh, M.S. A hybrid storage-wind virtual power plant (VPP) participation in the electricity markets: A self-scheduling optimization considering price, renewable generation, and electric vehicles uncertainties. *J. Energy Storage* **2019**, *25*, 100812. [CrossRef]
42. Yang, Y.; Wei, B.; Qin, Z. Sequence-based differential evolution for solving economic dispatch considering virtual power plant. *IET Gener. Transm. Distrib.* **2019**, *13*, 3202–3215. [CrossRef]
43. Sikorski, T.; Jasiński, M.; Ropuszyńska-Surma, E.; Węglarz, M.; Kaczorowska, D.; Kostyła, P.; Leonowicz, Z.; Lis, R.; Rezmer, J.; Rojewski, W.; et al. A Case Study on Distributed Energy Resources and Energy-Storage Systems in a Virtual Power Plant Concept: Economic Aspects. *Energies* **2019**, *12*, 4447. [CrossRef]
44. Ringkjøb, H.-K.; Haugan, P.M.; Solbrekke, I.M. A review of modelling tools for energy and electricity systems with large shares of variable renewables. *Renew. Sustain. Energy Rev.* **2018**, *96*, 440–459. [CrossRef]
45. Zhang, C.; Chen, H.; Liang, Z.; Hua, D. An interval DC power flow analysis through the optimizing-scenarios method. In Proceedings of the 2017 IEEE Innovative Smart Grid Technologies Asia (ISGT-Asia), Auckland, New Zealand, 4–7 December 2017; pp. 1–5.
46. De Oliveira Gomes, A.; Meirelles Gouvea, M. DC power flow optimization with a parallel evolutionary algorithm. In Proceedings of the 2012 Sixth IEEE/PES Transmission and Distribution: Latin America Conference and Exposition (T&D-LA), Montevideo, Uruguay, 3–5 September 2012; pp. 1–6.
47. Simpson-Porco, J.W. Lossy DC Power Flow. *IEEE Trans. Power Syst.* **2018**, *33*, 2477–2485. [CrossRef]
48. Plass, M. Energy Storage System Standards and Test Types. Available online: https://www.dnvgl.com/services/energy-storage-system-standards-and-test-types-111331 (accessed on 16 December 2020).
49. International Electrotechnical Commission. *IEC 61427-1:2013 Secondary Cells and Batteries for Renewable Energy Storage—General Requirements and Methods of test—Part 1: Photovoltaic Off-Grid Application*; IEC: Geneva, Switzerland, 2013; pp. 1–37.

Publisher's Note: MDPI stays neutral with regard to jurisdictional claims in published maps and institutional affiliations.

Article

The Impact of Supply Voltage Waveform Distortion on Non-Intentional Emission in the Frequency Range 2–150 kHz: An Experimental Study with Power-Line Communication and Selected End-User Equipment

Marek Wasowski [1,*], Tomasz Sikorski [1], Grzegorz Wisniewski [1], Pawel Kostyla [1], Jaroslaw Szymanda [1], Marcin Habrych [1], Lukasz Gornicki [2], Jaroslaw Sokol [2] and Mariusz Jurczyk [2]

[1] Faculty of Electrical Engineering, Wroclaw University of Science and Technology, 50-370 Wrocław, Poland; tomasz.sikorski@pwr.edu.pl (T.S.); grzegorz.wisniewski@pwr.edu.pl (G.W.); pawel.kostyla@pwr.edu.pl (P.K.); jaroslaw.szymanda@pwr.edu.pl (J.S.); marcin.habrych@pwr.edu.pl (M.H.)
[2] TAURON Dystrybucja Pomiary Ltd., 33-100 Tarnow, Poland; Lukasz.Gornicki@tauron-dystrybucja.pl (L.G.); Jaroslaw.Sokol@tauron-dystrybucja.pl (J.S.); Mariusz.Jurczyk@tauron-dystrybucja.pl (M.J.)
* Correspondence: marek.wasowski@pwr.edu.pl

Abstract: Knowledge of the conducted emissions in the frequency range 2–150 kHz contains some gaps related to the impact of the harmonics in the supply voltage on the nature of these emissions. It can be noticed that the conducted emissions from non-sinusoidal power supplies have not been studied sufficiently, and that the impact of this distortion may be greater than the generally known results of emission tests carried out under standardized test conditions. This paper is aimed at investigating experimental cases of the influence of supply voltage waveform distortion on non-intentional emission in the range 2–150 kHz and the efficiency of power line communication based on selected PRIME (PoweRline Intelligent Metering Evolution) power line communication (PLC) technology. A series of experimental laboratory studies were investigated, representing the operation of the investigated PLC system with different types of end-user equipment (LED—Light Emitting Diode, CFL—Compact Fluorescent Lamp, induction motor with frequency converter) working under a distorted supply voltage condition obtained by the programmable power supply for different scenarios of the admissible harmonics contribution in the range 0–2 kHz. The scenarios included limits defined in standards EN 50160 and IEC 61000-4-13. The researchers used spectral analysis with a notation to emission limits, compatibility levels, and mains signalling, as well as statistics of the PLC communication. The obtained results provide important conclusions, which may be applied both in the development of the design of the appliances in question and the higher frequency emission testing methods.

Keywords: conducted disturbances; power quality; supraharmonics; 2–150 kHz; Power Line Communications (PLC); intentional emission; non-intentional emission; mains signalling

Citation: Wasowski, M.; Sikorski, T.; Wisniewski, G.; Kostyla, P.; Szymanda, J.; Habrych, M.; Gornicki, L.; Sokol, J.; Jurczyk, M. The Impact of Supply Voltage Waveform Distortion on Non-Intentional Emission in the Frequency Range 2–150 kHz: An Experimental Study with Power-Line Communication and Selected End-User Equipment. *Energies* **2021**, *14*, 777. https://doi.org/10.3390/en14030777

Academic Editor: Akhtar Kalam
Received: 28 December 2020
Accepted: 26 January 2021
Published: 2 February 2021

Publisher's Note: MDPI stays neutral with regard to jurisdictional claims in published maps and institutional affiliations.

1. Introduction

Smart meters, which are now present in the majority of households in Europe, often use power line communication (PLC) over the low-voltage (LV) grid. In countries belonging to the European Union countries, 123 million electricity meters are read remotely using PLC transmission, which accounts for 43 percent of all meters installed in the member states. The most advanced in implementing smart metering are the Swedes, who have already replaced 100% of their meters with smart meters, Finns (99.8%), Estonians (98.9%), Italians (98.5% of ~37 million meters) and the Spanish (93.8% of ~28 million meters) [1]. Given that the member states will continue to update the installation of the smart meter in line with their new planning and target periods (see Figure 1), it is estimated that 223 million meters (77%) will be remotely read by 2024, and 266 million meters by 2030 (92%).

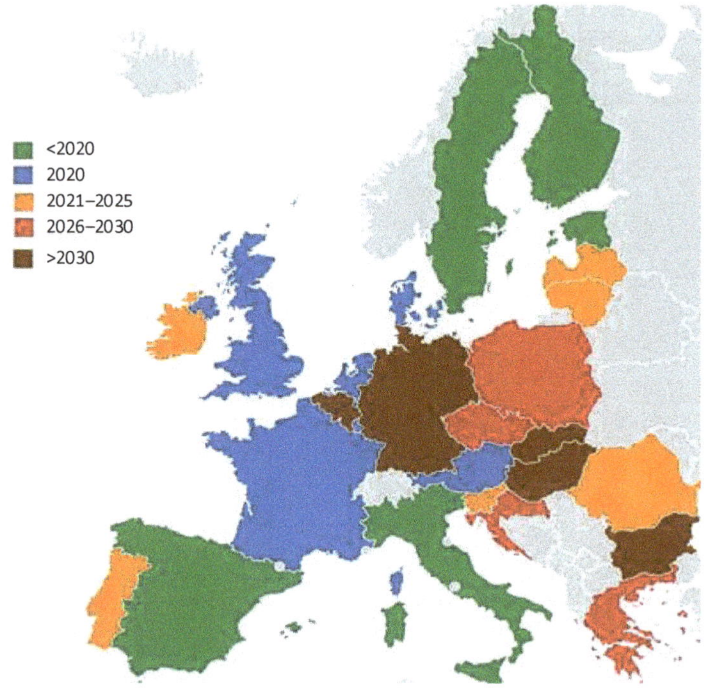

Figure 1. Review of the target completion date for smart electricity meters installation covering at least 80% of all consumers based on the plans of the European Union (EU) member states (based on [1]).

From the point of view of signal categorisation occurring in power networks, power line communication (hereinafter PLC transmission) is treated as mains signalling and is considered as intentional emission in the frequency range 2(3)–148.5 kHz [2,3]. However, intentional emission in LV networks is accompanied by the phenomenon of non-intentional emission, which is a growing problem in the operation of smart metering systems based on PLC transmission. All European Union (EU) members stated that, despite the formal separation of a dedicated 3–95 kHz frequency band for the transmission of metering data in low-voltage power networks, the phenomenon of harmful non-intentional emission in this band is increasing. This issue is confirmed by the experiences of power system operators and from many publications all over the world.

Several main issues have been formulated at this time, including:

- Classification and identification of supraharmonics, signalling on low-voltage electrical installations in the frequency range 2(3) kHz to 148.5 kHz [2–8].
- Accurate assessment of waveform distortion in the presence of supraharmonics [9–15].
- Source of non-intentional emissions and propagation of supraharmonics, interaction between devices [16–20].
- Potential problems with communication via the power grid [1,20–25].

The analysis of big data from smart electricity meters can contribute to significant results in terms of grid management. There are many alternative approaches to analysing data from such systems [26]. In the advanced analysis of the collected measurement data, there is potential for:

- Improving the security of the system by increasing its observability.
- The use of demand side management (DSM) tools, which is carried out by dynamic billing of electricity consumers and prosumers.

- Forecasting the amount of energy released to the grid by micro-installations.

The primary motivation for this work is the relatively small number of research studies concerning the influence of power supply conditions on supraharmonics emissions. The direction of the investigation was formulated on the basis of observation of the different levels of non-intentional emission generated by the same class of devices in a different node of the low-voltage network, which may be characterized by the respective quality of the voltage or impedance condition. This study led to the formulation of two hypotheses: (a) the non-intentional emission generated by particular devices may depend on the condition of the supply voltage; and (b) non-intentional emission generated by particular devices may depend on the impedances of the supplying circuits in the point of the connection. This paper is focused on the first hypothesis related to voltage supply waveform distortion. Designated research directions can be initially confirmed by the results of similar investigations related to harmonic current emissions (0–2 kHz) under the flat-top supply waveform distortion [27], or detected differences in the performance between compact fluorescent lamps (CFLs) and light emitting diode (LED) light sources under different voltage distortions [17], as well as the effect of supply voltage harmonics on the input current of a single-phase diode bridge rectifier load [28]. However, although the above-mentioned works highlight the issue of harmonic current emissions, they do not provide information about emissions in the range of higher frequencies up to 150 kHz, nor do they express the influence of the supply voltage distortion on mains signalling (usually expressed in dBμV). The unit decibel microvolt (dBμV) is directly associated with volt (V) as twenty decimal logarithms of volt divided by one microvolt. Since the subject of the research is the phenomenon of primary emission, the observation results can be expressed in volts or dBμV. It might be interpreted as a coupling coefficient in the point of the connection as a result of current emission flowing by the network impedances.

Technical measures to improve transmission—that are more software-based than hardware-based—should also be considered. It is worth paying attention to the context of using machine learning techniques to solve data transmission problems. For example, in [29], the authors designed the mechanism based on the matching rule, and further traffic collision avoidance, channel occupancy, power consumption, and delay in wireless networks.

This paper aims to extend the current state-of-the-art by investigating experimental cases of the influence of supply voltage waveform distortion on non-intentional emission in the range 2–150 kHz and the efficiency of selected power line communication systems. In order to achieve the proposed aim, in Section 2, the gap in the current studies related to the influence of the supply voltage distortion on supraharmonics emission and efficiency of power line communications is identified. Furthermore, in order to characterize the background of the investigation, a critical review of the emission limits, compatibility levels, and immunity tests for conducted disturbances and mains signalling in the frequency range 2–150 kHz was performed. Section 3 describes in detail the laboratory setup implemented for the experimental studies including electrical connections, the performance of programmable alternating current (AC) supply, selected devices, applied power-line communication technology as well as measurement setup for supraharmonics. Additionally, the software for frequency spectrum analysis is described in relation to current testing and measurement techniques. Section 4 presents a collection of the experimental results. First, the methodology of the investigation is presented and then the set of results of non-intentional emission for a different range of supply waveform distortion and different end-user equipment in the presence of selected power-line communications is analysed. The results describe the spectra of intentional and non-intentional emissions and the evaluation of the quality of operation of the PRIME PLC system. Distorted supply voltage conditions consist of different scenarios of the admissible harmonics contribution, in the range 0–2 kHz, including limits defined in standards EN 50160:2010 [3] and IEC 61000-4-13 [30]. Section 5 highlights crucial results and constitutes elements of the discussion.

2. Literature Review

2.1. Gap in the Research Related to Influence of Supply Voltage Distortion on Supraharmonics Emission

The phenomenon of electromagnetic emission occurring in the frequency range 2–150 kHz was termed "supraharmonics" in the literature (hereinafter SH) [2–8]. The lower limit of the SH range (2 kHz) is taken as the upper limit of traditional energy quality standards covering emission up to 40th harmonic order. The upper limit of the SH (150 kHz) range is the lower limit of the electromagnetic compatibility (EMC). This division indicates that there is an area of insufficiently regulated standards. Hence, several standardisation committees are currently working on a description of emission and immunity limits and methods for emission testing in this frequency range.

The scientific reports on the SH problem show that there are three lines of research in this area. The first concerns research into the impact of emissions generated by different types of loads on the level of current emissions [8,22,25], which should be considered as primary emissions. This group of research provides an overview of the types of noises that affect smart metering using power line communications, including noises coming from common electronic devices (compact fluorescent lamp (CFL) and light emitting diode (LED) light sources) and noises coming from photovoltaic inverters or electric vehicle charging spots, measured in a controlled environment. The previously mentioned studies, however, ignore the secondary emission in load currents, which is understood as the part of the current of a tested load that does not originate from its primary emission, but is transferred through parallel connections from another load. In order to describe the primary and secondary emission, a simplified model in a system of two loads is presented in Figure 2. Load 1 is represented by a constant current source I_{h1} and impedance z_1. The primary emission from load 1 is denoted as I_1. Impedances z_1 and z_2 are the internal impedances of load 1 and 2 respectively, and z_g is the network impedance. The primary emission (I_1) flows partially into the grid and partially through the impedance of load 2. Thus, the total emission of load 2 consists of its primary emission I_2 superimposed on the secondary emission caused by load 1 ($I_{1,2}$). Described relations are true, assuming the sinusoidal rated voltage. In practice, such conditions are hard to achieve.

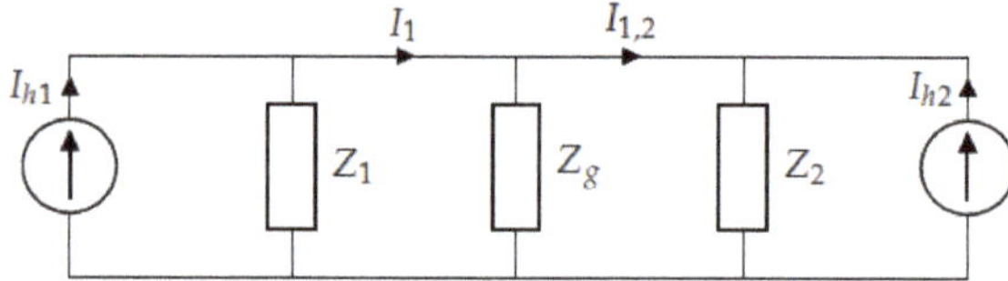

Figure 2. A simplified model, describing primary and secondary emissions in a system of two loads, assuming a sinusoidal rated voltage.

It is worth noting that, in [24], the authors noticed the influence of harmonics on the operating of electrical energy meters in a network with nonlinear loads. It is shown that electronic static meters of active energy are tested in the presence of distortions, and electronic static meters of reactive energy accuracy requirements do not take into account the possible presence of harmonics.

The second line of research focuses on the so-called resultant emission of loads, resulting from the simultaneous primary emission of multiple appliances connected to the same subnetwork. In paper [16], among others, the authors presented interesting results of research on the effect of gradual incorporating of 1, 10, and 48 compact fluorescent lamps into one connection point, which shows that the level of current emission at the connection point does not increase linearly with the growth of loads of the same type. The authors observed the effect of decreasing the share of low frequencies 0–2 kHz in the total current while increasing the number of fluorescent lamps. It was observed that in the case of SH emissions above 2 kHz, some frequency components of currents close through input circuits of adjacent loads and some frequency components of currents close through supply circuits and neutral wires. However, these tests also ignore the issue of low-frequency emissions from the distorted supply voltage.

In turn, in article [24], the authors outlined the third line of considerations. Based on tests carried out with the use of LED light sources and compact fluorescent lamps, the authors confirmed that the share of non-intentional emission introduced into the network by these loads might be greater than the results of emission tests carried out under the conditions of emission tests of loads following EN 61000-3-2 standard. Although the authors observed emission in the frequency band 2–40 kHz, it should be assumed that this phenomenon might also apply to other frequency ranges.

In each of the three lines of the problem consideration presented above, it can be shown that the aspect of the harmonic content in the supply voltage, in the range 0–2 kHz, as the potential origin of the increased level of non-intentional emission of the electrical devices, in the range 2–150 kHz, has not been (as of yet) sufficiently investigated. Many works investigate the impact of selected supply voltage parameters on current harmonics up to 2 kHz. Many works also present current supraharmonic emissions of different electrical devices, but are usually investigated under sinusoidal supply voltage. This work is focused on the non-intentional emission of the selected electrical devices under non-sinusoidal supply voltage conditions and their impact on power line communications.

In [27,31], authors attempted to assess the influence of the voltage curve distortion in the form of a flattening of the peak of a sine wave, which corresponds to an increase in the harmonic content factor. The authors of these works, using the example of lighting devices, assessed the harmonic content during parallel operating the devices. The authors demonstrated, among other findings, that new electronic equipment based on active power factor correction (PFC) shows a qualitatively different dependence of harmonic emissions on the degree of distortion (flattening) of the supply voltage concerning loads without PFC.

The influence of voltage distortion on the differences in work efficiency between compact fluorescent lamps (CFL) and light emitting diode (LED) light sources in the low and high-frequency range was described in [17]. The tests were carried out for three scenarios: sinusoidal voltage, flat-top distorted voltage (total harmonic distortion in voltage THDV about 3%), and pointed-top distorted voltage (overswing) (THDV about 4% in accordance with IEC 61000-4-13), which were used to power a group of 142 lamps manufactured in 2009–2016 (69 CFL sources and 73 LED sources). These lamps worked in various ranges of rated power (CFL range up to 46 W, LED range up to 17 W).

A frequent phenomenon is also the secondary emission of the load, which is understood as the part of the current of the tested load that is not derived from its primary emission, but is transferred by parallel connections from another load. The research results presented in [9,17,18,21] indicate that, due to the use of various input systems of loads, including EMC filters, and due to the phenomenon of secondary emission from the parallel operation of loads, the share of higher frequencies in the supply current at the point of common coupling (PCC) point might be smaller than would result from the emission of individual loads. It has been observed that in the case of supraharmonic emissions above 2 kHz, some frequency components of the currents are closed by the input circuits of neighbouring loads and some of the frequency components of the currents are closed by the supply and neutral circuits. The issue of secondary emission is also present in the considerations concerning the influence of loads' operation on the narrowband PLC transmission. In [23], authors indicate selected disturbances originating from loads connected to the LV network, which may affect the transmission efficiency in the range of 0–2 kHz and 2–150 kHz. The frequency range of these disturbances coincides with the PLC transmission frequencies; as a result, the useful signal of PLC transmission is "covered" by the spectrum of unintentional disturbances.

The authors of [17] also analysed the operation of light sources in the band above 2 kHz. The analyses were performed in three frequency sub-ranges: 9–30 kHz, 30–95 kHz, and 95–150 kHz. Tests of a representative set of lamps, both compact fluorescent lamps and LED lamps, showed clear differences in the current emission not only between the two types of lamps, but also between lamps within the same type. The discussed results indicate the influence of the quality of the supply voltage on the content of higher frequencies in the current of LED lamps in the wide 30–95 kHz range, used by some narrowband and broadband PLC technologies. Hence, the conclusion that in the case of operation of these loads in conditions of supply other than sinusoidal, the share of unintentional disturbances introduced to the network by these loads may be greater than the results of emissivity obtained during standardized tests.

This paper presents the results of the research devoted to the issue of the impact of harmonic content in the supply voltage, or more generally, the degree of distortion of the supply voltage waveform on non-intentional emission in the 2–150 kHz band by loads, which, according to literature, are responsible for most of the emission of the intentional PLC transmission. The methodology of the investigation is aimed to cover two issues: characterisation of the non-intentional emission under the different quality of the supply voltage as well as the assessment of its influence on continuity of the transmission in the selected PLC system.

2.2. The Background of the Investigations—Identification of the Emission Limits, Compatibility Levels, and Immunity Levels for Conducted Disturbances and Signalling in the Frequency Range 2–150 kHz

Due to the manner of interaction between the transmission systems (intentional emission, mains signalling) and the power grid including loads (non-intentional emission), it is worth analysing the contents of the standards in terms of permissible levels relating to:

- Levels of non-intentional emission referring to individual equipment and the power grid (non-intentional emission).
- Compatibility levels in the power grids (compatibility levels, environment characteristics).
- Intentional emission (transmission levels and mains signalling).
- Immunity test levels.

For this reason, in Figure 3, several characteristics in the range 2–150 kHz were collected. Presented curves express limits defined in applicable standards. Details of selected standards are given in the legend of the figure. The prominent curve is related to intentional emission (i.e., mains signalling and power-line communication denoted in the figure by the lines described using the letter "S") and non-intentional emission (i.e., network distortions, denoted in the figure by the line described using the letter "E"). Additionally the compatibility level in an electrical network is also represented (denoted in the figure, using letter "C"). The boundary conditions are represented by the curve related to the immunity test of the communication systems in the range 2–150 kHz (denoted in the figure by the line described using the letter "I"). Additionally, the figure shows the frequency area 42–89 kHz, which represents the PRIME broadband power-line communication technology used in the research.

Legend of the curves in Figure 3:

Non-intentional emission referring to IEC 61000-3-8:1997 [4] adopted in the standard for PLC transmission EN 50065:2012 [2]—it can be treated as background emission, potentially affecting the communication by the power grid (solid line, green colour, letter "E").

- Compatibility levels referring to IEC 61000-2-2: AMD1:2017 [5] and AMD2:2018 (dashed line, pink colour, letter "C").
- Permissible level of intentional emission, mains signalling, referring to IEC 61000-3-8:1997 [4] interpreted as the permissible level of transmission in power line communication systems adopted in EN 50065:2012 [2] (solid line, red colour, letter "S").
- Permissible level of intentional emission, mains signalling, referring to IEC 61000-2-2: AMD1:2017 [5] (dashed line, red colour, letter "S") and referring to EN 50160:2012 [3] (dashed-dot line, red colour, letter "S").
- Levels of tests for immunity to conducted, differential mode disturbances, and signalling in the frequency range 2 kHz to 150 kHz at AC power ports referring to

PN–EN 61000-4–19:2014 [6] (solid line, black colour, notation "P4"—test level with the reference to environment class 4: severe industrial environment; dashed line, colour black—notation "P3"—test level with the reference to environment class 3: typical residential, commercial, and light industrial environment).

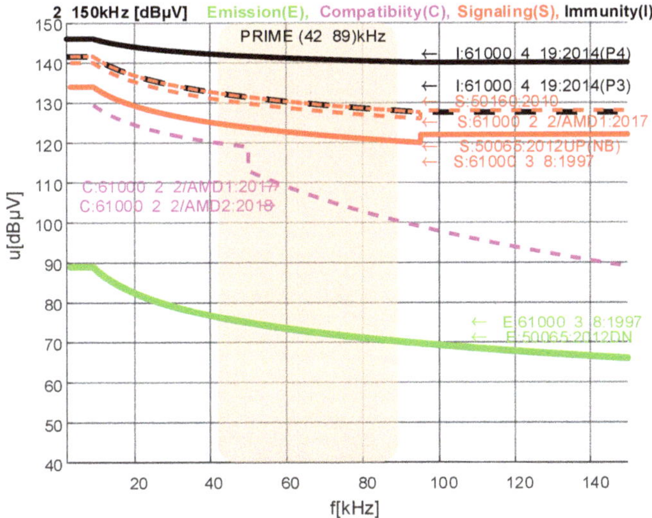

Figure 3. Comparison of the emission (E), compatibility (C), mains signalling (S), and immunity (I) levels for conducted disturbances and signalling in the frequency range 2–150 kHz (based on standards [16–21].

3. Characteristic of the Laboratory Hardware and Software Setup

The laboratory setup for the analysis of the direct impact of harmonic content in the supply voltage on the emission of supraharmonics consists of three 3-phase programmable power supplies KIKUSUI PCR500LA [32], which, together with the standardised reference impedances (for the phase wire $z_A = (0.24 + j0.15)\ \Omega$ and the neutral wire $z_N = (0.16 + j0.1)\ \Omega$) formed the equivalent of a standard low-voltage grid connection point (Figure 4a). The reference impedances follow the standard IEC 61000-3-3 [33] and are recommended to realize the equivalent of the power grid used in the calculation and measurement of the directly measured parameters of voltage changes, voltage fluctuations, and flicker. The materials used in constructions of the impedances preserve the reference value of the impedances in a 50Hz system with a nominal current up to 5A. The operation of power supplies in the three-phase source mode was obtained using 3P03-PCR-LA three-phase output driver cards [34]. The programmable power supplies were controlled by the Quick Immunity Sequencer [35] installed on a PC that was used to define power supply voltage waveform and subsequent test series. Communication between PC and the programmable AC supply used general purpose interface bus (GPIB interface IB03-PCR-LA) [36]. Using the authorised software different scenarios were created for the input signal with pre-sets of harmonics contribution in the range 0–2 kHz. This application realized communication with the programmable power supply and configuration sets for the harmonics contribution in the supply voltage. It used a waveform bank and the sequence operation function, which is responsible for transfer, execution, and output control of the waveform. In this paper, some selected pre-sets of harmonic contributions were considered: (a) pure sinusoid, (b) limits of harmonics in the low-voltage public network formulated in EN 50160:2010 [3], (c) low-frequency immunity test levels of the equipment defined in IEC 61000-4-13 [30]. These pre-sets were defined in the waveform bank.

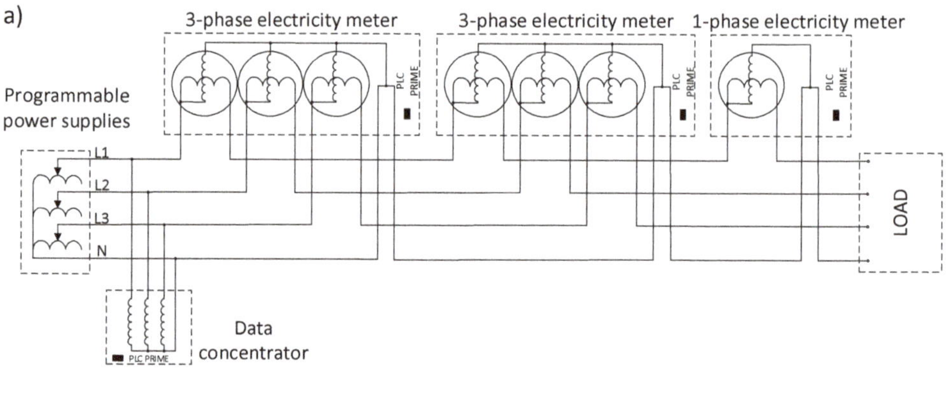

Figure 4. The laboratory setup: (**a**) electrical diagram of the test stand, (**b**) view of the power supply system in the form of programmable AC power supplies and model with power line communication (PLC) meters.

The PLC transmission was provided by a compact model containing three electricity meters equipped with PRIME PLC communication modules (two 3-phase and one 1-phase meter) and a data concentrator working in the same technology. PRIME is a specification for narrowband power line communication. The PRIME physical layer is based on Orthogonal Frequency Division Multiplexing (OFDM). The transmission band is 42–89 kHz. In the course of individual research steps, using functions and reports implemented in the PLC concentrator software, the statistics of PLC transmission between the concentrator and the meters were observed and recorded. The view of the constructed laboratory setup is presented in (Figure 4b).

In terms of information and communication technologies, the test stand was separate, both physically and logically, local area network (LAN) computer network connecting computers (in this case, meters, concentrator and the controlling computer). The functional diagram of the test stand as a local area network (LAN) is presented in Figure 5.

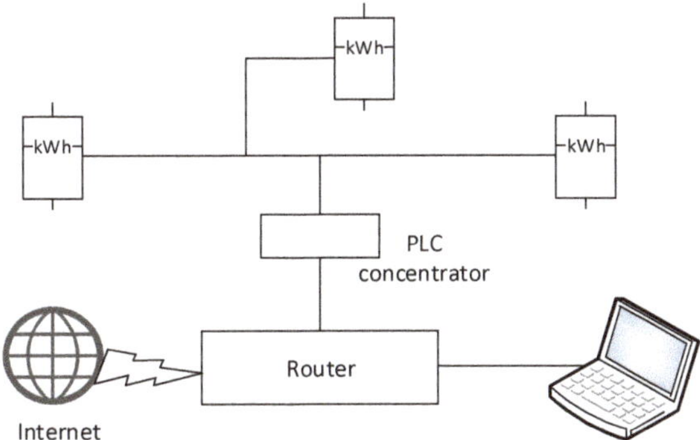

Figure 5. Information and communications technology (ICT) functional diagram of the test stand as a local area network (LAN).

The recording of measurement data was carried out by the authorised university hardware setup based on National Instrument cRIO technology (Figure 6). In detail, the voltage measurement acquisition is realized by National Instrument C series $\pm$10V voltage input module with 1 MS/s sampling rate and 16-Bit resolution [37]. In order to adapt the range of the measured voltage to the voltage input module a voltage transducer provided by Life Energy Motion Company (LEM) model CV 3-1000 was implemented with a voltage ratio of 1000:10 and frequency bandwidth to 500 kHz [38]. The obtained measurement setup allows recording simultaneously four voltage and four current waveforms with a maximum frequency of 1 MHz per channel. The investigations were focused on voltage measurements. For the measurements in the frequency range 2–150 kHz, the voltage units (instead of volts) are expressed in dBμV. Thanks to the use of cables between the concentrator and individual meters, with a cross-section of 2.5 mm^2 and their small length (1–2 m), the impact of power supply cables on the attenuation of the PLC signal in the voltage line should be considered insignificant. The accuracy of measurements was validated using commercial spectrum analysers used in the power line communication investigation: MFA 500 Spectrum Analyser 3–500 kHz by Swemet [39] and USB-SA44B Spectrum Analyser 1 Hz–4.4 GHz by Signal Hound [40]. The validation was mainly based on the comparison of the spectrum of the simultaneously recorded signal. The frequencies and magnitudes of spectrum components were compared. The obtained results were comparable. The advantage of using an authorized solution based on the cRIO platform is the open possibility to define the acquisition process e.g., time recording of the waveform.

Figure 6. View of 1 MHz voltage and current recorder components.

Specification of the hardware components of the laboratory test circuits is collected in Table 1.

Table 1. Specification of the hardware components of the laboratory test circuits.

Programmable AC Source/Kikusui PCR500LA	Specification
Power capacity	500 VA
Input voltage variation (with respect to changes in the rated range)	±0.1%
Output current variation (with respect to 0% to 100% changes in the rating)	Within ±0.1 V/±0.2 V (output voltage range 100 V/200 V)
Output frequency variation in AC mode (40–999.9 Hz)	Within ±0.3%
Ripple noise: DC mode (5 Hz to 1 MHz component)	0.1 Vrms or less
Output frequency stability (with respect to changes in the rated range)	Within $\pm 5 \times 10^{-5}$
Output voltage waveform distortion	0.3% or less
Output voltage response speed	30 μs
Compliant standards	EMC Directive 89/336/EEC EN61326:1997/A2:2001 Emission Class A IEC 61000-3-2:2000 IEC 61000-3-3:1995/A:2001
Reference impedances/authorised university solution	**Specification**
Nominal frequency	50 Hz
Nominal RMS current	5 A
Nominal impedance	Phase wire z_A = (0.24 + j0.15)Ω for 50 Hz R_A = 240 mΩ ± 0.2%, L_A = 477.5 μH ± 1% Neutral wire z_N = (0.16 + j0.10) Ω for 50 Hz R_N = 160 mΩ ± 0.2%, L_N = 318.3 μH ± 1%
Signal recorder/authorised university solution	**Specification**
FPGA platform	National Instrument cRIO FPGA 1 MS/s
Voltage input module	National Instrument NI-9223 C Series ±10 V, 1 MS/s, 16-Bit, Simultaneous Input, 4-Differential Channel
Voltage Transducer	LEM CV 3–1000 Primary voltage ±1000 V Secondary voltage ±10 V Frequency bandwidth DC ÷ 500 kHz Accuracy 0.2%

cRIO–compact reconfigurable input output platform, CV–voltage transducer, product of LEM, DC–direct current, EMC–electromagnetic compatibility, FPGA–field-programmable gate array, IEC–International Electrotechnical Commission, LEM–Life Energy Motion Company, NI–National Instruments, PCR–programmable power supply, product of Kikusui, RMS–root mean square.

Recently, a wide discussion has taken place in literature referring to the measurements method for the frequency range 2–150 kHz [9,12,13]. Some consideration is also provided by the informative annex in IEC 61000-4-30 [15]. One of the suggested approaches is to implement a discrete Fourier transform (DFT) defined in IEC 61000-4-7 [14] for the frequency range of 2 kHz and extend it above 2 kHz. Traditionally DFT implementation uses a rectangle data acquisition window with a width of 200 ms, corresponding to approximately 10 (12) periods of power system frequency 50 Hz (60 Hz). Consequently, spectrum resolution is 5 Hz. Before 2009, the informative annex of this standard suggested implementation of DFT in the range 2–9 kHz using DFT technique with 10 Hz resolution, realized by a time windowed signal with a window width equals to 100 ms. The current form of the annex recommends preserving 5 Hz spectrum resolution also in the range 2–9 kHz, obtained by using a 200 ms window. Additionally, a grouping concept is also proposed. For the frequency range 2–9 kHz, it is realized using 200 Hz band centred at frequencies being a multiplication of 100 Hz. The first group is centred at 2100 Hz. The grouped spectrum has a final resolution of 100 Hz.

The intentional transmission signal used by the power line communication systems has usually time-variant nature. The narrow or broadband signals are transmitted in time packages. In order to investigate the coexistence of these time-variant spectrum components and non-intentional emission of the equipment, the short-Fourier transform (STFT) in the range 2–150 kHz was implemented. The rectangle data acquisition window without overlapping was used. The window width was 100 ms and the DFT output was grouped using a 200 Hz band centralized at frequencies being a multiplication of 100 Hz starting from 2000 Hz. The selection of the shorter window width was dictated by two conditions. First, the time resolution of the STFT is enhanced by using a short window. Second, one of the commercial spectrum analyser used for the validation used a 10 Hz resolution spectrum. Additionally, the inherent effect of short window width is smoothed by grouping concept with 100 Hz resolution. The recording time was 20 s. Therefore, it was possible to observe changes in the spectrum and capture the voltage spectra on the load both with and with communication represented by intentional emission in the 42–89 kHz frequency band used by the PRIME PLC system.

A set of loads representing end-user equipment was based on:

- LED light sources (the three-phase circuit was made using three single-phase LED with a power of 8 W connected in a star, the LED have implemented active power factor correction by an active power filter (APF) in order to keep $\cos\varphi = 0.9$).
- CFL compact fluorescent lamps (the three-phase circuit was made of three single-phase compact fluorescent lamps with the power of 11 W, 18 W and 20 W connected in a star).
- Three-phase asynchronous induction motor 300 W/400 V/50 Hz rated speed 1390 rpm, powered by a single-phase frequency converter with input voltage 230 V, output voltage 3 × 230 V, power 400 W.

4. Results of an Experimental Study with Power-Line Communication and Selected End-User Equipment

In the conducted experiments, the waveform of supply voltage was shaped by the superposition of harmonics with set amplitudes. Using programmable power supplies and dedicated software, the supply voltage was shaped in three scenarios as follows:

- Supply voltage scenario 1: pure sinusoidal voltage.
- Supply voltage scenario 2: distorted voltage with harmonics according to permissible limits for public networks defined in EN 50160 [3].
- Supply voltage scenario 3: distorted voltage with harmonics used for immunity tests of load according to IEC 61000-4-13 [30]—equipment with the third class of immunity.

Selected harmonics contributions represent margin waveform distortion. The total harmonic distortion index (THD) achieved as a sum of permissible harmonics levels defined in EN 50160 equals 11.62%; however, the normal operating condition of the public grid is usually 2–8%.The study aimed to assess the impact of supply voltage distortion on the operation of a commercial PRIME PLC system with the simultaneous presence of non-intentional emission generated by the loads. In point of the load contribution, several load conditions were considered:

- Load condition 0: no load.
- Load condition 1: LED light source.
- Load condition 2: compact fluorescent lamp (CFL) light source.
- Load condition 3: induction motor powered by a frequency converter.

The PRIME PLC system is based on recommendation G.9904 [41,42]. The technology represents narrowband orthogonal frequency division multiplexing power line communication that uses 97 subcarriers, 96 of which are used for data transmission, where: first carrier frequency is 41.99 kHz (41,992.18750 Hz), last carrier frequency is 88.87 kHz (88,867.18750 Hz), the distance between carriers—0.488 kHz (488.28125 Hz). For each scenario of the supply waveform distortion and each group of loads, the investigation cases were performed in order to express the variability of non-intentional emission affecting intentional emission of the PRIME PLC transmission system. First, characteristic spectra of the background signal and the transmission signal obtained among 20 s of the STFT observation were compared with the normative curves identified in Section 2.2 referring to standards [2–6]. Second, selected spectrum signal parameters were derived. Over the given frequency interval in the frequency range of the investigated PRIME PLC system, a single local maximum magnitude of a frequency component of the background signal was identified. It can be treated as a maximum noise contribution (N_{max}). Then, corresponding to the N_{max} a magnitude of the transmission signal ($S(N_{max})$) related to N_{max} was selected. The parameters are used to derive a local signal-to-noise ratio (SNR_{local}), which expresses the local minimum SNR in the transmission band due to reference to the maximum noise frequency component:

$$SNR_{local[dB]} = 20 \quad log\frac{S(N_{max})_{[V]}}{N_{max[V]}} = 20logS(N_{max})_{[V]} - 20logN_{max[V]} = S(N_{max})_{[dB]} - N_{max[dB]}$$
$$= S(N_{max})_{[dB\mu V]} - N_{max[dB\mu V]} \tag{1}$$

Additionally, the classical signal-to-noise ratio in the full frequency range of the PRIME PLC band (SNR_{band}) was calculated using the power of the transmission signal (P_{singal}) and power of the background signal (P_{noise}) calculated based on M frequency components belonging to the transmission band:

$$SNR_{band[dB]} = 10log\frac{P_{signal}}{P_{noise}} = 10log\frac{\frac{1}{M}\sum_{m=1}^{M} S^2_{m[V]}}{\frac{1}{M}\sum_{m=1}^{M} N^2_{m[V]}} = P_{signal[dB]} - P_{noise[dB]} \tag{2}$$

Proposed parametrization using the local and broadband SNR aims to represent the variability of transmission conditions in the transmission band.

In order to extend the assessment of the communication conditions, the communication statistics were also analysed, considering the intensification of the connection attempts. In detail, for particular scenarios of the investigation, the statistics were monitored based on the total number of the connections, the number of failed connections, the number of successful connections, and the duration of failed connections in one hour. The statistics represent 2 h of continuous work of the PLC system. The data were reported using a dedicated ADDAX software solution for residential metering [43]. Following interpretations of the connection statistics may be defined:

- Increasing the number of attempted connections is the preventive action of the PLC system to keep the transmission successful and can be interpreted as a symptom of deteriorated transmission condition.
- Percentage contribution of the number of failed connections is not directly representative quantity indicating the deterioration of the transmission condition due to increasing number of attempted connections.
- Increasing the duration of failed connections is the direct symptom of the problems with continuity of the communication (referring to ADDAX recommendation the threshold value for the duration of failed connections indicating communication problem is 180 s in one hour).

4.1. Case 1 (Supply Voltage Scenario 1, Load Condition 0): Operation of the Tested PRIME PLC System without Loads during Sinusoidal Supply Voltage—Reference Analysis

As the system itself makes use of the supply voltage (power supplies for meters, meter displays, concentrator, etc.), it was decided to initially study the impact of voltage distortion without the loads, first of all. This testing stage serves, in a way, as an immunity test of transmission efficiency at a different degree of distortion of the supply voltage, but without the participation of higher frequencies introduced by the loads. At this level of testing, the transmission efficiency statistics available in the master system of the tested PLC are mainly used. With the use of PCR-LA-500 generators, a perfectly sinusoidal waveform of the supply voltage was defined and the first cycle of observation of the operation of the tested PLC system was performed. The results of spectral analysis for the considered case of PRIME PLC system operation in scenario 1 (pure sinusoid supply voltage) without loads are shown in Figure 7. Figure 7a. presents voltage in the point of the equipment connection, which is also the point of energy meter connection with PRIME PLC receiver (denoted in figures as line voltage "uL"). The only impact on the primary voltage distortion, i.e., before the PLC communication is switched on, are the currents of the power supplies of the concentrator and the power supplies of the systems and the meter displays. It should be noted that the power supply itself is not an ideal appliance. According to the technical specification, the error of the forced waveform could be up to 0.3%. Figure 7b presents the time-varying nature of the PRIME PLC communication during 20 s of observation. The figure was obtained using a sliding 100 ms window of DFT analysis using grouping in the 200 Hz band in 100 Hz central frequency. It can be visible that the investigated technology uses several packages of broadband signal in the range of 42–89 kHz. In order to identify details of the communication signal and the frequency spectrum of the reference analysis (pure sinusoid supply voltage and no load), the maximum and minimum spectra from the 200 local spectra (20 s of observation with 100 ms window) have been selected. Figure 7c identifies spectra representing voltage, with and without PRIME PLC communication. As shown in Figure 7c, a supply voltage spectrum in investigated cases is characterised by a practically constant value (grey colour in the diagram) not exceeding 60 dBμV with narrow-band components of higher frequencies. This waveform was obtained under the conditions of supplying the system with a nominal voltage of 400 V, 50 Hz, $THD_U \approx 0\%$. This characteristic can be treated as a background for the transmission system (noise). The green line in Figure 7c represents the non-intentional emission limit values, according to EN 50065 [3]. It can be noted that using the proposed laboratory setup, it is possible to obtain conditions representing the permissible range of non-intentional emission in the public grid. The black line in Figure 7c represents the identified spectrum of intentional emission introduced by PRIME PLC transmission. From the graph, the following PLC transmission parameters in the 42–89 kHz band can be identified:

- Local maximum of the non-intentional emission level (noise, background): N_{max} = 65.72 dBμV.
- The intentional transmission signal level related to N_{max}: $S(N_{max})$ = 87.51 dBμV.
- Local minimum signal-to-noise ratio: SNR_{local} = 21.79 dB.
- Signal-to-noise ratio in the full transmission band: SNR_{band} = 31.21 dB.

Such conditions guarantee the correct operation of the PRIME PLC system in the 42–89 kHz band, which has been confirmed by the observation of meter-concentrator communication PLC statistics. Communication statistics have been also checked during a continuous system operation for 2 h. The correct transmission between the concentrator and the energy meters was confirmed.

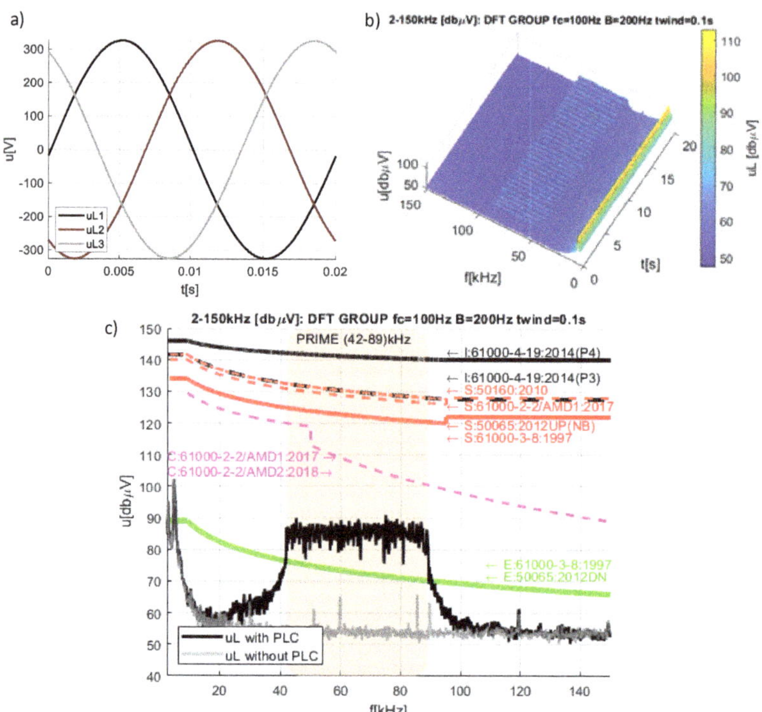

Figure 7. Case 1: Results of the analysis of PRIME PLC system operation without loads at pure sinusoidal supply voltage (reference case); (**a**) supply voltage waveform; (**b**) time-frequency representation in the range 2–150 kHz band using 100 ms window during the 20 s observation; (**c**) comparison of the spectrum of the corresponding PRIME PLC transmission (black) and the spectrum of the voltage in the connection point of the load and energy meter (grey) with relation to the normative curves (uL—line voltage in the point of the equipment connection, which is also the point of energy meter connection with PRIME PLC).

4.2. Case 2 (Supply Voltage Scenario 1, Load Condition 1): Operation of the Tested PRIME PLC System with LED Light Sources during Sinusoidal Supply Voltage

The measurements were then repeated for the condition when the sinusoidal voltage supplied the LED light sources. Figure 8 represents: (a) investigated voltage, (b) time-frequency plane of 20 s observation of the transmission including non-intentional component, (c) comparison of the spectrum of the transmission and non-intentional emission signals (noise, background for the transmission). In comparison to the no-load condition, the spectrum of voltage measured in the connection point of the LED and the energy meter is characterised by an additional narrowband non-intentional component around 60–80 kHz. However, the magnitude of this component did not affect the transmission signal. Details of the characteristic parameters of the spectrum component in the PRIME PLC transmission band 42–89 kHz are as follows:

- Local maximum of the non-intentional emission level (noise, background): N_{max} = 75.63 dBµV.
- The intentional transmission signal level related to N_{max}: $S(N_{max})$ = 92.67 dBµV.
- Local minimum signal-to-noise ratio: SNR_{local} = 17.04 dB.
- Signal-to-noise ratio in the full transmission band: SNR_{band} = 30.63 dB.

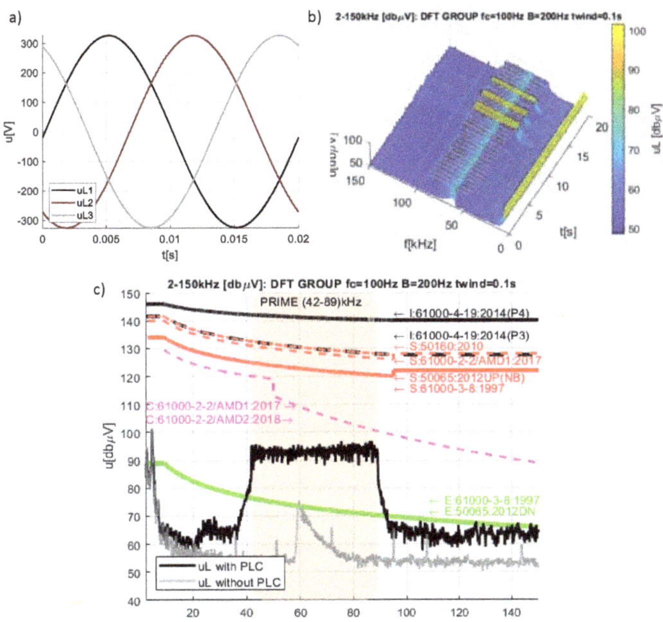

Figure 8. Case 2: results of the analysis of PRIME PLC system operation with LED light sources at pure sinusoidal voltage supply; (**a**) supply voltage waveform; (**b**) time-frequency representation in the range 2–150 kHz using 100 ms window during the 20 s observation; (**c**) comparison of the spectrum of the corresponding PRIME PLC transmission (black) and the spectrum of the voltage in the connection point of the load and energy meter (grey) with the relation to the normative curves (uL—line voltage in the point of the equipment connection, which is also the point of energy meter connection with PRIME PLC).

Communication statistics have been also simultaneously checked. During a continuous system operation for 2 h, correct transmission between the concentrator and the energy meters was confirmed.

- Statistics of the connection in the presented case are expressed by:
- Total number of the connections: 53 (including 44 successful and 9 failed connections).
- Duration of failed connections in one hour: 2 s.

4.3. Case 3 (Supply Voltage Scenario 2, Load Condition 1): Operation of the Tested PRIME PLC System with LED Light Sources during Distorted Supply Voltage Referring to Permissible Limits for Public Grid EN 50160

The tests presented in the previous section have been repeated under operating conditions with a distorted supply voltage with the maximum harmonic content allowed by EN 50160:2015 [3] (see Table 2). The supply voltage waveform distortion was achieved using the programmable AC source and authorised software for harmonic superposition. The created waveform of the distorted supply voltage is illustrated in Figure 9a. The waveform is characterised by the total harmonic distortion index equals to 11.62%.

Table 2. The magnitude of particular harmonics of the supply voltage according to the permissible values of EN 50160:2015 [3].

Odd Harmonics				Even Harmonics	
Undividable by 3		Dividable by 3			
Order	Amplitude	Order	Amplitude	Order	Amplitude
5	6.0%	3	5.0%	2	2.0%
7	5.0%	9	1.5%	4	1.0%
11	3.5%	15	0.5%	6 … 24	0.5%
13	3.0%	21	0.5%	>24	0.5%
17	2.0%	>21	0.5%		
19	1.5%				
23	1.5%				
25	1.5%				
>25	1%				

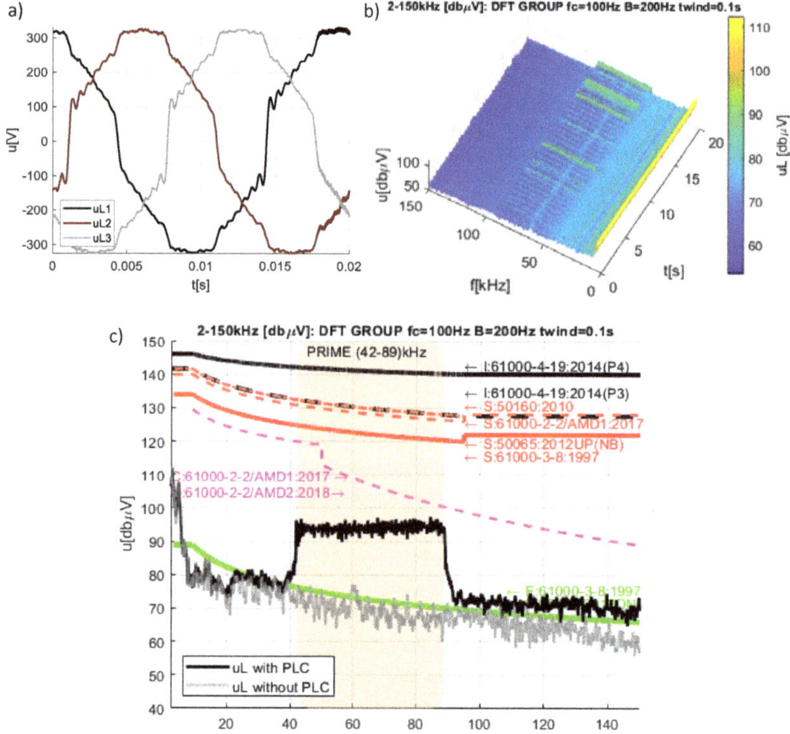

Figure 9. Case 3: results of the analysis of PRIME PLC system operation with LED light sources at distorted supply voltage referring to permissible limits for public grid defined in EN 50160:2015 [3]; (**a**) supply voltage waveform; (**b**) time-frequency representation in the range 2–150 kHz using 100 ms window during the 20 s observation; (**c**) comparison of the spectrum of the corresponding PRIME PLC transmission (black) and the spectrum of the voltage in the connection point of the load (grey) with the relation to the normative curves (uL—line voltage in the point of the equipment connection, which is also the point of energy meter connection with PRIME PLC).

Spectrum analysis using 200 of 100 ms windows (Figure 9b) allowed identifying the characteristic spectra of non-intentional emission and PRIME PLC transmission. Observing Figure 9c it can be concluded that as it results from the distorted voltage supply waveform the voltage spectrum of the non-intentional emission in the connection point of the load and energy meter resulting from the distorted voltage supply waveform is higher than in the case of operation under reference conditions (sinusoidal rated voltage). From the graph in Figure 9c, one can read the following signal parameters in the point of the 42–89 kHz transmission band are expressed by:

- Local maximum of the non-intentional emission level (noise, background): N_{max} = 77.08 dBμV.
- The intentional transmission signal level related to N_{max}: $S(N_{max})$ = 93.55 dBμV.
- Local minimum signal-to-noise ratio: SNR_{local} = 16.47 dB.
- Signal-to-noise ratio in the full transmission band: SNR_{band} = 23.57 dB.

The SNR value in the PRIME PLC communication band proved to be sufficient for the transmission of signals in the majority of trials to be effective, as evidenced by the PLC system statistics. However, going into the details of the statistics it has to be emphasized that in order protect against the potential loss of the transmission the investigated PRIME PLC system increased the number of the connection two times in comparison to reference conditions. The magnitude of the transmission signal was also adapted to worse connection conditions and increased from 90 dBμV (reference condition) to 95–97 dBμV.

Statistics of the connection in the presented case are expressed by:

- Total number of the connections: 103 (including 83 successful and 20 failed connections).
- Duration of failed connections in one hour: 187 s.

4.4. Case 4 (Supply Voltage Scenario 4, Load Condition 1) Operation of the Tested PRIME PLC System with LED Light Sources during Distorted Supply Voltage Referring to Immunity Test Limits

The final scenario of the supply voltage distortion represents harmonic contribution used in low-frequency immunity tests, according to IEC 61000-4-13 [30] addressed to the equipment of the third class of the immunity. Details of the harmonic contents are presented in Table 3. The distorted supply voltage waveform obtained using the programmable source is depicted in Figure 10a. The waveform is characterised by the total harmonic distortion index equals 25.69%. This scenario can be treated as a case that exceeds the normal operating condition of the equipment significantly.

Table 3. The magnitude of particular harmonics of the supply voltage used in low-frequency immunity tests according to IEC 61000-4-13 [30] addressed to the equipment of third class of the immunity.

The Harmonic Order h = 3n + 1	Class 3 Test Levels in %U1	The Harmonic Order h = 3n	Class 3 Test Levels in %U1
5	12	3	9
7	10	9	4
11	7	15	3
13	7	21	2
17	6	27	2
19	6	33	2
23	6	39	2
25	6		
29	5		
31	3		
35	3		
37	3		

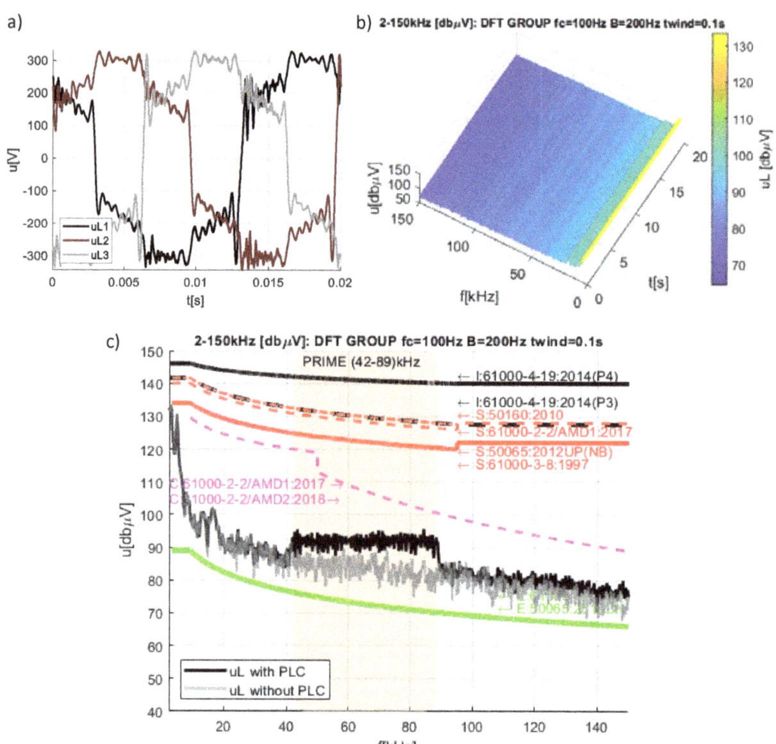

Figure 10. Case 4: results of the analysis of PRIME PLC system operation with LED light sources at distorted supply voltage referring to permissible limits for low-frequency immunity tests defined in IEC 61000-4-13 [30]; (**a**) supply voltage waveform; (**b**) time-frequency representation in the range 2–150 kHz using 100 ms window during the 20 s observation; (**c**) comparison of the spectrum of the corresponding PRIME PLC transmission (black) and the spectrum of the voltage in the connection point of the load (grey) with the relation to the normative curves (uL—line voltage in the point of the equipment connection, which is also the point of energy meter connection with PRIME PLC).

In contrast to the previously described scenarios, Figure 10b shows a significant impact of the distorted supply voltage on the non-intentional emission at the point of the connection of the LED and energy meter. The characteristic time packages of the broadband transmission signal in the range 42–89 kHz are not recognized now. It can be generally concluded that the non-intentional emission "covered" the intentional transmission signal. A comparison of the characteristic spectra of the intentional and non-intentional emission is presented in Figure 10c. Following signal parameters in the point of the 42–89 kHz transmission band can be identified as:

- Local maximum of the non-intentional emission level (noise, background): N_{max} = 89.82 dBμV.
- The intentional transmission signal level related to N_{max}: $S(N_{max})$ = 92.96 dBμV.
- Local minimum signal-to-noise ratio: SNR_{local} = 3.14 dB.
- Signal-to-noise ratio in the full transmission band: SNR_{band} = 7.61 dB.

In point of the transmission statistics, the transmission was effective in the majority of trials. However, similar to the previous case, when the distorted supply voltage referring to permissible levels of harmonics in a public grid was considered, using supply waveform distortion basing on IEC 61000-4-13 kept increasing number of the connection between concentrator and energy meters. The maximum time of failed connection extends the acceptable 180 s in one hour.

Statistics of the connection in the presented case are expressed by:

- Total number of the connections: 106 (including 86 successful and 20 failed connections).
- Duration of failed connections in one hour: 371 s.

4.5. Comparative Analysis: Operation of the Tested PRIME PLC System under the Different Scenario of Supply Voltage Distortion and Variant Types of Loads

Previously presented results described in detail cases with LED light sources and consequently increased level of supply voltage distortion. The presented results aimed to express the methodology of the investigation with special consideration of the identification and analysis of the characteristic spectra of voltage in the point of the connection of the load and energy meter. As already mentioned, the loads responsible for the majority of data transmission interference in systems based on PLC transmission are LED and CFL light sources and induction motors powered by a pulse width modulation (PWM) frequency inverter. Therefore, in order to assess the impact of the power supply voltage distortion on the operation of the tested PRIME PLC system, with the simultaneous presence of emissions of these loads, extended investigations were performed using several cases formulated as mixtures of the different scenario of supply voltage distortion and variant types of load. It resulted in ten investigation cases for deliberate comparative analysis:

- No-load and sinusoidal supply voltage (reference condition)—1 case.
- LED light sources (sinusoidal, EN 50160, IEC 61000 4-13)—3 cases.
- CFL light sources (sinusoidal, EN 50160, IEC 61000 4-13)—3 cases.
- Induction motor powered by PWM (sinusoidal, EN 50160, IEC 61000 4-13)—3 cases.

Spectral analysis in the range of 2–150 kHz was carried out for the mentioned ten investigation cases. Figure 11 presents spectra of the investigated voltage in the point of the connection of the load and energy meter with PRIME PLC communication for: (a) LED, (b) CFL, and (c) induction motor with PWM converter, respectively. For better representation, every figure consists of margin characteristic spectra reflecting the background of the transmission obtained for the sinusoidal supply and no-load denoted as reference condition (grey) as well as spectrum of the signal with PRIME PLC transmission (black). The next three curves express the investigated voltage spectra for the consequently deteriorating supply voltage condition (brown—sinusoidal, yellow—EN 50160, orange—IEC 61000-4-13).

The first comment to the obtained results presented in Figure 11 might be addressed to the comparison of the non-intentional emission of the investigated loads for the sinusoidal supply voltage (brown) and normative curves representing emission in a power grid. It can be noticed that, for normative supply conditions, the spectra differ significantly and are characterised by different frequency components in the observed range 2–150 kHz. Moreover, temporarily the obtained spectra exceed the curves representing the normative level of non-intentional emission in the public grid defined in IEC 61000-3-8 [4] adopted in EN 50065:2012 [2] (green normative curve).

The second comment might be addressed to the influence of the distorted voltage condition on non-intentional emission in the range 2–150 kHz. As it was shown in Figure 11, increasing the level of the contribution of the low-frequency component 0–2 kHz in the supply voltage affected the non-intentional emission in the observed range 2–150 kHz in the connection point of the loads and energy meter with PRIME PLC transmission (yellow and orange lines). For the margin condition representing low-frequency contribution used in the immunity test, according to IEC 61000-4-13, the investigated spectra were very close to the intentional transmission signal of the PRIME PLC system (black).

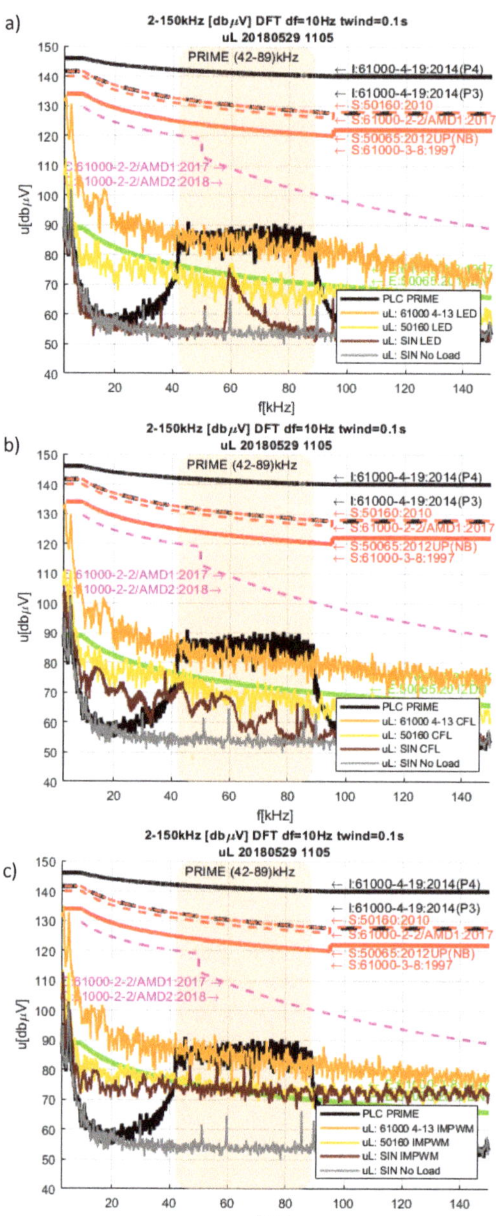

Figure 11. Summary: spectra of the voltage in the connection point of the load and energy meter for mixture cases representing results of the analysis of PRIME PLC system operation with (**a**) LED; (**b**) CFL; and (**c**) induction motor with PWM converter; under the different contribution of low-frequency distortion in the supply voltage (pure sinusoid (brown), EN 50160:2015 [3] (yellow), IEC 61000-4-13 [30] (orange)) with relation to no-load condition (grey) and PRIME transmission (black) (uL—line voltage in the point of the equipment connection, which is also the point of energy meter connection with PRIME PLC).

Details of the representative parameters of the transmission signal and the non-intentional emission in the transmission band 42–89 kHz of the investigated PRIME PLC system were collected in Table 4. It can be noticed that the condition of the transmission was consequently worse when the distortion of the supply voltage was increased.

Table 4. Characteristic parameters of the signal in the transmission band 42–89 kHz of PRIME PLC during the different scenarios of supply voltage distortion and variant types of loads.

Sinusoidal Supply Voltage		**No-Load**	**LED**	**CFL**	**Induction Motor with PWM**
$S(N_{max})$	(dBµV)	87.51	92.67	94.09	93.43
N_{max}	(dBµV)	65.72	75.63	73.47	83.32
SNR_{local}	(dB)	21.79	17.04	20.61	10.11
P_{signal}	(dB)	−31.22	−23.63	−25.37	−24.38
P_{noise}	(dB)	−62.42	−54.26	−51.68	−43.35
SNR_{band}	(dB)	31.21	30.63	26.31	18.97
Distorted Supply Voltage EN 50160 (Table 2)		**No-Load**	**LED**	**CFL**	**Induction Motor with PWM**
$S(N_{max})$	(dBµV)	87.51	93.55	92.46	92.92
N_{max}	(dBµV)	65.72	77.08	78.12	81.22
SNR_{local}	(dB)	21.79	16.47	14.35	11.70
P_{signal}	(dB)	−31.22	−22.84	−23.53	−23.99
P_{noise}	(dB)	−62.42	−46.41	−46.55	−42.42
SNR_{band}	(dB)	31.21	23.57	23.02	18.43
Distorted Supply Voltage IEC 61000-4-13 Class 3 (Table 3)		**No-Load**	**LED**	**CFL**	**Induction Motor with PWM**
$S(N_{max})$	(dBµV)	87.51	92.96	91.80	91.11
N_{max}	(dBµV)	65.72	89.82	88.86	91.90
SNR_{local}	(dB)	21.79	3.14	2.94	−0.79
P_{signal}	(dB)	−31.22	−24.92	−24.48	−29.45
P_{noise}	(dB)	−62.42	−32.53	−34.22	−31.94
SNR_{band}	(dB)	31.21	7.61	9.73	2.49

N_{max}—local maximum magnitude of the frequency component of the background signal; $S(N_{max})$—local magnitude of the frequency component of the transmission signal related to N_{max}; SNR_{local}—signal-to-noise ratio derived on the basis of local values of $S(N_{max})$ and N_{max}; P_{signal}—power of the transmission signal in the transmission band; P_{noise}—power of the background signal (noise) in the transmission band; SNR_{band}—signal-to-noise ration derived on the basis of P_{signal} and P_{noise}.

The impact of the supply voltage distortion on the signal-to-noise ratio (SNR) in the transmission band 42–89 kHz of the investigated PRIME PLC system is presented in Table 4 and illustrated in Figure 12. In particular:

- With LED load the SNR coefficient decreased from 30.06 dB (SNR_{band})/17.04 dB (SNR_{local}) for the sinusoidal normal condition to 7.61 dB(SNR_{band})/3.14 dB (SNR_{local}) for margin distorted condition, which made PLC transmission more difficult;
- With CFL load the SNR coefficient decreased from 26.31 dB (SNR_{band})/20.61 dB (SNR_{local}) for the sinusoidal normal condition to 9.73 dB (SNR_{band})/2.94 dB (SNR_{local}) for margin distorted condition, which made PLC transmission more difficult;
- With induction motor powered by PWM converter, the SNR coefficient decreased from 18.97 dB (SNR_{band})/10.11 dB (SNR_{local}) for the sinusoidal normal condition to 2.49 dB (SNR_{band})/−0.79dB (SNR_{local}) for margin distorted condition, which made PLC transmissions impossible.

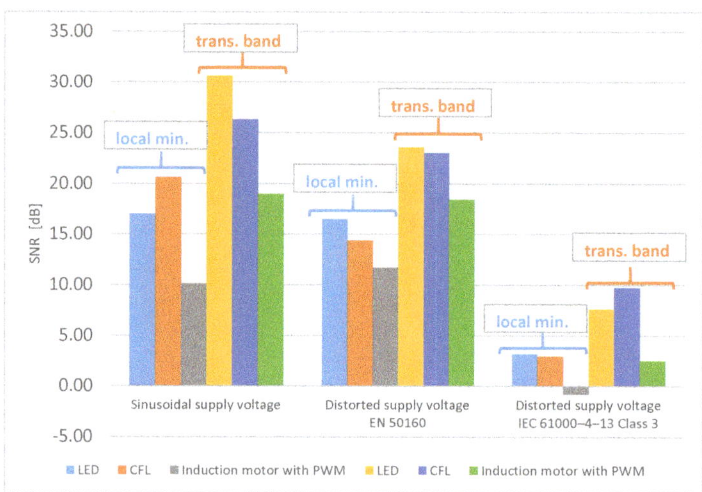

Figure 12. Impact of the supply voltage distortion grade on the signal to noise ratio (SNR) in the transmission band 42–89 kHz of the investigated PRIME PLC system.

Direct investigation of a bit error rate (BER) was not performed in the presented study. However, based on [44,45] it can be roughly estimated that for the achieved range of SNR reduction from 30.63 dB to 2.94 dB possible increase of BER could be in the range from 10^{-5} to 10^{-1}.

Details of the connections statistics in the investigated cases are collected in Table 5. It can be generally noticed that increasing the level of the supply voltage distortion resulted in increasing the total number of the attempted connections as well as a long duration of the failed connections during one hour. Figure 13 depicts the impact of the supply voltage distortion on the duration of the failed connections in one hour of the investigated PRIME PLC system concerning the different types of the investigated equipment. A prominent case is represented by waveform distortion used in the immunity test of the equipment. This case indicates the duration of the failed connections in one hour longer than level 180 s specified by the producer as a margin condition for proper connection. In particular:

- With LED load the number of attempted connections increased two times and the duration of failed connections increased from 2 s to 371 s when the distortion of the supply voltage was increasing;
- With CFL load the number of attempted connections were relatively higher than in the case of LED load and the duration of failed connections increased from 186 s to 366 s;
- With induction motor powered by PWM converter the duration of the failed connections obtained the highest value of 551 s noticed under the distorted supply condition. Additionally, in the case of the margin-investigated supply voltage distortion, the PLC system was not able to retry the connections and the number of the attempted connection was relatively small.

Table 5. The statistics of the transmission of the investigated PRIME PLC during the different scenarios of supply voltage distortion and variant types of loads (2 h continuous work).

Sinusoidal Supply Voltage	LED	CFL	Induction Motor with PWM
total number of attempted connections	53	95	110
number of successful connections	44	76	90
number of failed connections	9	19	20
duration of failed connections in one hour (s)	2	186	186
Distorted supply voltage EN 50160 (Table 2)	**LED**	**CFL**	**Induction motor with PWM**
total number of connections	103	54	93
number of successful connections	83	44	83
number of failed connections	20	10	10
duration of failed connections in one hour (s)	187	187	551
Distorted supply voltage IEC 61000-4-13 Class 3 (Table 3)	**LED**	**CFL**	**Induction motor with PWM**
total number of connections	106	104	28
number of successful connections	80	84	24
number of failed connections	20	20	4
duration of failed connections in one hour (s)	371	366	192

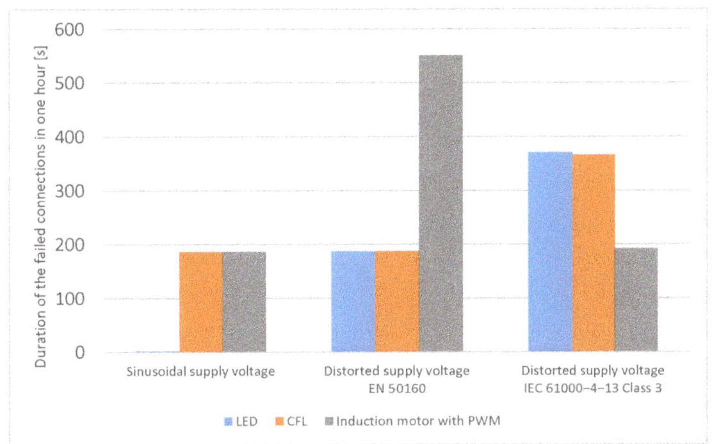

Figure 13. Impact of the supply voltage distortion grade on the duration of the failed connections in one hour of the investigated PRIME PLC system.

5. Discussion

Taking into account the results of the research presented in this paper, it can be stated that the harmonic content of the supply voltage can be considered as one of the origins of the increasing level of supraharmonics non-intentional emission. As it was shown in the paper, successively increased deterioration grade of the supply voltage waveform, representing respectively harmonic content of permissible level defined for the public grid as well as for the immunity test, had a direct influence on non-intentional emission in the range 2–150 kHz, that in consequence affected the transmission of the investigated PRIME PLC system.

The obtained results allow us to draw several conclusions, namely:

- For all investigated loads (CFL, LED, the motor with PWM) it was observed that increasing the content of higher harmonics in the supply voltage from the sinusoidal condition, by the waveform distortion representing permissible content of harmonic in a public grid, to the waveform distortion defined in the immunity test of the equipment, resulted in a higher level of non-intentional emission in the range 2–150 kHz. As a consequence, the SNR derived from the local maximum magnitude of the background signal and associated with it the local magnitude of the transmission signal of the investigated PRIME PLC system was consistently decreasing, from 17.04 dB to 3.14 dB (for LED), from 20.61 dB to 2.94 dB (for CFL) and from 10.11 dB to −0.79 dB (for the motor with PWM). The SNR calculated on the basis of power of the transmission signal and power of the background signal in the transmission band was also decreasing, from 30.06 dB to 7.61 dB (for LED), from 26.31 dB to 9.73 dB (for CFL), and from 18.97 dB to 2.49 dB (for the motor with PWM), respectively.
- An indirect result of the supply voltage distortion is an increasing number of the connections activated by the investigated PLC system in order to prevent loss of the connection. However, in case of a high level of supply voltage distortions, the duration of the failed connections increased significantly. The duration of failed connections decreased from 2 s to 371 s for the LED load, and from 186 s to 366 s for the CFL load. The highest value of the duration of failed connections, 551 s, was achieved in the case of the induction motor powered by PWM.
- The scenarios of the supply voltage distortion used in the investigation represent a relatively high level of deterioration (direct sum of harmonics permissible harmonic for public grid is represented by THD equals 11.62%; voltage distortion used in the immunity test equipment of the third class of immunity is represented by THD equals 25.69%). However, the formulated relation between the condition of the supply voltage and non-intentional emission suggests extending the discussion about the condition of the tests of non-intentional emissions, which currently are performed under sinusoidal conditions. A proposition for an extended test using deteriorated supply voltage, referring to the permissible level of THD in a low-voltage public network, gradually increasing from a few percentage points to 8%, can be considered.
- The proposition of extended testing of non-intentional emissions under a distorted supply voltage might also be valuable for a more effective filter specification and selection for particular power line communication technologies.

Author Contributions: Conceptualization: M.W., T.S., M.H.; methodology: M.W., T.S., M.H.; software: T.S., J.S. (Jaroslaw Szymanda), G.W.; validation: M.H., P.K., L.G., J.S. (Jaroslaw Sokol), M.J.; formal analysis: M.W., T.S., M.H., P.K., L.G., J.S. (Jaroslaw Sokol), M.J.; investigation: M.W., T.S., P.K.; resources: M.H., P.K., L.G., J.S. (Jaroslaw Sokol), M.J.; data curation: P.K., J.S. (Jaroslaw Szymanda), G.W.; writing—original draft: M.W., T.S.; writing—review and editing: M.H., P.K., G.W.; visualization: M.W., T.S., G.W., P.K.; supervision: T.S., M.H. All authors have read and agreed to the published version of the manuscript.

Funding: This research was founded by a subsidy of the Polish Ministry of Science and Higher Education for research activities at the Faculty of Electrical Engineering, Wrocław University of Science and Technology.

Data Availability Statement: The data presented in this study are available on request from the corresponding author.

Acknowledgments: In the research, commercial measurement equipment and software for supraharmonics analysis was used to validate the university laboratory setup. This part of the investigations was supported by TAURON Dystrybucja Pomiary Ltd., Poland.

Conflicts of Interest: The authors declare no conflict of interest.

References

1. Benchmarking smart metering deployment in the EU-28. In *Final Report, European Commission, Directorate-General for Energy, Directorate B—Internal Energy Market Unit B.3—Retail Markets; Coal & Oil*; Publications Office of the European Union: Luxembourg, Luxembourg, 2020. [CrossRef]
2. EN 50065-1:2012. *Signalling on Low-Voltage Electrical Installations in the Frequency Range 3 kHz to 148.5 kHz—Part 1: General Requirements; Frequency Bands and Electromagnetic Disturbances*; CEN-CENELEC Management Centre: Brussels, Belgium; BSI: London, UK, 2011.
3. EN 50160: 2010. *Voltage Characteristics of Electricity Supplied by Public Distribution Systems*; CEN-CENELEC Management Centre: Brussels, Belgium, 2010.
4. IEC 61000-3-8: 1997. *Part 3: Limits—Section 8: Signalling on Low-Voltage Electrical Installations—Emission Levels, Frequency Bands and Electromagnetic Disturbance Levels*; International Electrotechnical Commission: Geneva, Switzerland; BSI: London, UK, 1997.
5. IEC 61000-2-2: AMD1:2017. *Part 2-2: Environment—Compatibility Levels for Low-Frequency Conducted Disturbances and Signalling in Public Low-Voltage Power Supply Systems*; International Electrotechnical Commission: Geneva, Switzerland, 2017.
6. IEC 61000-4-19:2014. *Part 4-19: Testing and Measurement Techniques—Test for Immunity to Conducted, Differential Mode Disturbances and Signalling in the Frequency Range 2 kHz to 150 kHz at a.c. Power Ports*; International Electrotechnical Commission: Geneva, Switzerland, 2014.
7. Meyer, J.; Khokhlov, V.; Klatt, M.; Blum, J.; Waniek, C.; Wohlfahrt, T.; Myrzik, J. Overview and Classification of Interferences in the Frequency Range 2–150 kHz (Supraharmonics). In Proceedings of the International Symposium on Power Electronics; Electrical Drives; Automation and Motion (SPEEDAM), Amalfi, Italy, 20–22 June 2018; pp. 165–170.
8. Rönnberg, S.K.; Bollen, M.H.; Amaris, H.; Chang, G.W.; Gu, I.Y.; Kocewiak, Ł.H.; Meyer, J.; Olofsson, M.; Ribeiro, P.F.; Desmet, J. On waveform distortion in the frequency range of 2–150 kHz. Review and research challenges. *Electr. Power Syst. Res.* **2017**, *150*, 1–10. [CrossRef]
9. Alfieri, L.; Bracale, A.; Carpinelli, G.; Larsson, A. Accurate assessment of waveform distortions up to 150 kHz due to fluorescent lamps. In Proceedings of the 6th International Conference on Clean Electrical Power (ICCEP), Santa Margherita Ligure, Italy, 27–29 June 2017.
10. Alfieri, L.; Bracale, A.; Varilone, P.; Leonowicz, Z.; Kostyla, P.; Sikorski, T.; Wasowski, M. Methods for Assessment of Supraharmonics in Power Systems. Part I: Theoretical Issues. In Proceedings of the 7th International Conference on Clean Electrical Power (ICCEP), Otranto, Italy, 2–4 July 2019; pp. 117–122.
11. Alfieri, L.; Bracale, A.; Varilone, P.; Leonowicz, Z.; Kostyla, P.; Sikorski, T.; Wasowski, M. Methods for Assessment of Supraharmonics in Power Systems. Part II: Numerical Applications. In Proceedings of the 7th International Conference on Clean Electrical Power (ICCEP), Otranto, Italy, 2–4 July 2019; pp. 123–128.
12. Angulo, I.; Arrinda, A.; Fernandez, I.; Uribe-Perez, N.; Arehalde, I.; Hernandez, L. A review of measurement technique for non-intentional emissions above 2 kHz. In Proceedings of the IEEE International Energy Conference, Leuven, Belgium, 4–8 April 2016.
13. Grevener, A.; Meyer, J.; Ronberg, S. Comparison of measurement methods for the frequency range 2–150 kHz (Supraharmonics). In Proceedings of the IEEE 9th International Workshop on Applied Measurements for Power Systems, Bologna, Italy, 26–28 September 2018.
14. IEC 61000-4-7/A1:2009. *General Guide on Harmonics and Interharmonics Measurements and Instrumentation*; International Electrotechnical Commission: Geneva, Switzerland, 2009.
15. IEC 61000-4-30:2015. *Part 4-30: Testing and Measurement Techniques—Power Quality Measurement Methods*; International Electrotechnical Commission: Geneva, Switzerland, 2015.
16. Bollen, M.; Rönnberg, S. Propagation of Supraharmonics in the Low Voltage Grid. In Proceedings of the Energiforsk, Stockholm, Sweden, 25 January 2017.
17. Gil-de-Castro, A.; Medina-Gracia, R.; Rönnberg, S.K.; Blanco, A.M.; Meyer, J. Differences in the performance between CFL and LED lamps under different voltage distortions. In Proceedings of the 18th International Conference on Harmonics and Quality of Power (ICHQP), Ljubljana, Slovenia, 13–16 May 2018. [CrossRef]
18. Gil-de-Castro, A.; Rönnberg, S.K.; Bollen, M.H.J. Harmonic interaction between an electric vehicle and different domestic equipment. In Proceedings of the International Symposium on Electromagnetic Compatibility, Chiyoda, Tokyo, Japan, 12–16 May 2014.
19. Gil-de-Castro, A.; Rönnberg, S.K.; Bollen, M.H.J. A study about harmonic interaction between devices. In Proceedings of the 16th International Conference on Harmonics and Quality of Power (ICHQP), Bucharest, Romania, 25–28 May 2014.
20. IEC CLC/TR 50669:2017. *Technical Report. Investigation Results on Electromagnetic Interference in the Frequency Range below 150 kHz*; CEN-CENELEC Management Centre: Brussels, Belgium, 2017.
21. Andersson, M.O.J.; Rönnberg, S.K.; Lundmark, C.M.; Larsson, E.O.A.; Wahlberg, M.; Bollen, M.H.J. Interfering signals and attenuation—potential problems with communication via the power grid. In Proceedings of the 19th International Conference on Electricity Distribution, Vienna, Austria, 21–24 May 2007.
22. López, G.; Moreno, J.I.; Sánchez, E.; Martínez, C.; Martín, F. Noise Sources, Effects and Countermeasures in Narrowband Power-Line Communications Networks: A Practical Approach. *Energies* **2017**, *10*, 1238. [CrossRef]

23. Rönnberg, S.K.; Bollen, H.H.J.; Wahlberg, M. Interaction between Narrowband Power-Line Communication and End-User Equipment. *IEEE Trans. Power Deliv.* **2011**, *26*, 20134–22039. [CrossRef]

24. Shklyarskiy, Y.; Hanzelka, Z.; Skamyin, A. Experimental Study of Harmonic Influence on Electrical Energy Metering. *Energies* **2020**, *13*, 5536. [CrossRef]

25. Uribe-Pérez, N.; Angulo, I.; Hernández-Callejo, L.; Arzuaga, T.; De la Vega, D.; Arrinda, A. Study of Unwanted Emissions in the CENELEC-A Band Generated by Distributed Energy Resources and Their Influence over Narrow Band Power Line Communications. *Energies* **2016**, *9*, 1007. [CrossRef]

26. Thai, M.T.; Wu, W.; Xiong, H. (Eds.) *Big Data in Complex and Social Networks*, 1st ed.; CRC Press: Boca Raton, FL, USA, 2016.

27. Blanco, A.M.; Gupta, M.; de Castro, A.G.; Rönnberg, S.K.; Meyer, J. Impact of flat-top voltage waveform distortion on harmonic current emission and summation of electronic household appliances. In Proceedings of the International Conference on Renewable Energies and Power Quality ICREPQ'18, Salamanca, Spain, 21–23 March 2018.

28. Mansoor, A.; Grady, W.M.; Thallam, R.S.; Doyle, M.T.; Krein, S.S.; Samotyj, M.J. Effect of Supply Voltage Harmonics on the Input Current of Single-phase Diode Bridge Rectifier Loads. *IEEE Trans. Power Deliv* **1995**, *10*, 1416–1422. [CrossRef]

29. Huang, X.-L.; Ma, X.; Hu, F. Machine learning and intelligent communications. *Mob. Netw. Appl.* **2018**, *23*, 68–70. [CrossRef]

30. IEC 61000-4-13:2002 & A1:2009 & A2:2015. *Electromagnetic Compatibility (EMC): Testing and Measurement Techniques. Harmonics and Interharmonics Including Mains Signalling at a.c. Power Port; Low-Frequency Immunity Tests;* International Electrotechnical Commission: Geneva, Switzerland, 2002.

31. Yanchenko, S.; Meyer, J. Harmonic Emission of Household Devices in Presence of Typical Voltage Distortions. In Proceedings of the 2015 IEEE PowerTech Conference, Eindhoven, The Netherlands, 29 June 2015.

32. KIKUSUI AC Power Supply PRC-LA-Series. Available online: https://www.kikusui.co.jp/common/product/pdf/pcr-la.pdf (accessed on 20 March 2020).

33. IEC 61000-3-3:2013. *Part 3: Limits—Section 3: Limitation of Voltage Changes, Voltage Fluctuations and Flicker in Public Low-Voltage Supply Systems, for Equipment with Rated Current ≤ 16 A per Phase and Not Subject to Conditional Connection;* International Electrotechnical Commission: Geneva, Switzerland; BSI: London, UK, 2013.

34. KIKUSUI Three-phase Output Driver for PCR-LA Series. Available online: https://https://manual.kikusui.co.jp/I/IB03_PCR_LA_E2.pdf (accessed on 5 January 2021).

35. KIKUSUI Quick Immunity Testing Software. Available online: https://www.kikusui.co.jp/en/download/en/?fn=sd003-pcrla_AP (accessed on 5 January 2021).

36. KIKUSUI GPIB Interface for PCR-LA Series. Available online: https://manual.kikusui.co.jp/NUMERIC/3P03_PCR_LA_E3.pdf (accessed on 5 January 2021).

37. National Instruments NI 9223 Voltage Input Module Datasheet. Available online: https://www.ni.com/pdf/manuals/374223a_02.pdf (accessed on 5 January 2021).

38. LEM Voltage Transducer CV 3-1000 Datasheet. Available online: https://www.lem.com/sites/default/files/products_datasheets/cv_3-1000.pdf (accessed on 5 January 2021).

39. Swemet MFA 500, Spectrum Analyser for PLC 3–500 kHz. Available online: http://en.swemet.se/files/Produkter/produktblad_mfa500_ver1.pdf (accessed on 5 January 2021).

40. Signal Hound USB-SA44B Spectrum Analyser Measuring Receiver 1 Hz–4.4 GHz Datasheet. Available online: https://signalhound.com/products/usb-sa44b/ (accessed on 5 January 2021).

41. ITU-T G.9904 (10/2012) Narrowband Orthogonal Frequency Division Multiplexing Power Line Communication Transceivers for PRIME Networks, International Telecommunication Union. Available online: https://www.itu.int/rec/T-REC-G.9904/en (accessed on 17 January 2020).

42. PRIME Alliance Specification for PoweRline Intelligent Metering Evolution, PRIME Alliance Technical Working Group. Available online: https://www.prime-alliance.org/wp-content/uploads/2020/04/PRIME-Spec_v1.4-20141031.pdf (accessed on 17 January 2021).

43. SIMS—Smart Integrated Metering System—ADDAX Software Solution Based on up-to-Date Addax Technology for Residential Metering—Datasheet. Available online: www.addgrup.com (accessed on 18 January 2021).

44. Korki, M.; Hosseinzadeh, N.; Moazzeni, T. Performance Evaluation of a Narrowband Power Line Communication for Smart Grid with Noise Reduction. *IEEE Trans. Consum. Electron.* **2011**, *57*, 4. [CrossRef]

45. Seijo, M.; López, G.; Matanza, J.; Moreno, J.I. Planning and Performance Challenges in Power Line Communications Networks for Smart Grids. *Int. J. Distrib. Sens. Netw.* **2016**, *2016*, 17. [CrossRef]

Article

A Case Study on a Hierarchical Clustering Application in a Virtual Power Plant: Detection of Specific Working Conditions from Power Quality Data

Michał Jasiński [1,*], Tomasz Sikorski [1], Dominika Kaczorowska [1,*], Jacek Rezmer [1], Vishnu Suresh [1], Zbigniew Leonowicz [1], Paweł Kostyła [1], Jarosław Szymańda [1], Przemysław Janik [2], Jacek Bieńkowski [2] and Przemysław Prus [2]

[1] Faculty of Electrical Engineering, Wroclaw University of Science and Technology, 50-370 Wroclaw, Poland; tomasz.sikorski@pwr.edu.pl (T.S.); jacek.rezmer@pwr.edu.pl (J.R.); vishnu.suresh@pwr.edu.pl (V.S.); zbigniew.leonowicz@pwr.edu.pl (Z.L.); pawel.kostyla@pwr.edu.pl (P.K.); jaroslaw.szymanda@pwr.edu.pl (J.S.)

[2] TAURON Ekoenergia Ltd., 58-500 Jelenia Góra, Poland; przemyslaw.janik@tauron-ekoenergia.pl (P.J.); jacek.bienkowski@tauron-ekoenergia.pl (J.B.); przemyslaw.prus@tauron-ekoenergia.pl (P.P.)

* Correspondence: michal.jasinski@pwr.edu.pl (M.J.); dominika.kaczorowska@pwr.edu.pl (D.K.); Tel.: +48-713-202-022 (M.J.); +48-713-202-901 (D.K.)

Citation: Jasiński, M.; Sikorski, T.; Kaczorowska, D.; Rezmer, J.; Suresh, V.; Leonowicz, Z.; Kostyła, P.; Szymańda, J.; Janik, P.; Bieńkowski, J.; et al. A Case Study on a Hierarchical Clustering Application in a Virtual Power Plant: Detection of Specific Working Conditions from Power Quality Data. *Energies* **2021**, *14*, 907. https://doi.org/10.3390/en14040907

Academic Editor: Pedro Faria

Received: 6 January 2021

Accepted: 6 February 2021

Published: 9 February 2021

Publisher's Note: MDPI stays neutral with regard to jurisdictional claims in published maps and institutional affiliations.

Abstract: The integration of virtual power plants (VPP) has become more popular. Thus, research on VPP for different issues is highly desirable. This article addresses power quality issues. The presented investigation is based on multipoint, synchronic measurements obtained from five points that are related to the VPP. This article provides a proposition and discussion of using one global index in place of the classical power quality (PQ) parameters. Furthermore, in the article, one new global power quality index was proposed. Then the PQ measurements, as well as global indexes, were used to prepare input databases for cluster analysis. The mentioned cluster analysis aimed to detect the short-term working conditions of VPP that were specific from the point of view of power quality. To realize this the hierarchical clustering using the Ward algorithm was realized. The article also presents the application of the cubic clustering criterion to support cluster analysis. Then the assessment of the obtained condition was realized using the global index to assure the general information of the cause of its occurrence. Furthermore, the article noticed that the application of the global index, assured reduction of database size to around 74%, without losing the features of the data.

Keywords: virtual power plant; power quality; data mining; clustering; distributed energy resources; energy storage systems; short term conditions

1. Introduction

The concept of renewable energy sources (RES) and energy storage systems (ESS) integration into virtual power plants (VPP) includes different areas. This investigation concerns power quality (PQ) and data mining (DM) issues in VPP.

The article [1] is concerned with standard IEC 61850. It proposed an extension to this standard as a step to enhance the interaction between the controller of RESs and VPPs. As one of the elements, the power quality recorders' issues were included. The demands for them in point of IEC 61850 are discussed. Finally, the proposed methodology was verified in the virtual power plant that consists of HPP, PV, and wind power plant as well as storage systems. The indicated VPP operates on a medium voltage (MV) level. In Pudjianto et al. [2], the virtual power plant is treated as an instrument to enable a cost efficient integration of RES with the present power systems. The article includes the performance analysis of a VPP system from the point of view of different indicators such as energy efficiency, power quality, and security. The analyzed case consists of fuel cells, a wind microturbine PV system, a fly wheel, a combined heat and power plant, and a storage system. Caldon et al. [3] propose the applicable framework for harmonizing operations of different VPP units. The authors indicated that decisions and profits although complying with power quality requirements and real network constraints must be a significant part of VPP operation strategies. Zhang et al. [4] considered the coordinated operation of VPPs. The bi objective dispatch model was applied for the performance optimization of multi energy VPP in terms of both economic and power quality issues. The article concerned both simulations based on a 118-node IEEE test system and a real case from Hongfeng Eco-town in China. Gong et al. [5] proposed the optimization issues concerning VPP management strategies. The customers' satisfaction, system stability, and PQ were used together with the economic objective to formulate a multi-objective optimization problem. The fuzzy multiple objective optimizations were applied to solve optimization problems. The proposed approach was verified in a test system. Beguin et al. [6] presented simulations for an islanded grid model of the virtual power plant. The VPP integrates a 0.2 GW wind power plant, a 0.1 GW PV power plant, and a 0.25 GW pumped storage system. The outcomes of the research were control strategies of the storage plant. The control strategies were investigated to highlight VPP on PQ.

Another element of the investigation is using data mining (DM) techniques to obtain knowledge from dataset. In Ref. [7], one of the DM techniques available is cluster analysis (CA). The current research connecting CA and VPP are as follows. Luo et al. [8] proposed demand response schemes based on data mining for electricity trading between VPPs and their participants. Participants' bid offers were used to categorize them using ordering points to identify the clustering structure (OPTICS) algorithm. Yi et al. [9] presented multi time scale scheduling for VPP. The scheduling contains both a day ahead bidding and real-time operations. The proposed models were based on the deferrable loads' aggregation connected with the k-mean cluster analysis. The proposed strategy enables efficient management of massive deferrable loads. Then it results in a reduction of the energy management complexity, and increases the general economics. This approach seems promising for s efficient scheduling of VPPs. Kong et al. [10] presented the robust stochastic optimal dispatching approach to solve scheduling issues. K-medoids cluster analysis was used to define typical scenarios of different units, that were integrated into VPP. Ai et al. [11] proposed a VPP load curve cluster analysis approach. It uses the principal-component dimension-reduction analysis, aggregation level clustering, and fc-means clustering. The principal component analysis is applied to determine different loads, that are aggregated into a VPP. Then hierarchical aggregation and fc-means cluster analysis was applied to divide all load output curves participating in the aggregation process. The last step of the investigation was the analysis of clustering results to establish an evaluation system. The article [12] proposes the methodology concerning distributed energy sources management. It concerns resource scheduling, aggregation, as well as remuneration. To realize the aggregation process, the k-means algorithm was proposed. The calculations of the clusters optimum number were used to determine the best number of demand response programs to be implemented by the VPP units. The research was based on 20,000 consumers and 500 DG units. Then, in article [13], the proposed methodology was extended to investigate other 2592 operation scenarios.

Under this investigation, a VPP that operates at both low voltage (LV) and medium voltage (MV) distribution networks in Poland was selected. The investigated part of the VPP consists of a 1250 kW hydropower plant (HPP) and associated 500 kWh battery energy storage system (BESS) as well as low voltage loads. The investigation is concerned with power quality issues in the selected part of the VPP. The PQ measurements were performed synchronically in five measurement points of the VPP area. The measurement points include the HPP, the BESS, the associated MV line, and two LV loads. The measurements were conducted for 182 days: from 1 May 2020 to 28 October 2020. Therefore, they are 26 weeks long from the point of view of classical PQ assessment [14].

The single parameter analysis of each measurement point for such a period of time would be very time-consuming. Thus, the concept of the global index was introduced. Such an approach, in the literature, is known under different names e.g., global power quality index [15,16]; unified power quality index [17,18]; total power quality index [19,20]; or synthetic power quality index [21,22].

This article applied the ADI index proposed in [16], and the newly proposed power quality pollution index (PQPI). Then both the classic PQ parameters and the global indexes were used to define datasets for cluster analysis. Global indexes were applied to reduce the size of the input database, without losing features of the PQ data. Then for those databases, hierarchical clustering was performed. Cluster analysis aimed to detect short term specific working conditions of the VPP from the point of power quality. As a tool to realize this, the cubic clustering criterion was applied for results assessment of hierarchical cluster analysis with the Ward algorithm for indicated databases. Finally, the application of PQPI was presented to highlight the difference from the point of PQ level for clusters.

The contributions of this research are as follows:

- The source of the data was multipoint, synchronic, and long-term power quality measurements, that were obtained from a real VPP.
- The global index approach for PQ issues was discussed, and a new index is proposed.
- The proposed input databases to cluster analysis are concerned with raw PQ data and global indexes. Global indexes were proposed to reduce the size of the input databases but the reduction has been realized while maintaining existing features of the PQ data.
- The cubic clustering criterium for hierarchical cluster analysis results was used to detect short-term working conditions of VPP, that were specific in point of power quality.
- The global index was used for comparative assessment between clusters.

To realize those contributions, the article is organized into five sections. In Section 2 the virtual power plant description, global index proposition, clustering methods, and input databases are presented. Section 3 presents the results of specific working conditions detection in view of power quality using hierarchical clustering. Section 4 presents a discussion of the results. Section 5 concludes the article.

2. Methodology and Research Object Description

This section is based on four elements. The first element is a description of the real VPP, that operates in Poland. This VPP became a source for long term, synchronic power quality measurements. Then, the global index approach was discussed. Next the long-term measurements and the global index were combined to obtain different datasets. These datasets consisted of classic power quality parameters and global power quality indexes. Then those datasets were used as an input for hierarchical clustering. The assessment of cluster assignment for those measurement data was realized using cubic clustering criterion (CCC). The CCC were applied to select the adequate number of clusters, that will indicate short-term specific working conditions from a power quality point of view. To summarise the proposed approach, the simplified methodology scheme was presented in Figure 1.

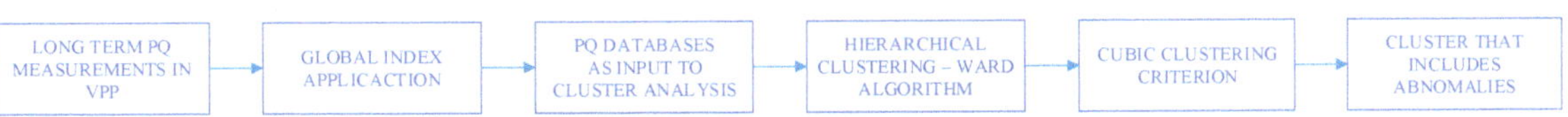

Figure 1. Simplified methodology scheme.

2.1. Virtual Power Plant That Operates in Poland as a Source of Power Quality Measurements

This article deals with a real VPP that operates in Poland, in a region called Lower Silesia. The virtual power plant consists of a fragment of the distribution network on both medium voltage (MV) and low voltage (LV) [23]. The two substations 110/20 kV are used as points of connection to the 110 kV polish grid. However, under this investigation one MV network was selected. The network fed from the 110/20 kV station is an overhead cable network. The selected network has earth fault current compensation [24]. The main distributed energy resources that are integrated into the virtual power plant are a 1.25 MW HPP and a 0.5 MW battery ESS. Both are connected to a medium voltage level.

The scheme of the investigated fragment of the VPP is presented in Figure 1. The analyzed fragment consists of a 20 kV distribution network with a 1.25 MW hydropower plant (HPP_MV) and an 0.5 MW battery ESS (ESS_MV). Those energy sources are connected with the HV/MV substation by a 20 kV line (Line_MV). Additionally, representatives of low voltage loads are indicated: LOAD1_LV and LOAD2_LV. LOAD1_LV is connected with the indicated medium voltage line (LINE_MV). LOAD2_LV is connected with the node of the HPP_MV and ESS_MV. This fragment of the VPP is monitored by power quality recorders. Power quality recorders are indicated as "R". The location of these recorders is also included in Figure 2. The HPP_MV and ESS_MV are connected to one node and their PQ recorders use the same voltage transformer. Thus, in this research, they are treated as one point for the PQ level (HPP and ESS_MV) and another for the active power level (HPP_MV and ESS_MV).

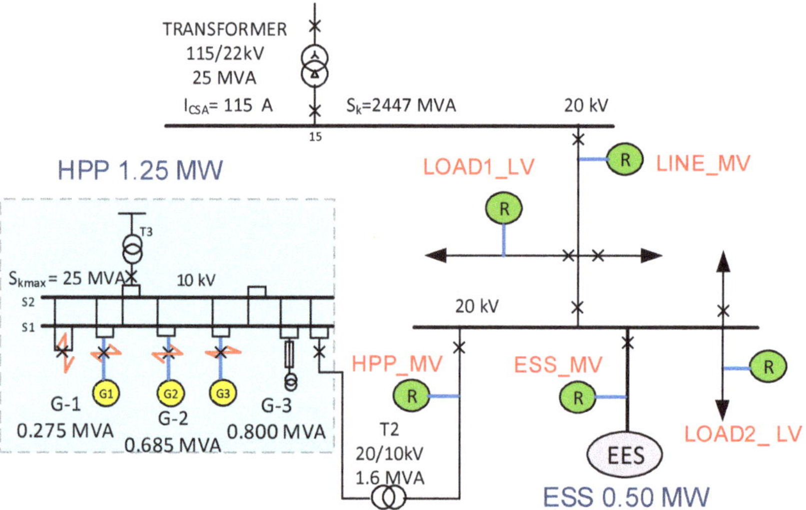

Figure 2. The virtual power plant with the power quality (PQ) recorders placements. Where: LINE_MV: medium voltage line that connects the virtual power plant (VPP) to 110/20 substation; HPP_MW: 1.25 MW hydropower plant; ESS_MW: 0.5 MW electric energy storage; LOAD1_LV: low voltage load related with LINE_MV; LOAD2_LV: low voltage load related with the hydropower plant (HPP) and ESS substation.

2.2. Global Power Quality Index

Recent research on power quality considers different areas. One of these areas is a simplification of the assessment using global values. Thus this article is concerned with the application of the global index-aggregated data index (ADI) [16], as well as the proposition of a new power quality pollution index (PQPI). ADI consists of seven 10-min PQ parameters–frequency (f), voltage (U), an envelope of voltage, short term flicker severity (Pst), unbalance (ku2), total harmonic distortion in voltage (THDu), and maximum harmonic distortion.

However, it was decided to exclude frequency as a customization step of the index to VPP issues as proposed in [16]. Thus, the ADI index corresponds to:

- voltage indicator,
- an envelope of voltage deviation obtained by the difference between the maximum and minimum of 200-ms U values identified during the 10-min aggregation interval,
- short term flicker severity indicator,
- asymmetry indicator,
- total harmonic distortion in voltage indicator,
- a maximum of the 200-ms value of total harmonic distortion of voltage indicator, identified in the 10-min aggregation interval [16].

The mentioned indicators are in response to a 10-min aggregation interval proposed by standard IEC 61000-4-30 [25]. They use the mean value from three phase values to calculate one as a representative. Those factors of ADI are based on the differences between the measured 10-min data and the recommended limit as a division. The limits may be taken differently and based on the object. For VPP that operates in the distribution network standard EN 50,160 [26] was selected. The applied limits based on EN 50,160 [26] are:

- voltage—10% of the declared voltage,
- short term flicker severity—1.0,
- unbalance—2%,
- total harmonic distortion in voltage—8%.

ADI index responds to the voltage level as well as an envelope of voltage. The same situation is with total harmonic distortion and maximum total harmonic distortion. So in view of data features, they are similar. Thus, in this article, the power quality pollution index (PQPI) is representative of the ADI factor but with the reduction of voltage and harmonic distortion indicators. This reduction would retain the data features of those parameters using the envelope of voltage and maximal THDu. Thus, PQPI includes following indicators:

- voltage distortion that responds to the envelope of voltage,
- unbalance distortion that responds to the asymmetry of voltage,
- flicker distortion that responds to short term flicker severity,
- harmonic distortion that responds to maximal total harmonic distortion in voltage.

2.3. Input Databases Description

During the investigation, three different databases were analyzed. The data for each parameter or indicator were used in a 10-min aggregation interval. Generally, the applied variables represented classical PQ parameters, global indexes as well as the active power level (P). The indicated databases are presented in Table 1.

Table 1. Description of input databases.

Database	Parameters	Number of Variables for Each Measurement Point That Describe Every 10-Min Data
Database A	PQ parameters data + P: consisting of classical PQ parameters and active power levels	20
Database B	ADI indicators + P: consists of ADI components and active power level	7
Database C	PQPI indicators + P: consists of PQPI components and active power level	5

Database A (PQ parameters + P) consists of classic power quality parameters and active power level:

- 3 values of U,
- 3 values of 200-ms minimum values of U,
- 3 values of 200-ms maximum values of U,
- 3 values of Pst,
- 1 value of ku2,
- 3 values of THDu,
- 3 values of 200-ms maximum values of THDu,
- 1 value of active power level.

The database B (ADI + P) consists of ADI components and active power

- 1 value that represents U,
- 1 value that represents 200-ms minimum and maximum values of U,
- 1 value that represents Pst,
- 1 value that represents ku2,
- 1 value that represents to THDu
- 1 value that represents 200-ms maximum values of THDu,
- 1 value that active power level.

The third database (PQPI + P) consists of PQPI and active power:

- 1 value that represents 200-ms minimum and maximum values-an envelope of U,
- 1 value that represents Pst,
- 1 value that represents ku2,
- 1 value that represents 200-ms maximum values of THDu,
- 1 value that active power level.

To summarize the database construction a simplified scheme is presented in Figure 3.

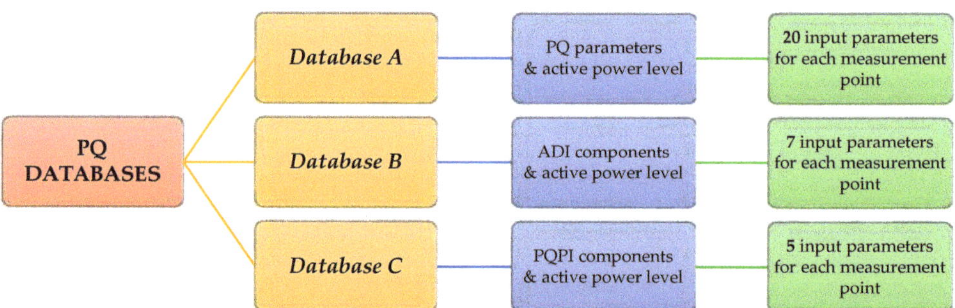

Figure 3. The comparison between investigated databases for hierarchical clustering.

2.4. Hierarchical Clustering

Clustering is one of the data mining (DM) techniques, that aims to divide data into groups that represent similar features [27]. Clustering may be realized in two approaches: a hierarchical or a nonhierarchical [28].

The nonhierarchical algorithms aim toward assigning all observations to the earlier known number of clusters [29]. The most commonly used methods are e.g., K-mean, K-median, or expectation maximization [30].

The hierarchical methods constitute x classes of y observations. Hierarchical methods are also realized by two approaches: agglomerative or divisive. In this research, the agglomerative approach was selected. Generally, the agglomerative approach represents a set of observations, when each piece of data is treated as a separate cluster during the first step. Then, the data are connected into new clusters until one single cluster is established.

This single cluster contains all the data [31]. The agglomerative methods to obtain clusters are single linkage, complete linkage, average linkage, weighted pair-group average linkage, unweighted pair-group centroid linkage, and the Ward method [31,32].

In this article, hierarchical clustering was selected. It determines if the connection is better realized by a single data point or by a group of similar data (achieved in the previous agglomeration) to get a final classification. In this article hierarchical clustering is realized using the Ward algorithm. Ward algorithm cluster analysis is carried out to connect data concentrated in an average value until the data has a similar value (range). The hierarchical CA algorithm, that uses the Ward method of minimal variance, is based on six steps [31,33].

- step 1: Initiate an agglomeration clustering -> divide into x clusters from x data -> calculate the distance between each pair of clusters -> create symmetrical Dis matrix, that consists of distances.
- step 2: find the one pair of clusters that has the smallest squares sum of the distances between adequate object and the related cluster center of the object.
- step 3: create the new cluster, that connects those indicated clusters.
- step 4: update the matrix Dis with the distance between a new cluster and other clusters.
- step 5: check if the number of clusters is equal to 1? YES–go to step 6; NO–back to step 2.
- step 6: final classification when all data are connected to one cluster.

In the Ward method, the indicated "finding pair of clusters which have the smallest sum of squares distance between the object and the cluster center to which this object belongs" is obtained using Equation (1) [31,33].

$$\text{Dis}_{ik} = \frac{n_i + n_k}{n_i + n_j + n_k} * \text{dis}_{ik} + \frac{n_j + n_k}{n_i + n_j + n_k} * \text{dis}_{jk} + \frac{-n_k}{n_i + n_j + n_k} * \text{dis}_{ij}, \tag{1}$$

where: [31,33]

Dis_{pr}: the distance of a new cluster to cluster of number "k",
k: the proceed numbers of cluster from "i" to "j",
dis_{ik}: the distance of a primary cluster "i" from cluster "k",
dis_{jk}: the distance of a primary cluster "j" from cluster "k",
dis_{ij}: the common distance of primary clusters "i" and "j",
n: number of a single object inside each object.

The big problem in cluster analysis, irrespective of the method, is to determine the final number of clusters [34]. The solution in literature for the Ward method is the cubic clustering criterion (CCC). This criterion is obtained by comparing an observed coefficient of determination (R^2) to the approximate expected R^2. It is realized using an approximate variance-stabilizing transformation [35]. The positive values of the cubic clustering criterion inform us that the obtained coefficient of determination is greater than would be expected if sampled from a uniform distribution and, therefore, indicate the possible presence of clusters. The features of the cubic clustering criterion are [35]:

- Extremum of CCC for cluster number greater than two or three indicate good clustering.
- CCC can have several local extremums if the data have a hierarchical structure.
- If CCC values are negative for at least 2 clusters, the distribution is probably unimodal or long tailed.
- Very negative values of the CCC (e.g., −30), could be caused by the outliers.

The last feature was a contribution to apply CCC criteria to detect short-term working conditions of VPP that are specific (outliners) in view of PQ.

3. Hierarchical Clustering of Power Quality Measurement Obtained from the Virtual Power Plant

In this section, the comparison between different databases was performed on the basis of hierarchical clustering with the Ward algorithm. The assessment of cluster assignment was realized using the cubic clustering criterion. Then for the selected number of the cluster and the PQ, comparison between clusters was performed using PQPI.

3.1. Comparison between Databases Using Cubic Clustering Criterion

The PQ measurement time was from 1 May 2020 to 29 October 2020. The analyzed number of PQ points were five: Line_MV, HPP_MV, ESS_MV, Load1_LV, Load2_LV. However, due to the fact that the HPP_MV and ESS_MV PQ records were obtained from the same voltage transformer they were treated as the same point (HPP and ESS_MV). The only difference was the active power level. So, for the analysis of HPP and ESS_MV, there is one more active power parameter than the number of measurement points. For the observed time period, 26 weeks, there is an analysis of 26,208 single 10-min data points. However, the coverage of data point is equal to 97.7% due to measurement device problems. Thus 25,069 10-min data points were accessible [24]. However, as a preprocessing of PQ data, the voltage events connecting data were also excluded as suggested in [36]. Thus, the final number of 10-min data points was 24,612.

Then, for such defined measurement dataset, the input databases for hierarchical clustering were prepared. The database was a matrix that has 24,612 rows (10-min aggregated data) and a different number of columns. The number of columns was connected with the measurement points and the number of variables for each database (check Figure 2). The size of each database set is as follows:

- Database A: matrix 24,612 × 81 consisting of 1,993,572 single cells,
- Database B: matrix 24,612 × 29 consisting of 713,748 single cells,
- Database C: matrix 24,612 × 21 consisting of 516,852 single cells.

The first aim was the verification of using global indexes (ADI and PQPI) in place of classic PQ parameters. These global indexes aimed to minimize the size of the database and retrain the features of data in point of power quality.

The investigated datasets concern long term measurements. During this measurement time, the specific working conditions (like high/low harmonic content, high low voltage level or asymmetry), are indicated. Such states are mainly, short. Thus, it may be treated as an anomaly in view of the general assessment. The selection of such anomalies (specific short-term working conditions) by analyzing every 10-min data of each PQ parameter separately may be time-consuming. Thus, in this article the application of the cubic clustering criterion (CCC) is proposed.

The CCC was used on the databases (A, B, C) for hierarchical clustering with the Ward algorithm. Under the investigation, the minimum number of clusters was equal to two and maximum was equal to ten. Ten, as the maximum value, was selected on the basis of justification presented in e.g., [37] or [38]. The results were presented in Table 2.

Table 2. The results of the cubic clustering criterion for different databases.

Number of Clusters	Cubic Clustering Criterion		
	Database A (PQ Parameters + P)	Database B (ADI Indicators + P)	Database C (PQPI Indicators + P)
2	−52.14	−41.34	−80.81
3	−81.65	−66.43	−84.94
4	−84.51	−71.51	−86.20
5	−78.08	−69.19	−77.45
6	−67.57	−64.51	−59.78
7	−67.47	−50.28	−40.24
8	−58.11	−35.86	−30.65
9	−57.84	−25.89	−19.91
10	−52.43	−15.95	−8.67

The results in point of CCC values were different for each database for different clusters, but for each database the extremal value was indicated for a final number of clusters equal to 4. It means that this division assured clusters that represent data which are very different from one another. Thus, four clusters were selected as the most appropriate to detect the specific short-term working conditions for the investigated measurement. Furthermore, as for all databases that indicated extremum for CCC the least numerous input database was selected to further investigate (database C), which uses PQPI indicators and an active power level.

3.2. Results of Hierarchical Clustering

As it was indicated in the previous subsection the optimal selection in view of the size of the input database is those that consist of PQPI indicators and active power level (database C). Thus, in this section, that database was used for hierarchical clustering with the Ward algorithm.

The main result of the hierarchical clustering is a dendrogram. The dendrogram for the selected database for 26 measurement weeks is presented in Figure 4. On the vertical line, the connection distance between clusters is presented. On a horizontal line, each of 24,612 single 10-min data points is indicated.

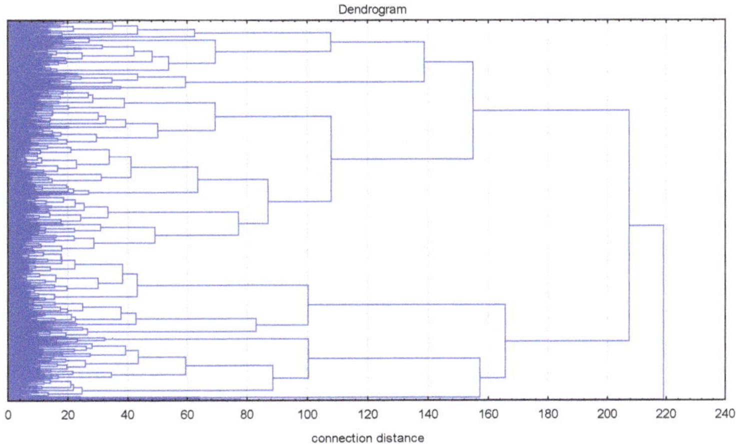

Figure 4. Dendrogram of the hierarchical clustering.

As it was indicated in the previous subsection, division into four clusters has an extremum value for cubic clustering criteria. Based on the dendrogram, it was indicated that for four clusters one is with a small number of 10-min data. The numbers for each cluster are presented and the cluster assignment in the time-domain is presented in Figure 5. It can be observed that cluster 4 is a short-term condition that is represented by only 165 10-min data points. Those 165 10-min data points represent around 0.7% of the measurement time.

Cluster	1	2	3	4
Number of 10 minute data	14,912	5 580	3 955	165

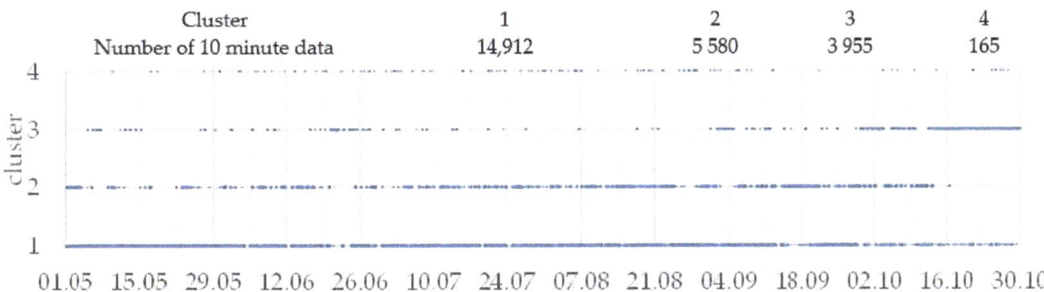

Figure 5. Clustering results in the point of cluster size and time domain.

3.3. Qualitative Assessment of Hierarchical Clustering Results Using the Global Index

In this subsection, the qualitative assessment for indicated clusters is performed. The assessment is realized using PQPI. Thus, the comparison for voltage indicator (a), unbalance indicator (b), flicker indicator (c), and harmonic distortion indicator (d) was presented in Figure 6. The assessment goes towards obtaining knowledge about what PQ parameters and for which measurement point there was a reason for indicating this short-term working condition of VPP. Based on this comparison it may be concluded that cluster 4 represents:

- a problem with voltage indicator in all measurement points. However, a relatively higher value of the indicator is observed for LINE_MV and HPP and ESS_MV;
- a problem unbalance indicator for HPP and ESS_MV, LOAD1_LV, LOAD2_LV;
- a problem flicker indicator for LINE_MV.

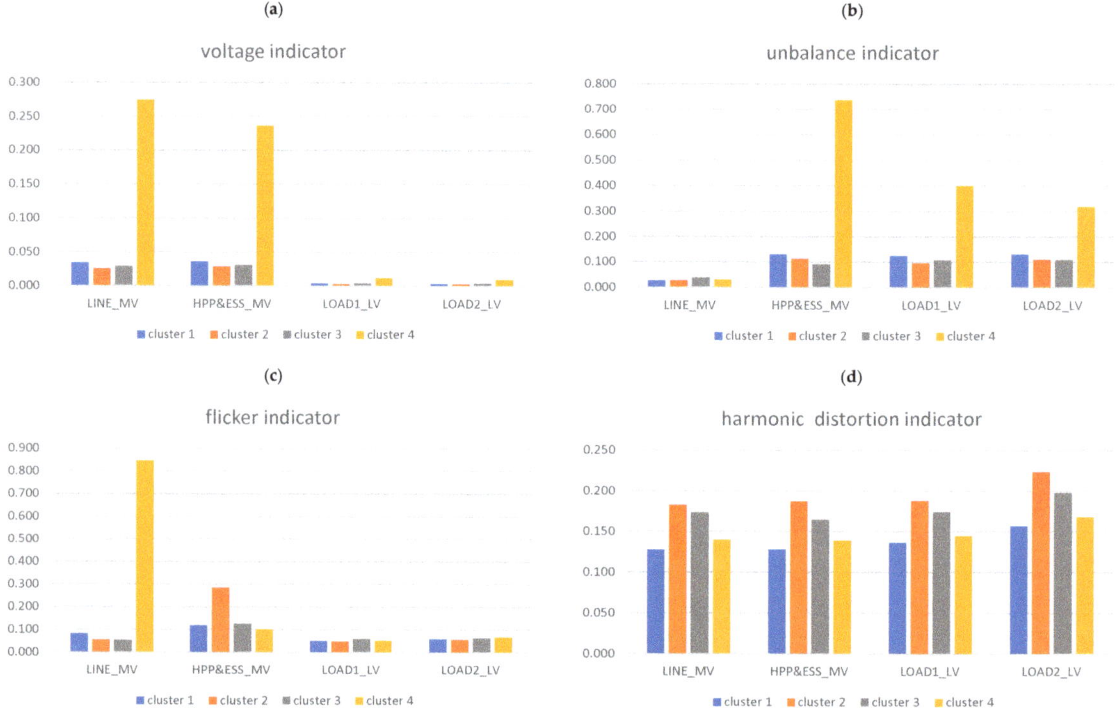

Figure 6. The qualitative comparison between clusters for power quality pollution index (PQPI) indicators: (**a**) voltage indicator; (**b**) unbalance indicator; (**c**) flicker indicator; (**d**) harmonic distortion indicator.

4. Discussion

This article is concerned with a virtual power plant that operates in Poland. The presented research is based on synchronic measurements from five PQ recorders. The PQ measurements were performed at both medium and low voltage levels. The PQ data consists of 182 days (26 weeks) (from 1 May 2020 to 28 October 2020). Thus, these data represent long term data during which different working conditions may occur. This dataset, in work [24], was used to compare the working conditions that were defined a-priory. The states were connected with HPP and ESS working condition, and level of active power. Thus in this work, the conditions were obtained without any predefinition but based on data features.

Those long-term measurements were used as the input to define different PQ databases. The proposed databases were based on classical PQ parameters, PQ global indices as well as the level of active power. Database A represents classical PQ parameters and active power. It concerns 20 variables for each measurement point. Database B uses the ADI components separately and active power level. It concerns seven variables for each measurement point. Database C consists of PQPI and active power level. It concerns five variables for each measurement point. It is important to notice that both global indexes (ADI and PQPI) aim to reduce the amount of data that are analyzed. However, this minimization should not cause one to lose the data features.

The research aimed to detect short-term working conditions in view of power quality. This task was realized using hierarchical clustering with the Ward algorithm. For those three databases, the clustering was realized using the cubic clustering criterion to realize the qualitative assessment of the clustering process. As it is known from the literature if CCC has a negative value it may mean that clustered data has anomalies. For all databases, the CCC was negative for final clusters equal to the range of 2 to 10. Additionally, the extremum for each database was the same and equal to four clusters. Thus, for further investigation, the database with the smallest number of variables (database C) was selected. Furthermore, the final number of clusters was selected as four.

Then, for the indicated circumstances, the comparison between clusters for each measurement point was performed. The PQPI indicators such as voltage indicator, unbalance indicator, flicker indicator and harmonic distortion indicator was used. Based on a comparison of the above mentioned parameters it was concluded that this short term condition (anomaly) was connected with problems of voltage, unbalance, and flicker. The harmonic distortion does not have a significant impact on this condition. It is very important to notice that this short-term condition was not connected with the voltage event. All data that contained voltage events were excluded during the preprocessing of measurements.

The appliances of PQPI index and hierarchical clustering indicated the short-term condition and measurement points, for which it occurs. Thus, the general information about the outcome in PQ problems was also indicated. However, using all this information, it is impossible to define the reason for this condition. Thus, the proposed solution seems essential as a first step of the investigation to define the time period of specific working condition occurrence.

Furthermore, to generalize the discussion of the results, it is important to notice that:

- The investigation was realized in real VPP that operates in Poland, but it also may be applied to other VPP, only if long term power quality data would be available.
- The investigation was based on four measurement points but it may be conducted also for other numbers of points. The minimum number is one, and maximum is limited to the computer computing capabilities.
- In the investigation the extremal value for CCC was for four clusters. However, if the other measurement data would be applied to this methodology, another number would be obtained. However, the most crucial aspect is to select the division when CCC has an extremum. So, the results should be treated in view of proposition of the methodology as well as investigation of the real case study.
- The proposed global index was directed only to selected voltage issues (voltage level, flicker, unbalance and harmonic distortion) and active power level. However, there is a possibility to add other parameters to the global index to make the division more sensitive to other phenomes like current parameters or reactive power.

5. Conclusions

The article proposes the application of clustering to long term power quality data obtained from a virtual power plant. The synchronic, multipoint measurements were used as common input to prepare different databases. The databases were based on both classic PQ parameters and global indexes as well as active power level. The selected PQ global indexes (ADI and PQPI) enabled reduction of the size of the input databases and retrain the features of the PQ data.

The application of the global indexes for clustering the input dataset reduced the size by around 74%. The results for 26 weeks clustering in view of cubic clustering criterion had an extremum for the same number of clusters and indicated specific short-term working conditions of the virtual power plant in view of PQ.

Additionally, using the proposed global index PQPI helped decide which measurement points and which group of parameters had caused the specific working condition. However, it is not possible to define the reason for such a situation. Thus, the single parameter for a single measurement point assessment is still needed even though the time period of this short condition is strictly defined. So, the application of global index and hierarchical clustering may be treated as a first step for deeper analysis.

Author Contributions: Conceptualization, M.J. and T.S.; methodology, M.J. and T.S.; software, M.J. and T.S.; validation, D.K., J.R. and V.S.; formal analysis, M.J., T.S., D.K., J.R.; investigation, M.J. and T.S.; resources, P.K., P.J., J.B. and P.P.; data curation, V.S. and J.S.; writing—original draft preparation M.J. and T.S.; writing—review and editing, D.K. and J.R.; visualization, D.K. and J.R.; supervision, T.S., J.R., Z.L.; project administration, T.S. and P.J.; funding acquisition, T.S. and P.J. All authors have read and agreed to the published version of the manuscript.

Funding: This research was funded by the National Center of Research and Development in Poland, the project "Developing a platform for aggregating generation and regulatory potential of dispersed renewable energy sources, power retention devices and selected categories of controllable load" supported by European Union Operational Programme Smart Growth 2014–2020, Priority Axis I: Supporting R&D carried out by enterprises, Measure 1.2: Sectoral R&D Programmes, POIR.01.02.00-00-0221/16, performed by TAURON Ekoenergia Ltd.

Institutional Review Board Statement: Not applicable.

Informed Consent Statement: Not applicable.

Data Availability Statement: The data are not publicly available due to policy of associate company.

Conflicts of Interest: The authors declare no conflict of interest.

References

1. Etherden, N.; Vyatkin, V.; Bollen, M.H.J.J. Virtual Power Plant for Grid Services Using IEC 61850. *IEEE Trans. Ind. Inform.* **2016**, *12*, 437–447. [CrossRef]
2. Pudjianto, D.; Ramsay, C.; Strbac, G. Microgrids and virtual power plants: Concepts to support the integration of distributed energy resources. *Proc. Inst. Mech. Eng. Part A J. Power Energy* **2008**, *222*, 731–741. [CrossRef]
3. Caldon, R.; Patria, A.; Turri, R. Optimal Control of a Distribution System with a Virtual Power Plant. In Proceedings of the Bulk Power System Dynamics and Control-VI, Cortina d'Ampezzo, Italy, 22–27 August 2004; pp. 278–284.
4. Zhang, J.; Xu, Z.; Xu, W.; Zhu, F.; Lyu, X.; Fu, M. Bi-Objective Dispatch of Multi-Energy Virtual Power Plant: Deep-Learning-Based Prediction and Particle Swarm Optimization. *Appl. Sci.* **2019**, *9*, 292. [CrossRef]
5. Gong, J.; Xie, D.; Jiang, C.; Zhang, Y. Multiple Objective Compromised Method for Power Management in Virtual Power Plants. *Energies* **2011**, *4*, 700–716. [CrossRef]
6. Beguin, A.; Nicolet, C.; Kawkabani, B.; Avellan, F. Virtual power plant with pumped storage power plant for renewable energy integration. In Proceedings of the 2014 International Conference on Electrical Machines (ICEM), Berlin, Germany, 2–5 September 2014; pp. 1736–1742. [CrossRef]
7. CIGRE. *Broshure 292: Data Mining Techniques and Applications in the Power Transmission Field*; CIGRE: Paris, France, 2006.
8. Luo, Z.; Hong, S.; Ding, Y. A data mining-driven incentive-based demand response scheme for a virtual power plant. *Appl. Energy* **2019**, *239*, 549–559. [CrossRef]
9. Yi, Z.; Xu, Y.; Gu, W.; Wu, W. A Multi-Time-Scale Economic Scheduling Strategy for Virtual Power Plant Based on Deferrable Loads Aggregation and Disaggregation. *IEEE Trans. Sustain. Energy* **2020**, *11*, 1332–1346. [CrossRef]
10. Kong, X.; Xiao, J.; Liu, D.; Wu, J.; Wang, C.; Shen, Y. Robust stochastic optimal dispatching method of multi-energy virtual power plant considering multiple uncertainties. *Appl. Energy* **2020**, *279*, 115707. [CrossRef]
11. Ai, X.; Yang, Z.; Hu, H.; Wang, Z.; Peng, D.; Zhao, Z. A load curve clustering method based on improved k-means algorithm for virtual power plant and its application. *Dianli Jianshe/Electr. Power Constr.* **2020**, *41*, 28–36. [CrossRef]
12. Silva, C.; Faria, P.; Vale, Z. Multi-Period Observation Clustering for Tariff Definition in a Weekly Basis Remuneration of Demand Response. *Energies* **2019**, *12*, 1248. [CrossRef]

13. Faria, P.; Spínola, J.; Vale, Z. Distributed Energy Resources Scheduling and Aggregation in the Context of Demand Response Programs. *Energies* **2018**, *11*, 1987. [CrossRef]
14. Klajn, A.; Bątkiewicz-Pantua, M. *Application Note–Standard EN 50 160: Voltage Characteristics of Electricity Supplied by Public Electricity Networks*; European Copper Institute: Brussels, Belgium, 2017; pp. 1–42.
15. Nourollah, S.; Moallem, M. A Data Mining Method for Obtaining Global Power Quality Index. In Proceedings of the 2011 2nd International Conference on Electric Power and Energy Conversion Systems (EPECS), Sharjah, UAE, 15–17 November 2011; pp. 1–7.
16. Jasinski, M.; Sikorski, T.; Kostyla, P.; Borkowski, K. Global power quality indices for assessment of multipoint Power quality measurements. In Proceedings of the 2018 10th International Conference on Electronics, Computers and Artificial Intelligence (ECAI), Iasi, Romania, 28–30 June 2018; pp. 1–6.
17. Serpak, M. A unified index and system indicator for global power quality assessment. *Sci. Int.* **2016**, *28*, 1131–1136.
18. Lee, B.; Sohn, D.; Kim, K.M. Development of Power Quality Index Using Ideal Analytic Hierarchy Process. In *Information Science and Applications (ICISA) 2016*; Springer: Singapore, 2016; pp. 783–793.
19. Raptis, T.E.; Vokas, G.A.; Langouranis, P.A.; Kaminaris, S.D. Total Power Quality Index for Electrical Networks Using Neural Networks. *Energy Procedia* **2015**, *74*, 1499–1507. [CrossRef]
20. Langouranis, P.A.; Kaminaris, S.D.; Vokas, G.A.; Raptis, T.E.; Ioannidis, G.C.; General, A. Fuzzy Total Power Quality Index for Electric Networks. In Proceedings of the MedPower 2014, Athens, Greece, 2–5 November 2014; pp. 1–6.
21. De Capua, C.; De Falco, S.; Liccardo, A.; Romeo, E. Imporvement of New Synthetic Power Quality Indexes: An Original Approach to Their Validation. In Proceedings of the 2005 IEEE Instrumentationand Measurement Technology Conference Proceedings, Ottawa, ON, Canada, 16–19 May 2005; Volume 2, pp. 819–822. [CrossRef]
22. Ge, B.; Pan, T.; Li, Z. Synthetic assessment of power quality using relative entropy theory. *J. Comput. Inf. Syst.* **2015**, *11*, 1323–1331. [CrossRef]
23. Sikorski, T.; Jasiński, M.; Ropuszyńska-Surma, E.; Węglarz, M.; Kaczorowska, D.; Kostyla, P.; Leonowicz, Z.; Lis, R.; Rezmer, J.; Rojewski, W.; et al. A Case Study on Distributed Energy Resources and Energy-Storage Systems in a Virtual Power Plant Concept: Technical Aspects. *Energies* **2020**, *13*, 3086. [CrossRef]
24. Jasiński, M.; Sikorski, T.; Kaczorowska, D.; Rezmer, J.; Suresh, V.; Leonowicz, Z.; Kostyla, P.; Szymańda, J.; Janik, P. A Case Study on Power Quality in a Virtual Power Plant: Long Term Assessment and Global Index Application. *Energies* **2020**, *13*, 6578. [CrossRef]
25. IEC 61000 4-30. *Electromagnetic Compatibility (EMC)–Part 4-30: Testing and Measurement Techniques–Power Quality Measurement Methods*; International Electrotechnical Commission: London, UK, 2003.
26. *EN 50160: Voltage Characteristics of Electricity Supplied by Public Distribution Network*; British Standards: London, UK, 2010; Available online: https://orgalim.eu/position-papers/en-50160-voltage-characteristics-electricity-supplied-public-distribution-system (accessed on 8 February 2021).
27. Vehkalahti, K.; Everitt, B.S. *Multivariate Analysis for the Behavioral Sciences*, 2nd ed.; CRC Press: Boca Raton, FL, USA, 2019; ISBN 9781351202275.
28. Roiger, R.J. *Data Mining*; Chapman and Hall/CRC: Boca Raton, FL, USA, 2017; ISBN 9781315382586.
29. Wierzchoń, S.; Kłopotek, M. *Modern Algorithms of Cluster Analysis*; Studies in Big Data; Springer International Publishing: Cham, Switzerland, 2018; Volume 34, ISBN 978-3-319-69307-1.
30. Jasiński, M.; Sikorski, T.; Borkowski, K. Clustering as a tool to support the assessment of power quality in electrical power networks with distributed generation in the mining industry. *Electr. Power Syst. Res.* **2019**, *166*, 52–60. [CrossRef]
31. Wierzchoń, S.; Kłopotek, M. *Algorithms of Cluster Analysis*; Institute of Computer Science Polish Academy of Sciences: Warsaw, Poland, 2015; Volume 3, ISBN 9789638759627.
32. Sneath, P.H.; Sokal, R.R. *Numerical Taxonomy*; Freeman: San Franciso, CA, USA, 1973; ISBN 9780716706977.
33. Statsoft Polska StatSoft Electronic Statistic Textbook. Available online: https://www.statsoft.pl/textbook/stathome.html (accessed on 15 December 2020).
34. Chowdhury, K.; Chaudhuri, D.; Pal, A.K. An entropy-based initialization method of K-means clustering on the optimal number of clusters. *Neural Comput. Appl.* **2020**. [CrossRef]
35. Sarle, W. Cubic clustering criteria. In *SAS Technical Report A-108*; SAS Institute Inc.: Cary, NC, USA, 1983; p. 51.
36. Jasiński, M.; Sikorski, T.; Leonowicz, Z.; Borkowski, K.; Jasińska, E. The Application of Hierarchical Clustering to Power Quality Measurements in an Electrical Power Network with Distributed Generation. *Energies* **2020**, *13*, 2407. [CrossRef]
37. Claeys, R.; Azaioud, H.; Cleenwerck, R.; Knockaert, J.; Desmet, J. A Novel Feature Set for Low-Voltage Consumers, Based on the Temporal Dependence of Consumption and Peak Demands. *Energies* **2020**, *14*, 139. [CrossRef]
38. Kang, J.; Lee, J.-H. Electricity Customer Clustering Following Experts' Principle for Demand Response Applications. *Energies* **2015**, *8*, 12242–12265. [CrossRef]

Article

A Case Study on Data Mining Application in a Virtual Power Plant: Cluster Analysis of Power Quality Measurements

Michał Jasiński [1,*], Tomasz Sikorski [1], Dominika Kaczorowska [1,*], Jacek Rezmer [1], Vishnu Suresh [1], Zbigniew Leonowicz [1], Paweł Kostyła [1], Jarosław Szymańda [1], Przemysław Janik [2], Jacek Bieńkowski [2,3] and Przemysław Prus [2]

[1] Faculty of Electrical Engineering, Wroclaw University of Science and Technology, 50-370 Wroclaw, Poland; tomasz.sikorski@pwr.edu.pl (T.S.); jacek.rezmer@pwr.edu.pl (J.R.); vishnu.suresh@pwr.edu.pl (V.S.); zbigniew.leonowicz@pwr.edu.pl (Z.L.); pawel.kostyla@pwr.edu.pl (P.K.); jaroslaw.szymanda@pwr.edu.pl (J.S.)

[2] TAURON Ekoenergia Ltd., 58-500 Jelenia Góra, Poland; przemyslaw.janik@tauron-ekoenergia.pl (P.J.); jacek.bienkowski@tauron-ekoenergia.pl (J.B.); przemyslaw.prus@tauron-ekoenergia.pl (P.P.)

[3] Faculty of Mechanical and Power Engineering, Wroclaw University of Science and Technology, 50-370 Wroclaw, Poland

[*] Correspondence: michal.jasinski@pwr.edu.pl (M.J.); dominika.kaczorowska@pwr.edu.pl (D.K.); Tel.: +48-71-320-2022 (M.J.); +48-71-320-2901 (D.K.)

Citation: Jasiński, M.; Sikorski, T.; Kaczorowska, D.; Rezmer, J.; Suresh, V.; Leonowicz, Z.; Kostyła, P.; Szymańda, J.; Janik, P.; Bieńkowski, J.; et al. A Case Study on Data Mining Application in a Virtual Power Plant: Cluster Analysis of Power Quality Measurements. *Energies* **2021**, *14*, 974. https://doi.org/10.3390/en14040974

Academic Editor: Andrea Mariscotti

Received: 2 January 2021

Accepted: 9 February 2021

Published: 12 February 2021

Publisher's Note: MDPI stays neutral with regard to jurisdictional claims in published maps and institutional affiliations.

Abstract: One of the recent trends that concern renewable energy sources and energy storage systems is the concept of virtual power plants (VPP). The majority of research now focuses on analyzing case studies of VPP in different issues. This article presents the investigation that is based on a real VPP. That VPP operates in Poland and consists of hydropower plants (HPP), as well as energy storage systems (ESS). For specific analysis, cluster analysis, as a representative technique of data mining, was selected for power quality (PQ) issues. The used data represents 26 weeks of PQ multipoint synchronic measurements for 5 related to VPP points. The investigation discusses different input databases for cluster analysis. Moreover, as an extension to using classical PQ parameters as an input, the application of the global index was proposed. This enables the reduction of the size of the input database with maintaining the data features for cluster analysis. Moreover, the problem of the optimal number of cluster selection is discussed. Finally, the assessment of clustering results was performed to assess the VPP impact on PQ level.

Keywords: virtual power plant (VPP); cluster analysis (CA); power quality (PQ); global index; distributed energy resources (DER); energy storage systems (ESS); power systems; smart grids

1. Introduction

Currently, in electrical power networks (EPN), the number of renewable energy sources (RES) and energy storage systems (ESS) are scientifically increasing [1]. Thus, different approaches that enable to control them are highly desirable [2,3]. Commonly known are microgrids and virtual power plants (VPP) [4]. Generally, virtual power plants are autonomous units, that are equipped with systems for effective power flow control. VPPs consist of different units that are integrated with the distribution network. The used elements are, e.g., generators, loads, as well as energy storage systems [5]. Thus, the current literature analyzes the real cases of VPPs. The international examples are in, e.g., United Kingdom [6], Denmark [7], Ireland [8], Greece [9], Germany [10], China [11], India [12], Australia [13], and South Korea [14]. The range of VPP functioning is wide. Thus, there is a possibility of indicating different research trends that concerns VPP, e.g., network voltage control by renewable energy sources integrated [15–18], power flow control and analysis [19–22], playing a role on the energy market [23–26], energy management in a VPP [27–30], optimal active and reactive power scheduling [31–34], and ESS sizing, localization, and management in VPP [35–38].

One of the elements that concerns VPP is power quality (PQ) issues. Reference [39] introduces the virtual power plant as vehicles to facilitate cost-efficient integration of distributed energy sources into the existing power system (PS). This research also includes a demonstration of the application of the test system. The analysis of the system performance concerns energy efficiency, power quality, and security. Reference [40] proposes the extensions of the standard International Electrotechnical Commission (IEC) 61850 as a solution to enhance the interaction between the virtual power plant controller and the distributed energy resources. This research concerns the implementation of virtual power plant communication and control architecture as a real case study. The presented case concerns the power quality recorders issues and demands presented in IEC 61850. Reference [41] presents virtual power plant management with the priority requirements optimized by the compromised methods. The model of operation optimization of a VPP was based on fuzzy multiple objective optimization problems. The indicated optimization problem concerns the satisfaction of both customers and suppliers, system stability, power quality, as well as the costs of operation limitations. The presented method was verified in a test system. Reference [42] attempts to assure the applicable framework for harmonizing operations of different units of the virtual power plant. The authors concluded that decisions and obtaining profits, although complying with the required power quality levels and real network constraints, are a crucial part of virtual power plant strategies. Reference [43] concerns simulations when a virtual power plant is in islanded grid mode with a thermal power plant for baseload. The presented virtual power plant consists of 0.2 GW wind power, 0.1 GW photovoltaic power, and ±0.25 GW pumped storage. The research results include different control strategies of the storage plant to highlight the impact of a virtual power plant on PQ. Reference [44] presents the coordinative operation problem of multi-energy virtual power plants. The bi-objective dispatch model was proposed for the optimization of the performance of a multi-energy virtual power plant in terms of economic issues and power quality. The presented case study was based on Hongfeng Eco-town in China.

Additionally, literature from 2019 and 2020 also present recent applications of clustering methods to virtual power plant issues. Reference [45] proposes the application of a robust stochastic optimal dispatching method for solving the scheduling problem. In this research, the K-medoids clustering is used to obtain typical scenes of different units integrated into VPP. Reference [46] proposes the virtual power plant multi-timescale scheduling that contains day-ahead bidding and real-time operation. The models are based on the deferrable loads' aggregation model with the k-mean clustering approach. The introduced strategy helps to obtain efficient management of massive deferrable loads to reduce the energy management complexity as well as to increase the general economics, which seems a promising application in the virtual power plant economic scheduling. Reference [47] presents a data-mining-driven incentive-based demand response scheme to model electricity trading between a virtual power plant and its participants. In this article, cluster analysis is used to divide consumers into different categories by their bid-offers. The algorithm applied in this article was based on ordering points to identify the clustering structure (OPTICS) algorithm. Reference [48] presents a virtual power plant load curve cluster analysis approach based on principal-component dimension-reduced analysis, aggregation level clustering, and fc-means clustering. The principal component analysis method is used to define the specific different loads in the virtual power plant aggregation. Then aggregation hierarchical clustering and fc-means clustering are used to realize the division of all load output curves participating in the aggregation. As a result, load curve clusters of the same class are obtained. Finally, the cluster analysis results are analyzed to establish an evaluation system.

Also, the concept of "global/synthetic/unified/total" power quality indices are indicated in the literature, such as total power quality index [49,50], synthetic power quality index [51,52], unified power quality index [53,54], or global power quality index [55,56]. All of these approaches are aimed toward reducing the number of analyzing power quality data. However, in the literature, there is a lack of research that uses the global indexes as an input to data-mining techniques to reduce the size of the input database. This investigation therefore verifies whether this approach is possible. The verification is based on real long-term PQ data obtained from VPP.

This research concerns a case study of a VPP that operates in Poland. The analyzed VPP is a part of both low-voltage (LV) and medium-voltage (MV) distribution networks. The main units of the VPP are a hydropower plant (HPP), a photovoltaic system (PV), and energy storage systems (ESS). This article aims to analyze a fragment of this virtual power plant that consists of 1.25 MW HPP and associates 0.5 MW ESS as well as LV loads. The presented research is based on PQ measurements. These measurements were realized synchronically in five measurement points. The observed measurement points are HPP, ESS, associated MV line, and two LV loads. The measurements were conducted from 1 May to 28 October 2020. Thus, the measurement period of time was 26 weeks (182 days). The analysis for full weeks is used as in classical PQ assessment [57]. From such a big amount of data, different datasets may be prepared. Thus, in this article, different databases were obtained. Those databases consist of classic PQ parameters as well as an application of the global index. Then, these databases were used as an input to data-mining algorithms. The selected algorithm was k-mean with Euclidean distance. A comparison of the results and effectiveness was performed. Additionally, the solution for defining the optimal number of clusters was performed using the v-fold cross-validation test. Finally, the qualitative assessment of clustering results was presented to respond to the different working conditions of VPP. All calculations were performed in Statistica 13 software (StatSoft Polska, Kraków, Poland).

To summarize the contributions of this article:

- The data that were used in this article were based on multipoint, synchronic, and long-term measurement performed in real VPP.
- The different input databases in point of power quality were proposed. The databases consisted of classic PQ parameters as well as global index values.
- The application of PQ global index enabled the reduction of the size of the input database while maintaining a similar division of data.
- The article investigated different PQ datasets and proposed a solution to define the optimal number of clusters selection.
- Application of CA enabled the definition of the different working conditions of the VPP based on data features. Additionally, the assessment of these working conditions was realized using the PQ global index.

To realize those contributions, the article is organized into 6 sections. In Section 2, the clustering methods, global index proposition, and virtual power plant description are presented. Section 3 presents the construction of different datasets and a comparison between them. Section 4 presents the optimal number of cluster selection and a qualitative assessment of the final classification. Section 5 is a discussion of the obtained results. Section 6 presents the conclusions.

2. Methodology and Research Object Description

This section is based on three main elements of this article. In Section 2.1, the cluster analysis algorithm is introduced as a representant of the data-mining technique. The justification of the k-mean algorithm for long-term data is included. Also, methods for obtaining the optimal number of clusters are presented. Then, the definition of the global index is proposed. The global index is proposed to minimize the number of parameters that represent power quality while retaining their features. Finally, the description of the real virtual power plant is presented. That VPP was used as a source to obtain multipoint synchronic power quality data.

2.1. Cluster Analysis

Cluster analysis (CA) is one of the data-mining techniques [58]. Generally, the aim of cluster analysis is to divide data in the point of their features [59]. Clustering may be realized in a hierarchical or nonhierarchical approach. The hierarchical methods constitute x classes of y observations. The nonhierarchical approach is based on assigning all observations to the earlier known number of clusters. The non-hierarchical clustering is not a tree as in the case of the hierarchical clustering. It assures the division into groups of the data in order to maximize/minimize some evaluation criteria [60]. The nonhierarchical methods most used in the literature are based on, e.g., k-mean algorithm, k-median algorithm, or expectation maximization (EM) algorithm [61,62]. In this paper, the authors suggest using the nonhierarchical with the k-mean algorithm with Euclidean distance. The Euclidean distance was selected because it is not sensitive to the changeability of a single measurement [63]. It assures the general information about the data difference, especially for multiparameter input [64].

The k-mean algorithm's goal is to find the extremum of the objective function. The k-mean algorithm function with Euclidean distance is defined as [65]:

$$J\,(U,\,M) = \sum_{i=1}^{m}\sum_{j=1}^{k} u_{ij} * \sqrt{\sum_{i}(x_i - y_i)^2}, \tag{1}$$

where:

J—objective function,
U—matrix of the object belonging to a cluster,
M—matrix in which a row vector represents the centroids of clusters,
$i = 1,2,3, \ldots ,m$—number of objects,
$j = 1,2,3 \ldots ,k$—number of classes (clusters),
u_{ij}—element indicating the fact of assignment of i-th object to the j-th class (cluster),
x_i—vector of observations belonging to cluster x,
y_i—vector of observations belonging to cluster y.

However, the main disadvantage is that for the nonhierarchical approach, there is a need to define the final number of clusters for which data may be divided. There are different approaches to realize this optimal number of cluster selection, e.g., an entropy-based initialization method [66], gap statistic [67], u-control chart [68], or k-fold cross-validation test [69]. In this research, the k-fold cross-validation test was selected. This type of cross-validation is useful for the situation where there is no known test sample: The user-specified 'v' value for v-fold cross-validation. Normally, the v is equal to 3. The v value determines the number of random subsamples that are used for the learning sample. Then, the tree of the specified size is computed v times. After each iteration, it leaves out one of the subsamples from the computations. Then, these subsamples are used as a test sample for cross-validation. The cross-validation cost is computed for each of the v test samples that are then averaged to give the v-fold estimate of the cross-validation costs [69]. Then, on the basis of cross-validation cost, the optimal number of clusters is selected.

2.2. Global Power Quality Index

One of the current trends in PQ is using global indexes. Thus, in this article, there is a proposition of using the synthetic (global) index. The selected index was called the aggregated data index (ADI) [55,70]. The selected index consists of five classic 10 min PQ parameters (frequency, voltage, flicker severity, asymmetry, harmonic distortion in voltage), and two extremal (envelope of voltage deviation and maximum harmonic distortion).

However, it was decided to exclude frequency as a customization step of the index to VPP issues. Thus, the index corresponds to:

- Voltage—U,
- An envelope of voltage deviation obtained by the difference between the maximum and minimum of 200 millisecond U values identified during the 10 min aggregation interval—ΔU,
- Short-term flicker severity—P_{st},
- Asymmetry factor—k_{u2},

- Total harmonic distortion in voltage—*THDu*,
- A maximum of the 200 millisecond value of *THDu*, identified in the 10 min aggregation interval—*THDumax* [71].

Selected parameters correspond to the demands of the standard IEC 61000-4-30 [72]. Additionally, the mean value of three phase values is used as a representative. All factors that are included in the ADI are based on the differences between the measured 10 min aggregated power quality data and the recommended limits of the selected standard. In this article, the European standard EN 50160 [73] was selected to obtain global values. The applied limits based on EN 50160 [73] are presented in Table 1.

Table 1. Acceptable values base on standard EN 50160 [73] used for the global index.

Parameter	Value
voltage	10% of declared voltage
short-term flicker severity	1.0
asymmetry	2%
total harmonic distortion in voltage	8%

2.3. Investigated VPP

The investigated VPP is based on a fragment of the distribution network in the Lower Silesia region in Poland [32]. It is supplied by two substations, 110/20 kV. The mentioned substations are connected to a 110 kV electrical power system of Poland [74]. However, in this research, only one substation area was selected. The investigated 20 kV network fed from the station is an overhead-cable network. The selected MV network fed from the second station is mainly an urban cable network. This network has earth fault current compensation. Connected distributed energy resources to the VPP are a 1.25 MW hydro power plant and a 0.5 MW battery energy storage system. Those resources are connected to a MV level.

The simplified scheme of the investigated fragment of the virtual power plant area is presented in Figure 1. It consists of a 20 kV distribution network with 1.25 MW hydropower plant and an 0.5 MW battery energy storage system (ESS) connected with the HV/MV substation by MV line (1—MV). HPP and ESS are connected to the same node of the network. Also, two additional low-voltage loads are indicated: 3—LV and 4—LV. 3—LV is connected with indicated MV-associated lined. 4—LV is connected with the node of the HPP and ESS. Power quality recorders are denoted as a "R" and are also presented in Figure 1. Due to that, the hydro power plant and electric energy storage are connected to one node and their PQ recorders use the same voltage transformer. In further research, they are treated as one point indicated as 2 for PQ issues but the active power level is treated separately as 2-HPP and 2-ESS.

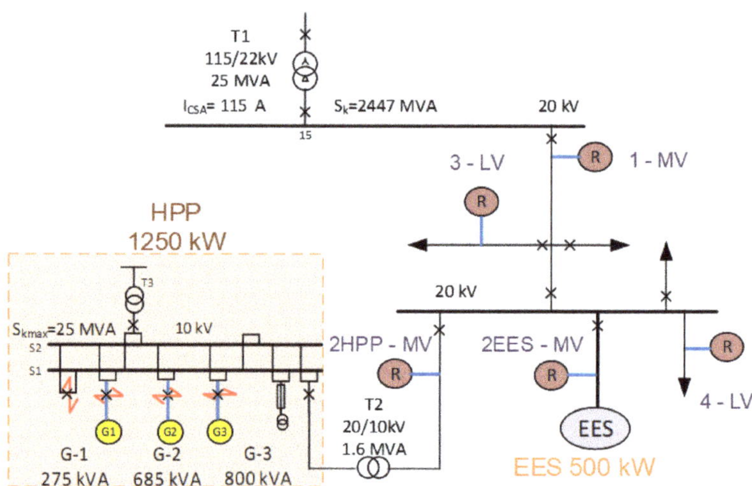

Figure 1. Investigated part of VPP with the placement of PQ recorders. Where: 1—MV: medium-voltage line; 2HPP—MW: medium-voltage hydro power plant; 2EES—MW: medium-voltage electric energy storage; 3—LV: low-voltage load related with MV line; 4—LV: low-voltage load related with hydro power plant substation.

3. Power Quality Data as an Input to Cluster Analysis Techniques

In this section, the power quality data from VPP were used to prepare three different input databases. The proposed databases consist of classical power quality measurement and active power level as well as global power quality index, which was proposed in Section 2.2. Then, in Section 3.2, the comparison of cluster analysis with the k-mean algorithm and Euclidean distance was performed for those different input databases.

3.1. Input Databases Describtion

Under this investigation, three different databases were included. Each of the parameters are represented as one value from a 10 min time interval. The indicated databases are:

- Database I—Raw PQ data + *Pphase*: consists of classical PQ parameters and active power level for each phase separately. This database consists of 22 variables that describe each 10 min data point.
- Database II—PQ Global Indicators + *Psum*: consists of ADI components and active power level as a sum of each phase. This database consists of 7 variables that describe each 10 min data point.
- Database III—Global PQ Index + *Psum*: consists of ADI and active power level as a sum of each phase. This database consists of 2 variables that describe each 10 min data point.

The first database (Raw PQ data + *Pphase*) consists of classic power quality parameters:

- Three phase values of voltage,
- Three phase values of 200 ms minimal values of voltage,
- Three phase values of 200 ms maximal values of voltage,
- Three phase values of short-term flicker severity,
- One value of voltage unbalance,
- Three phase values of total harmonic distortion in voltage,
- Three phase values of 200 ms maximal values of total harmonic distortion in voltage,
- Three phase values of active power level.

The second database (PQ Global Indicators + *Psum*) consists of ADI index components and sum of active power from all phases:

- One value that represents voltage,
- One value that represents 200 ms minimal and maximal values of voltage,
- One value that represents short-term flicker severity,
- One value that represents voltage unbalance,
- One value that represents total harmonic distortion in voltage,
- One value that represents 200 ms maximal values of total harmonic distortion in voltage,
- One value that represents active power level: sum from three phases.

The third database (Global PQ Index + *Psum*) consists of ADI index and sum of active power from all phases:

- One value that represents power quality: ADI,
- One value that represents active power level: sum from three phases.

To summarize database construction, the simplified scheme is presented in Figure 2.

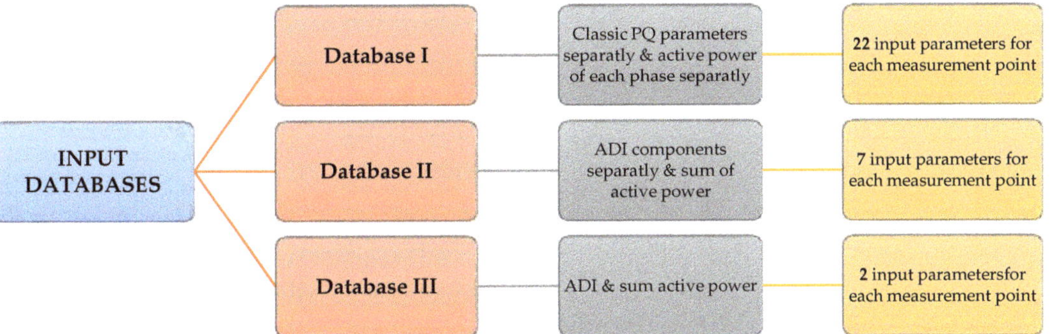

Figure 2. Simplified scheme of the proposed input databases.

3.2. Selection of Optimal Database—Results for 26-Week Measurements

In this section, the measurement from five PQ recorders localized as indicated in Section 2.2 was used. The observed time was selected from 1 May to 29 October 2020, which represents 26 weeks of the measurements. Thus, in point of PQ assessment, there is an analysis of 26,208 single 10 min data points. However, for such a long period of multipoint measurements, the coverage of data is equal to 97.7%, thus 25,069 data points were used [71]. Also, as a preprocessing of PQ data, the data that come from the time when a voltage event occurred were excluded, as suggested in Reference [62]. So, the final number of 10 min data points was equal to 24,612.

Then, for such a defined measurement dataset, the input databases for cluster analysis were prepared. The database was a matrix that has 24,612 rows and a different number of columns. The number of columns was connected with five measurements and features of the database (check Figure 2). The size of each database set is as follows:

- Dataset I: matrix 24,612 × 91, so concerns 2,421,692 single cells,
- Dataset II: matrix 24,612 × 29, so concerns 713,748 single cells,
- Dataset III: matrix 24,612 × 9, so concerns 221,508 single cells.

The next step of the investigation was to conduct a cluster analysis of the indicated power quality measurement. The aim was to verify if it is possible to use global indicators in place of classic PQ parameters to minimize the size of the dataset. As the main representant, the k-mean algorithm with Euclidean distance was selected. The final number of clusters was equal to 2. The results of this data-mining process in the time domain are presented in Figure 3.

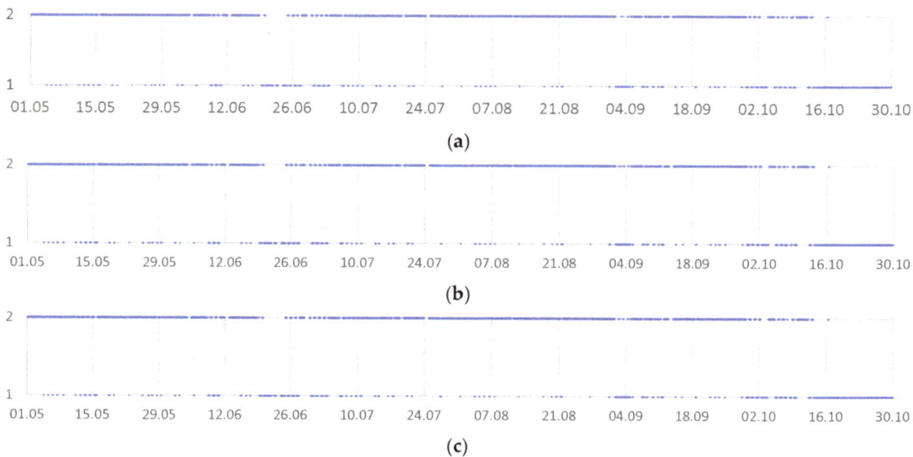

Figure 3. Results of cluster analysis for different input databases. (**a**) Database I: raw PQ data + *Pphase*, (**b**) Database II: ADI components separately + *Psum*, (**c**) Database III: ADI + *Psum*.

To summarize the results of cluster analysis for those three input databases, Table 2 was prepared. In this table, the number of data that were connected to each cluster is given. Additionally, to compare the results of the clustering for different databases, the number of single cells that were included in different clusters compared to database I were calculated. As it can be observed, generally, the final classification for the k-mean algorithm with Euclidean distance for each dataset is similar. The difference between classification is no more than 111 single data, which is only 0.5% of the difference. It is important to notice that the final classification was obtained with 99.5% similarity and the reduction of the dataset was at 90%. Thus, the application of the global index as an input to cluster analysis for general issues is desirable.

Table 2. The number of 10 min data points in each cluster for different input databases, and simplified comparison.

Database	Cluster 1	Cluster 2	Difference to Basic Database
Database I: raw PQ data + *Pphase*	6579	18,034	-
Database II: ADI components separately + *Psum*	6483	18,129	108
Database III: ADI + *Psum*	6661	17,951	111

4. Cluster Analysis for Identification of Different Working Conditions of Virtual Power Plant

The previous section indicated that database III is suitable for cluster analysis. The database uses a global index and sum of active power level for each observed measurement point. However, the division into two clusters of such a big database may not be sufficient. Thus, in this section, the investigation of the selection of the optimal number of clusters using the v-fold cross-validation scheme is performed. Then, for the selected number of the cluster, the qualitative assessment is realized to compare different working conditions of VPP.

4.1. Optimal Number of Clusters

As it was indicated in Section 3.2, the selected input database for PQ issues was database III, which represents global indexes for measurement points and active power level. However, the cluster analysis for this dataset was realized for the previously determined number of clusters (2 clusters). But, to analyze the long-term data, the adequate number of final clusters is unknown at the beginning. Thus, there is a need for the introduction of other methods to obtain this. One of the known solutions is the application of the v-fold cross-validation.

Thus, the application of v-fold cross-validation was realized for dataset III. As a result, the chart of the cost sequence was obtained and is presented in Figure 4. The circumstances for this test were performed for the minimal number of clusters equal to 2 and maximal equal to 10. The maximal value equal to 10 is then justified and indicated, e.g., as in References [75,76].

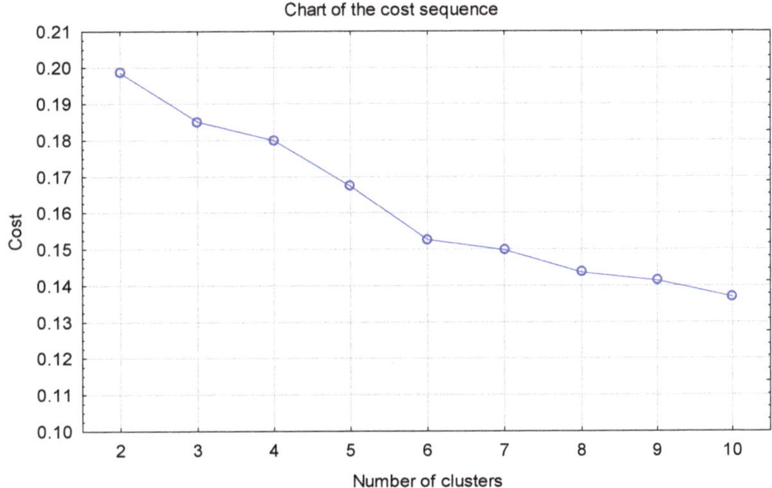

Figure 4. Cost sequence in relation for the number of clusters for v-fold cross-validation for database III and k-mean algorithm with Euclidean distance.

However, to analyze this cost chart, the minimal percentage decrease was calculated. The results of these calculations are indicated in Table 3. In the literature, the most commonly used value of the range of minimal decrease is equal to 5% [69]. Thus, for the observed measurement data applied for database III, the optimal number of clusters is 3.

Table 3. The calculations of minimal decrease for v-fold cross-validation test results.

Range of minimal decrease	1%	2%	3–6%	≤7%
Optimal number of clusters	9 clusters	6 clusters	3 clusters	2 clusters

4.2. Qualitative Assessment of Clusters

As indicated in the previous subsection, the optimal number of clusters using v-fold cross-validation is equal to 3. Thus, in this section, the qualitative comparison is realized for each cluster. The results are presented in Table 4. Additionally, using the knowledge about the operation schedule of HPP and ESS, the main feature for each cluster was indicated.

Generally, the cluster analysis indicated one major working condition (cluster 1). Cluster 1 presents a time when HPP is switched off and ESS is not working with high power. As it can be observed, these conditions last for around 72% of the time. Cluster 2 corresponds to a time when HPP is generally working with high power that takes around 22% of the time, and during this time, ESS is charged/discharged with low power. Cluster 3 represents a time when ESS is discharged with high power. In point of power quality level, the most positive working condition is cluster 3, which represents the lowest value for the global index in each measurement point.

Table 4. Qualitative assessment between clusters and main defining feature.

Cluster	Global Index				Main Feature of Dataset in Point of VPP	Number of 10 min Data Points
	1-G	2-G	3-G	4-G		
1	0.081	0.068	0.077	0.088	ESS and HPP are not working with high power	17,765
2	0.076	0.073	0.079	0.087	HPP is working with high power	5432
3	0.073	0.067	0.068	0.084	ESS is discharging with high power	1415

5. Discussion

This research demonstrated a case study of a virtual power plant that operates in the Lower Silesia in Poland. The presented virtual power plant operates at both low-voltage (LV) and medium-voltage (MV) levels of the distribution network. The investigation concerned synchronic measurements from 5 PQ recorders. The measurements were realized at both MV and LV levels. The measurements were based on 26 weeks (from 1 May to 28 October 2020). Thus, they represent long-term data when different working conditions may occur, but defining this working condition a priori would lead to missing some specific data. The solution for this is an application of cluster analysis that assures the division of data into clusters (that represent different working conditions) based on data features.

The indicated measurements were used as the sources for different databases. The used parameters were classical PQ parameters, proposed global indices, as well as the level of active power. Database I consists of classical PQ parameters and active power level for each phase, separately (22 variables for each measurement point). Database II consists of ADI components separately and active power level as a sum of each phase (7 variables for each measurement point). Database III consists of ADI and active power level as a sum of each phase (2 variables for each measurement point).

The cluster analysis for the k-mean algorithm with Euclidean distance for the three different databases generally indicated the same final classification for 2 clusters. The similarity of data clustering results was higher than 99.5%. So, applying the global index, which represents PQ level, reduced the size of the database by around 90% for the multipoint approach.

Another element of investigation was the proposition of selecting the optimal number of clusters. This step is always an important element of the research. In this article, the v-fold cross-validation was selected. Based on the cost sequence chart, the optimal number of clusters was equal to 3. However, the 6 and 9 clusters also were indicated if the cost sequence would be selected as 2% or 1%.

The final element of the investigation was the qualitative assessment of obtained cluster results. The qualitative comparison was realized for the number of clusters equal to 3, as indicated by the v-fold cross-validation test. The assessment was realized using the global index ADI [55,70]. The standard used to apply limits to the global value was EN 50160 [73]. The results of clustering were related to the working conditions of VPP units. The indicated results were described as a time when HPP and ESS are not working with high power (less than half of the nominal values), HPP is working with high power, and ESS is working with high power. The result of comparison in point of global index indicated that the most positive working condition was cluster 3 (ESS is working with high power). This result means that values of PQ parameter were closer to nominal (e.g., nominal voltage) or the smallest (e.g., asymmetry level).

The analysis of a single point for single PQ parameters seems more vulnerable. But for long-term assessment, such analysis is time-consuming. Additionally, the main purpose of the VPP is generally to achieve economic profits, and technical issues are mostly omitted. So, the proposition of using cluster analysis for the global index database seems interesting to extend the assessment of VPP in a long-term approach.

Finally, it is important to notice that the research was realized in a VPP that operates in Poland, but it may also be applied to other VPPs. Additionally, the proposed solution may also be applied only if long-term power quality data would be available and the different working conditions of monitored objects occur.

6. Conclusions

This article proposed the application of cluster analysis in a virtual power plant. Power quality measurements were used as input to prepare different databases. The databases consisted of classic parameters as well as the global indexes. The selected index (ADI) enabled simplifying the assessment from the number of classical PQ parameters to one global value as a representant of each measurement point. But, on the other hand, including extreme 200 ms values of voltage and harmonics parameters represents the extension to the classical methods. Thus, the global index approach simplifies the assessment and increases the range of used parameters during analysis.

The application of the global index as an input to the cluster analysis dataset enables to reduce the size to around 90% for clustering into 2 clusters. The results for 26 weeks of clustering were similar, with a level over 99.5%. Thus, such applications seem interesting.

The cluster analysis for the global index as an input database was indicated as useful for the identification of working conditions in point of PQ. Also, it enables to simplify the assessment between those working conditions, but using this global value does not enable to define the reason for this situation. However, it is a good first step for deep analysis.

Author Contributions: Conceptualization, M.J. and T.S.; methodology, M.J. and T.S.; software, M.J. and T.S.; validation, D.K., J.R. and V.S.; formal analysis, M.J., T.S., D.K. and J.R.; investigation, M.J. and T.S.; resources, P.K., P.J., J.B. and P.P.; data curation, V.S. and J.S.; writing—original draft preparation M.J. and T.S.; writing—review and editing, D.K. and J.R.; visualization, D.K. and J.R.; supervision, T.S., J.R. and Z.L.; project administration, T.S. and P.J.; funding acquisition, T.S. and P.J. All authors have read and agreed to the published version of the manuscript.

Funding: This research was funded by the National Center of Research and Development in Poland, the project "Developing a platform for aggregating generation and regulatory potential of dispersed renewable energy sources, power retention devices and selected categories of controllable load" supported by European Union Operational Programme Smart Growth 2014–2020, Priority Axis I: Supporting R & D carried out by enterprises, Measure 1.2: Sectoral R & D Programmes, POIR.01.02.00-00-0221/16, performed by TAURON Ekoenergia Ltd.

Institutional Review Board Statement: Not applicable.

Informed Consent Statement: Not applicable.

Data Availability Statement: The data are not publicly available due to policy of associate company.

Conflicts of Interest: The authors declare no conflict of interest.

References

1. Li, Y.; Gao, D.W.; Gao, W.; Zhang, H.; Zhou, J. Double-mode energy management for multi-energy system via distributed dynamic event-triggered Newton-Raphson algorithm. *IEEE Trans. Smart Grid* **2020**, *11*, 5339–5356. [CrossRef]
2. Li, Y.; Zhang, H.; Liang, X.; Huang, B. Event-triggered-based distributed cooperative energy management for multienergy systems. *IEEE Trans. Ind. Inform.* **2019**, *15*, 2008–2022. [CrossRef]
3. Yushuai, L.; Gao, W.; Gao, W.; Zhang, H.; Zhou, J. A distributed double-Newton descent algorithm for cooperative energy management of multiple energy bodies in energy internet. *IEEE Trans. Ind. Inform.* **2020**, *1*. [CrossRef]
4. Yavuz, L.; Önen, A.; Muyeen, S.; Kamwa, I. Transformation of microgrid to virtual power plant–a comprehensive review. *IET Gener. Transm. Distrib.* **2019**, *13*, 1994–2005. [CrossRef]
5. Lis, R.; Czechowski, R. Transformation of microgrid to virtual power plant. In *Variability, Scalability and Stability of Microgrids*; Institution of Engineering and Technology (IET): Stevenage, UK, 2019; pp. 99–142.
6. Jenkins, A.; Patsios, C.; Taylor, P.; Khayrullina, A.; Chirkin, V. Optimising virtual power plant response to grid service requests at newcastle science central by coordinating multiple flexible assets. *CIRED Workshop 2016* **2016**, *4*, 212. [CrossRef]
7. Gabderakhmanova, T.; Engelhardt, J.; Zepter, J.M.; Sorensen, T.M.; Boesgaard, K.; Ipsen, H.H.; Marinelli, M. Demonstrations of DC microgrid and virtual power plant technologies on the Danish island of Bornholm. In Proceedings of the 2020 55th International Universities Power Engineering Conference (UPEC), Torino, Italy, 1–4 September 2020; pp. 1–6.

8. Van Summeren, L.F.; Wieczorek, A.J.; Bombaerts, G.J.; Verbong, G.P. Community energy meets smart grids: Reviewing goals, structure, and roles in Virtual Power Plants in Ireland, Belgium and the Netherlands. *Energy Res. Soc. Sci.* **2020**, *63*, 101415. [CrossRef]
9. Nikolaou, T.; Stavrakakis, G.S.; Tsamoudalis, K. Modeling and optimal dimensioning of a pumped hydro energy storage system for the exploitation of the rejected wind energy in the non-interconnected electrical power system of the Crete island, Greece. *Energies* **2020**, *13*, 2705. [CrossRef]
10. Heimgaertner, F.; Schur, E.; Truckenmueller, F.; Menth, M. A Virtual power plant demonstration platform for multiple optimization and control systems. In Proceedings of the International ETG Congress 2017, Bonn, Germany, 28–29 November 2017; pp. 1–6.
11. Zhao, H.; Wang, B.; Pan, Z.; Sun, H.; Guo, Q.; Xue, Y. Aggregating additional flexibility from quick-start devices for multi-energy virtual power plants. *IEEE Trans. Sustain. Energy* **2020**, *1*, 646–658. [CrossRef]
12. Sharma, H.; Mishra, S. Techno-economic analysis of solar grid-based virtual power plant in Indian power sector: A case study. *Int. Trans. Electr. Energy Syst.* **2020**, *30*. [CrossRef]
13. Behi, B.; Baniasadi, A.; Arefi, A.; Gorjy, A.; Jennings, P.; Pivrikas, A. Cost–benefit analysis of a virtual power plant including solar PV, flow battery, heat pump, and demand management: A western australian case study. *Energies* **2020**, *13*, 2614. [CrossRef]
14. Jeon, W.; Cho, S.; Lee, S. Estimating the impact of electric vehicle demand response programs in a grid with varying levels of renewable energy sources: Time-of-use tariff versus smart charging. *Energies* **2020**, *13*, 4365. [CrossRef]
15. Moutis, P.; Georgilakis, P.S.; Hatziargyriou, N.D. Voltage regulation support along a distribution line by a virtual power plant based on a center of mass load modeling. *IEEE Trans. Smart Grid* **2018**, *9*, 3029–3038. [CrossRef]
16. Dall'Anese, E.; Guggilam, S.S.; Simonetto, A.; Chen, Y.C.; Dhople, S.V. Optimal regulation of virtual power plants. *IEEE Trans. Power Syst.* **2018**, *33*, 1868–1881. [CrossRef]
17. Paternina, J.L.; Contreras, L.; Trujillo, E.R. Study of voltage stability in a distribution network by integrating distributed energy resources into a virtual power plant. *Contemp. Eng. Sci.* **2017**, *10*, 1441–1455. [CrossRef]
18. Ishihara, H.; Nada, K.; Tanaka, M.; Inoue, S.; Kuwata, A.; Takano, T. A Voltage control method for power distribution lines utilizing dispersed customer resources. In Proceedings of the 2020 22nd European Conference on Power Electronics and Applications (EPE'20 ECCE Europe), Lyon, France, 7–11 September 2020; pp. 1–8.
19. Konara, K.; Kolhe, M.; Sharma, A. Power flow management controller within a grid connected photovoltaic based active generator as a finite state machine using hierarchical approach with droop characteristics. *Renew. Energy* **2020**, *155*, 1021–1031. [CrossRef]
20. Haque, M.M.; Wolfs, P.; Alahakoon, S. Active power flow control of three-port converter for virtual power plant applications. In Proceedings of the 2020 IEEE International Conference on Power Electronics, Smart Grid and Renewable Energy (PESGRE2020), Cochin, India, 20 April 2020; pp. 1–6.
21. Pudjianto, D.; Djapic, P.; Strbac, G.; Stojkovska, B.; Ahmadi, A.R.; Martinez, I. Integration of distributed reactive power sources through Virtual Power Plant to provide voltage control to transmission network. In Proceedings of the CIRED 2019 Conference, Madrid, Spain, 3–6 June 2019.
22. Kaczorowska, D.; Rezmer, J.; Sikorski, T.; Janik, P. Application of PSO algorithms for VPP operation optimization. *Renew. Energy Power Qual. J.* **2019**, *17*, 91–96. [CrossRef]
23. Candra, D.I.; Hartmann, K.; Nelles, M. Economic Optimal Implementation of Virtual Power Plants in the German Power Market. *Energies* **2018**, *11*, 2365. [CrossRef]
24. Moreno, B.; Díaz, G. The impact of virtual power plant technology composition on wholesale electricity prices: A comparative study of some European Union electricity markets. *Renew. Sustain. Energy Rev.* **2019**, *99*, 100–108. [CrossRef]
25. Sikorski, T.; Jasiński, M.; Sobierajski, M.; Szymańda, J.; Bejmert, D.; Janik, P.; Ropuszyńska-Surma, E.; Węglarz, M.; Kaczorowska, D.; Kostyła, P.; et al. A case study on distributed energy resources and energy-storage systems in a virtual power plant concept: Economic aspects. *Energies* **2019**, *12*, 4447. [CrossRef]
26. Foroughi, M.; Pasban, A.; Moeini-Aghtaie, M.; Fayaz-Heidari, A. A bi-level model for optimal bidding of a multi-carrier technical virtual power plant in energy markets. *Int. J. Electr. Power Energy Syst.* **2021**, *125*, 106397. [CrossRef]
27. Rahimiyan, M.; Baringo, L. Real-time energy management of a smart virtual power plant. *IET Gener. Transm. Distrib.* **2019**, *13*, 2015–2023. [CrossRef]
28. Othman, M.M.; Hegazy, Y.; Abdelaziz, A.Y. Electrical energy management in unbalanced distribution networks using virtual power plant concept. *Electr. Power Syst. Res.* **2017**, *145*, 157–165. [CrossRef]
29. Mears, A.; Martin, J. Fully flexible loads in distributed energy management: PV, batteries, loads, and value stacking in virtual power plants. *Engineering* **2020**, *6*, 736–738. [CrossRef]
30. Maanavi, M.; Najafi, A.; Godina, R.; Mahmoudian, M.; Rodrigues, E.M.G. Energy management of virtual power plant considering distributed generation sizing and pricing. *Appl. Sci.* **2019**, *9*, 2817. [CrossRef]
31. Jha, B.K.; Singh, A.; Kumar, A.; Misra, R.K.; Singh, D. Phase unbalance and PAR constrained optimal active and reactive power scheduling of Virtual Power Plants (VPPs). *Int. J. Electr. Power Energy Syst.* **2021**, *125*, 106443. [CrossRef]
32. Sikorski, T.; Jasiński, M.; Sobierajski, M.; Szymańda, J.; Bejmert, D.; Janik, P.; Solak, B.; Ropuszyńska-Surma, E.; Węglarz, M.; Kaczorowska, D.; et al. A case study on distributed energy resources and energy-storage systems in a virtual power plant concept: Technical aspects. *Energies* **2020**, *13*, 3086. [CrossRef]
33. Sun, H.; Meng, J.; Peng, C. Coordinated optimization scheduling of multi-region virtual power plant with wind-power/Photovoltaic/Hydropower/Carbon-capture units. *Dianwang Jishu/Power Syst. Technol.* **2019**, *43*, 4040–4049. [CrossRef]

34. Jiao, F.; Deng, Y.; Li, D.; Wei, B.; Yue, C.; Cheng, M.; Zhang, Y.; Zhang, J. A self-scheduling strategy of virtual power plant with electric vehicles considering margin indexes. *Arch. Electr. Eng.* **2020**, *69*, 907–920. [CrossRef]
35. Sun, J.; Li, X.; Ma, H. Study on optimal capacity of multi-type energy storage system for optimized operation of virtual power plants. In Proceedings of the 2018 China International Conference on Electricity Distribution (CICED), Tianjin, China, 31 December 2018; pp. 2989–2993.
36. Han, N.; Wang, X.; Chen, S.; Cheng, L.; Liu, H.; Liu, Z.; Mao, Y. Optimal configuration of energy storage systems in virtual power plants including large-scale distributed wind power. *IOP Conf. Series: Earth Environ. Sci.* **2019**, *295*, 042072. [CrossRef]
37. Sadeghian, O.; Oshnoei, A.; Khezri, R.; Muyeen, S. Risk-constrained stochastic optimal allocation of energy storage system in virtual power plants. *J. Energy Storage* **2020**, *31*, 101732. [CrossRef]
38. Kim, S.; Kwon, W.-H.; Kim, H.-J.; Jung, K.; Kim, G.S.; Shim, T.; Lee, D. Offer curve generation for the energy storage system as a member of the virtual power plant in the day-ahead market. *J. Electr. Eng. Technol.* **2019**, *14*, 2277–2287. [CrossRef]
39. Pudjianto, D.; Ramsay, C.; Strbac, G. Microgrids and virtual power plants: Concepts to support the integration of distributed energy resources. *Proc. Inst. Mech. Eng. Part. A: J. Power Energy* **2008**, *222*, 731–741. [CrossRef]
40. Etherden, N.; Vyatkin, V.; Bollen, M.H.J. Virtual power plant for grid services using IEC 61850. *IEEE Trans. Ind. Inform.* **2016**, *12*, 437–447. [CrossRef]
41. Gong, J.; Xie, D.; Jiang, C.; Zhang, Y. Multiple objective compromised method for power management in virtual power plants. *Energies* **2011**, *4*, 700–716. [CrossRef]
42. Caldon, R.; Patria, A.; Turri, R. Optimal control of a distribution system with a virtual power plant. In Proceedings of the Bulk Power System Dynamics and Control VI, Cortina d, 'AmpezzoItaly, 22–27 April 2004; pp. 278–284.
43. Beguin, A.; Nicolet, C.; Kawkabani, B.; Avellan, F. Virtual power plant with pumped storage power plant for renewable energy integration. In Proceedings of the 2014 International Conference on Electrical Machines (ICEM), Berlin, Germany, 2–5 September 2014; pp. 1736–1742.
44. Zhang, J.; Xu, Z.; Xu, W.; Zhu, F.; Lyu, X.; Fu, M. BI-objective dispatch of multi-energy virtual power plant: Deep-learning-based prediction and particle swarm optimization. *Appl. Sci.* **2019**, *9*, 292. [CrossRef]
45. Kong, X.; Xiao, J.; Liu, D.; Wu, J.; Wang, C.; Shen, Y. Robust stochastic optimal dispatching method of multi-energy virtual power plant considering multiple uncertainties. *Appl. Energy* **2020**, *279*, 115707. [CrossRef]
46. Yi, Z.; Xu, Y.; Gu, W.; Wu, W. A multi-time-scale economic scheduling strategy for virtual power plant based on deferrable loads aggregation and disaggregation. *IEEE Trans. Sustain. Energy* **2019**, *11*, 1332–1346. [CrossRef]
47. Luo, Z.; Hong, S.; Ding, Y. A data mining-driven incentive-based demand response scheme for a virtual power plant. *Appl. Energy* **2019**, *239*, 549–559. [CrossRef]
48. Ai, X.; Yang, Z.; Hu, H.; Wang, Z.; Peng, D.; Zhao, Z. A load curve clustering method based on improved k-means algorithm for virtual power plant and its application. *Dianli Jianshe/Electric Power Constr.* **2020**, *41*, 28–36. [CrossRef]
49. Raptis, T.; Vokas, G.; Langouranis, P.; Kaminaris, S. Total power quality index for electrical networks using neural networks. *Energy Procedia* **2015**, *74*, 1499–1507. [CrossRef]
50. Langouranis, P.; Kaminaris, S.; Vokas, G.; Raptis, T.; Ioannidis, G. Fuzzy total power quality index for electric networks. *MedPower* **2014**, 81. [CrossRef]
51. Ge, B.; Pan, T.; Li, Z. Synthetic assessment of power quality using relative entropy theory. *J. Comput. Inf. Syst.* **2015**, *11*, 1323–1331. [CrossRef]
52. De Capua, C.; De Falco, S.; Liccardo, A.; Romeo, E. Improvement of new synthetic power quality indexes: An original approach to their validation. In Proceedings of the 2005 IEEE Instrumentation and Measurement Technology Conference, Ottawa, ON, Canada, 13 March 2006; Volume 2, pp. 819–822.
53. Lee, B.; Sohn, D.; Kim, K.M. Development of power quality index using ideal analytic hierarchy process. *Lect. Notes Electr. Eng.* **2016**, *376*, 783–793. [CrossRef]
54. Serpak, M. A unified index and system indicator for global power quality assessment. *Sci. Int.* **2016**, *28*, 1131–1136.
55. Jasiński, M.; Sikorski, T.; Kostyła, P.; Leonowicz, Z.; Borkowski, K. Combined cluster analysis and global power quality indices for the qualitative assessment of the time-varying condition of power quality in an electrical power network with distributed generation. *Energies* **2020**, *13*, 2050. [CrossRef]
56. Nourollah, S.; Moallem, M. A data mining method for obtaining global power quality index. In Proceedings of the 2011 2nd International Conference on Electric Power and Energy Conversion Systems (EPECS), Sharjah, United Arab Emirates, 12 January 2011; pp. 1–7.
57. Klajn, A.; Bątkiewicz-Pantua, M. *Application Note–Standard EN 50 160: Voltage Characteristics of Electricity Supplied by Public Electricity Networks*; European Copper Institute: Woluwe-Saint-Pierre, Belgium, 2017.
58. Roiger, R.J. *Data Mining*; Chapman and Hall/CRC: London, UK, 2017; ISBN 9781315382586.
59. Vehkalahti, K.; Everitt, B.S. *Multivariate Analysis for the Behavioral Sciences*, 2nd ed.; CRC Press: Boca Raton, FL, USA, 2019; ISBN 9781351202275.
60. Filzmoser, P.; Hron, K.; Templ, M. *Applied Compositional Data Analysis*; Springer Series in Statistics; Springer International Publishing: Cham, China, 2018; ISBN 978-3-319-96420-1.
61. Wierzchoń, S.; Kłopotek, M. *Modern Algorithms of Cluster Analysis*; Studies in Big Data; Springer International Publishing: Cham, Switzerland,, 2018; Volume 34, ISBN 978-3-319-69307-1.

62. Jasiński, M.; Sikorski, T.; Borkowski, K. Clustering as a tool to support the assessment of power quality in electrical power networks with distributed generation in the mining industry. *Electr. Power Syst. Res.* **2019**, *166*, 52–60. [CrossRef]
63. Kapil, S.; Chawla, M. Performance evaluation of K-means clustering algorithm with various distance metrics. In Proceedings of the 2016 IEEE 1st International Conference on Power Electronics, Intelligent Control and Energy Systems (ICPEICES), Delhi, India, 16 February 2016; pp. 1–4.
64. Faber, P.; Fisher, R. Pros and cons of Euclidean fitting. *Computer Vis.* **2001**, *2191*, 414–420. [CrossRef]
65. Wierzchoń, S.; Kłopotek, M. *Algorithms of Cluster Analysis*; Institute of Computer Science Polish Academy of Sciences: Warsaw, Poland, 2015; Volume 3, ISBN 9789638759627.
66. Chowdhury, K.; Chaudhuri, D.; Pal, A.K. An entropy-based initialization method of K-means clustering on the optimal number of clusters. *Neural Comput. Appl.* **2020**, 1–18. [CrossRef]
67. Yang, J.; Lee, J.-Y.; Choi, M.; Joo, Y. A New Approach to Determine the Optimal Number of Clusters Based on the Gap Statistic. In *Machine Learning for Networking*; Springer International Publishing: New York, NY, USA, 2020; pp. 227–239. [CrossRef]
68. Silva, J.; Lezama, O.B.P.; Varela, N.; Guiliany, J.G.; Sanabria, E.S.; Otero, M.S.; Rojas, V. Álvarez U-control chart based differential evolution clustering for determining the number of cluster in k-means. *Computer Vis.* **2019**, 31–41. [CrossRef]
69. Statsoft Polska. StatSoft Electronic Statistic Textbook. Available online: http:www.statsoft.pl/textbook/stathome.html (accessed on 15 December 2020).
70. Jasinski, M.; Sikorski, T.; Kostvla, P.; Borkowski, K. Global power quality indices for assessment of multipoint Power quality measurements. In Proceedings of the 2018 10th International Conference on Electronics, Computers and Artificial Intelligence (ECAI), Iasi, Romania, 28–30 June 2018; pp. 1–6.
71. Jasiński, M.; Sikorski, T.; Kaczorowska, D.; Rezmer, J.; Suresh, V.; Leonowicz, Z.; Kostyla, P.; Szymańda, J.; Janik, P. A case study on power quality in a virtual power plant: Long term assessment and global index application. *Energies* **2020**, *13*, 6578. [CrossRef]
72. International Electrotechnical Commission (IEC) 61000 4-30. *Electromagnetic Compatibility (EMC)–Part 4-30: Testing and Measurement Techniques–Power Quality Measurement Methods*; IEC: Geneva, Switzerland, 2015.
73. European Standards EN 50160. Voltage Characteristics of Electricity Supplied by Public Distribution Network 2010. Available online: http://copperalliance.org.uk/uploads/2018/03/542-standard-en-50160-voltage-characteristics-in.pdf (accessed on 15 December 2020).
74. Jasinski, M.; Sikorski, T.; Kaczorowska, D.; Kostyla, P.; Leonowicz, Z.; Rezmer, J.; Janik, P.; Bejmert, D. Global power quality index application in virtual power plant. In Proceedings of the 2020 12th International Conference and Exhibition on Electrical Power Quality and Utilisation (EPQU), Kraków, Poland, 14–15 September 2020; pp. 1–6.
75. Kang, J.; Lee, J.-H. Electricity customer clustering following experts’ principle for demand response applications. *Energies* **2015**, *8*, 12242–12265. [CrossRef]
76. Claeys, R.; Azaioud, H.; Cleenwerck, R.; Knockaert, J.; Desmet, J. A novel feature set for low-voltage consumers, based on the temporal dependence of consumption and peak demands. *Energies* **2020**, *14*, 139. [CrossRef]

Article

Impact of Harmonic Currents of Nonlinear Loads on Power Quality of a Low Voltage Network–Review and Case Study

Łukasz Michalec, Michał Jasiński *, Tomasz Sikorski, Zbigniew Leonowicz, Łukasz Jasiński and Vishnu Suresh

Faculty of Electrical Engineering, Wroclaw University of Science and Technology, 50-370 Wroclaw, Poland; 253681@student.pwr.edu.pl (Ł.M.); tomasz.sikorski@pwr.edu.pl (T.S.); zbigniew.leonowicz@pwr.edu.pl (Z.L.); lukasz.jasinski@pwr.edu.pl (Ł.J.); vishnu.suresh@pwr.edu.pl (V.S.)
* Correspondence: michal.jasinski@pwr.edu.pl; Tel.: +48-71320202

Abstract: The paper presents a power-quality analysis in the utility low-voltage network focusing on harmonic currents' pollution. Usually, to forecast the modern electrical and electronic devices' contribution to increasing the current total harmonic distortion factor (THD_I) and exceeding the regulation limit, analyses based on tests and models of individual devices are conducted. In this article, a composite approach was applied. The performance of harmonic currents produced by sets of devices commonly used in commercial and residential facilities' nonlinear loads was investigated. The measurements were conducted with the class A PQ analyzer (FLUKE 435) and dedicated to the specialized PC software. The experimental tests show that the harmonic currents produced by multiple types of nonlinear loads tend to reduce the current total harmonic distortion factor (THD_I). The changes of harmonic content caused by summation and/or cancellation effects in total current drawn from the grid by nonlinear loads should be a key factor in harmonic currents' pollution study. Proper forecasting of the level of harmonic currents injected into the utility grid helps to maintain the quality of electricity at an appropriate level and reduce active power losses, which have a direct impact on the price of electricity generation.

Keywords: nonlinear loads; harmonics, cancellation, and attenuation of harmonics; waveform distortion; THDi; power quality; low-voltage networks

Citation: Michalec, Ł.; Jasiński, M.; Sikorski, T.; Leonowicz, Z.; Jasiński, Ł.; Suresh, V. Impact of Harmonic Currents of Nonlinear Loads on Power Quality of a Low Voltage Network–Review and Case Study. *Energies* **2021**, *14*, 3665. https://doi.org/10.3390/en14123665

Academic Editor: Abu-Siada Ahmed

Received: 10 May 2021
Accepted: 16 June 2021
Published: 19 June 2021

Publisher's Note: MDPI stays neutral with regard to jurisdictional claims in published maps and institutional affiliations.

1. Introduction

Power electronics components and microprocessors are the key entities of the evolution of nonlinear loads and smart devices/systems in commercial and household environments [1,2]. On one hand, due to their nonlinear V–I characteristics, these devices introduce power quality (PQ) problems in the electric power distribution system, and on the other hand they are sensitive to the PQ level. In the case of electrical energy billing, even the revenue meters that measure electrical quantities may be not immune enough to the impact generated by disturbing devices [3–5]. Since the intensified integration of nonlinear components, PQ monitoring is essential to improving the performance of both power system equipment and the end-user loads. [6,7] Thus, the PQ disturbances are the responsibility of not only utilities but also end-users, manufacturers as well as researchers, and engineers. [8,9] The remarkable growth in research and development work on identifying the avoidable energy losses caused by poor PQ has been observed in recent years [10–16]. The proper measurements and analysis of the distribution system and implementation of the evolving methods and techniques to improve the condition of the supply system are developing areas of present-day electrical engineering [17–19].

Due to the nonlinearity of the electrical devices, the major fault in PQ is observed as harmonic pollution. The severity of harmonic disturbances contained in the current signal can be specified collectively by the current total harmonic distortion factor (THD_I), which can be calculated according to the formula:

$$THD_I = \frac{\sqrt{\sum_{h=2}^{\infty} I_h^2}}{I_1} \cdot 100\%,$$

(1)

where,

I_h—RMS value of the individual harmonic component of order h for current,
I_1—RMS value of the fundamental harmonic (a component of order 1) for current.

To forecast how increased nonlinear loads will contribute to the THD_I level and when they may exceed the regulation limit, harmonic impact and propagation studies of the distribution network need to be performed for future scenarios [20–22]. Distribution system operators can use these analyses and simulations to take preventive steps to avoid deterioration and improve the PQ level [23].

Currently, the increase of new technologies and the changing of conventional appliances to power electronic-driven appliances can be observed in residential and commercial buildings feeding from the LV distribution networks [18,24]. Modern power electronic-driven appliances can be divided into four groups (Figure 1).

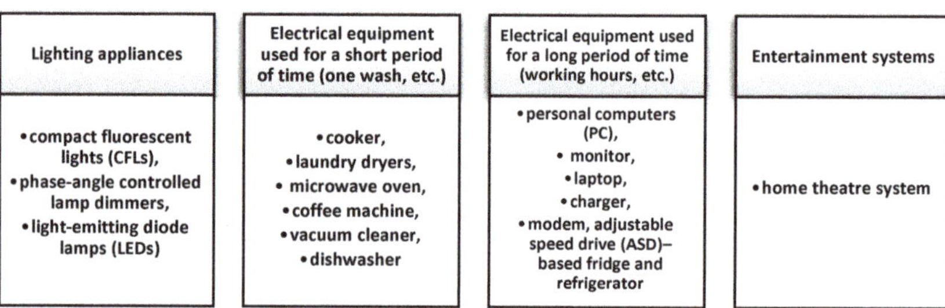

Lighting appliances	Electrical equipment used for a short period of time (one wash, etc.)	Electrical equipment used for a long period of time (working hours, etc.)	Entertainment systems
• compact fluorescent lights (CFLs), • phase-angle controlled lamp dimmers, • light-emitting diode lamps (LEDs)	• cooker, • laundry dryers, • microwave oven, • coffee machine, • vacuum cleaner, • dishwasher	• personal computers (PC), • monitor, • laptop, • charger, • modem, adjustable speed drive (ASD)–based fridge and refrigerator	• home theatre system

Figure 1. Groups of modern power electronic-driven appliances.

This nonlinear electrical equipment which, indeed, forms the power system is one of the main causes of harmonic distortion. Since its AC circuit topology draws, often in a pulsed manner, a distorted current whose waveform shape does not resemble the applied voltage waveform shape for the considered cycle, it is non–sinusoidal [21,25].

2. Review

2.1. Problems Caused by Harmonic Currents

The first and major problems caused by harmonic currents in the mains feeding residential and commercial buildings are those caused by voltage waveform distortion [18]. Due to the propagation of harmonic currents through a finite system impedance, the vector sum of all individual voltage drops results in the increased voltage total harmonic distortion factor (THD_V), which contributes to significant deterioration of power quality and affects the effective performance of consumers' devices connected to the same point of common coupling (PCC).

Figure 2 shows the power system where multiple customers are connected at PCC with different operating and load conditions. Because of the significantly distorted supply voltage waveform, all electrical devices (also linear ones) will generate nonlinear currents (in different degrees) [25].

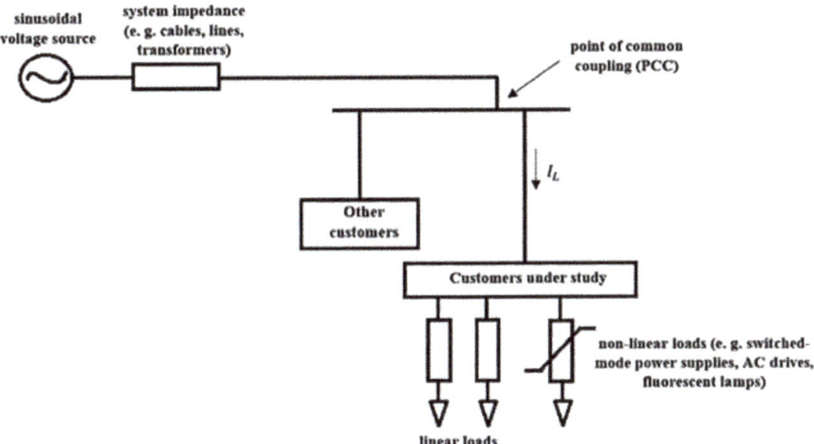

Figure 2. The typical model of the customer's loads supply system.

The levels of harmonic currents which the customers are allowed to inject into the power network at the PCC are limited by standards [26,27].

The second problem caused by harmonic currents is associated with the harmonic currents themselves [28]. The harmonic currents injected into distribution networks cause high disturbances [22,29], such as:

- overheating and failure of transformers and other power equipment;
- false tripping of protection relays;
- overcurrent on equipment-neutral connection wiring;
- errors in the operation and control of sensitive devices (e.g., microprocessors); and
- interference with communication signals (digital technology), and others.

2.2. Recent Research

The equipment within domestic as well as commercial installations is increasingly sensitive to some type of electromagnetic interference, both from internal and external sources [1,18]. Therefore, there is a need to control the electromagnetic environment, namely by limiting the harmonic pollutions caused by any type of electrical appliance. Thus, the IEC 61000–3–2 standard [30] classifies all electrical equipment into four categories, that are presented in Figure 3.

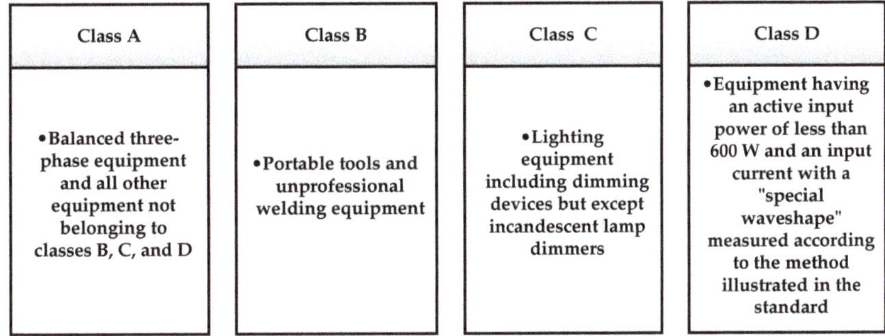

Class A	Class B	Class C	Class D
• Balanced three-phase equipment and all other equipment not belonging to classes B, C, and D	• Portable tools and unprofessional welding equipment	• Lighting equipment including dimming devices but except incandescent lamp dimmers	• Equipment having an active input power of less than 600 W and an input current with a "special waveshape" measured according to the method illustrated in the standard

Figure 3. Classification of electrical equipment to limit harmonic pollution.

Unlike IEC limits [30] that can be applied to all systems, irrespective of their stiffness, the IEEE standards (according to IEEE Std 519) [31] introduce the harmonics current distortion criteria limits (individual harmonic distortion as well as THD_I factors), which are a function of maximum short circuit current I_{SC} to maximum demand load current I_L at PCC [32,33].

From past literature reviews [34–36], it was found that many harmonic current impact studies were performed with individual usage of a specific type of nonlinear device or groups of devices used for the same purpose, often for sinusoidal conditions of the supply voltage [37].

The effects of harmonics current for fluorescent lamps (FLs) with different ballasts are studied in [38]. Various configurations of the fluorescent T8 and T5 tube lamps from different commercial brands were used. The results indicate that with the increased number of fluorescent T8 lamp (from 4 lamps to 10 lamps) with electromagnetic as well as electronic ballasts, the percentage of the total harmonic current distortion factor (THD_I) has little change, approximately 0.5–1%, while the consumed power increases significantly. For T8 FLs with electromagnetic ballast, the $THD_I \approx 9\%$, whereas for the T8 FLs with electronic ballasts the $THD_I \approx 10\%$. The newer technology, fluorescent T5 lamps with electronic ballasts, generates much more distorted current waveforms (selected in the experiment low-class lamps). Four T5 FLs show the $THD_I = 23.7\%$ and 10 T5 FLs demonstrate the $THD_I = 27.2\%$. The authors also present the high class T5 type lighting devices with electronic ballast, which indicate the current total harmonic distortion factor even less (3–5%) than the tested electromagnetic ballast for the T8 lamp.

The comprehensive overview conducted by several researchers investigating the harmonic performance of the utility grid when employing compact fluorescent lamps (CFLs) on a large sczzale is reviewed in [39]. The authors conclude that power quality is affected significantly by CFLs. For a typical CFL tested in [39], the THD_I is as high as about 75%. However, in [40] the lamps were tested in varying the RMS terminal voltage conditions. The results show that the THD_I for 12 CFLs of the different manufacturers is practically constant ($THD_I \approx 110\%$) in the operation range between 80% to 100% of the nominal voltage of the CFLs. The next research [41] signifies that when connecting subsequent CFLs, the harmonic content is expected to increase as the more distorted current is drawn. To anticipate the problems generated by a massive insertion of small nonlinear loads, several investigations focused on modeling, characterizing, and predicting the increase of harmonic currents due to the CFLs [40,41]. The results of the research on the cancellation of harmonics between groups of compact fluorescent lamps are demonstrated in [42].

Light-emitting diodes (LEDs) and another solid-state lighting (SSL) system's impact on power quality is studied in [12,43,44]. For these types of lighting, the individual current harmonic components as well as the THD_I values often exceed the limits set by the IEEE standard 519 [26]. According to [44], the LED and micro-LED lamps show moderated harmonic current pollutions. As indicated in [12], as a greater number of LED lamps are connected to the grid, their performance improves both in terms of power factor and harmonic distortion. In [45] it is shown that there is a relevant difference in harmonic pollution behavior of CFLs and LED lamps, not only between the two types of lamps but also within the same type. Thus, it seems to be sensible to utilize some lamp combinations to keep harmonic distortion within permissible limits.

Besides the lighting devices, the literature has documented measurements of other commercial as well as household appliances such as TV sets, personal computers, monitors, printers, laptops, chargers, etc. [18,41,46]. The conventional main interface of these electrical devices consists of a diode bridge rectifier followed by a large DC capacitor. In this case, the current drawn from the AC supply is relatively linear [21]. However, in the newer technology, switch-mode power supply (SMPS) systems, the rectifier is usually followed by a converter that can be controlled in different modes to achieve a smooth DC output. The input current waveform is highly nonlinear and often comes to the load in very short pulses [47]. Today, single-phase electronic equipment almost universally employs SMPSs [21,47,48]. Although individual consumption is small, the collective effect can be significant because many of them are often connected to the same PCC. Additionally, to satisfy harmonic standards' requirements, the latest SMPSs utilize complex circuits with active power factor correction (PFC) and usually change the equipment classification to Class A according to the mentioned IEC standard [30]. In [49,50], three main circuit topologies of these devices are proposed: without PFC, with passive PFC, and with active PFC. The result is low pollution in the case of sinusoidal supply voltage but often also increased harmonic currents' pollution, which then increases the distortion of the supply voltage as already described.

The experimental measurements of harmonic currents' generation by a cluster of desktop personal computers (20 PCs during different operating modes) are described in [41,51,52]. Generally, the connecting subsequent number of PCs decrease the current total harmonic distortion factor (THD_I). The main reason is the attenuation phenomenon for the 3rd, 5th, and 7th harmonics [52]. Interestingly, the obtained result shows that when the number of PCs is increased from 1 to 124, the 3rd current harmonic is reduced from 76% to 24% [41]. Other results [51] provide insight into PC harmonic behavior in different conditions. Due to voltage change, the THD_I of 20 PCs is about 7% lower than that of a single PC. When a group of 20 PCs is fed from the same supply, the THD_I decreases by more than 15% (in comparison to a single PC) for maximum system impedance. Even with a group of PCs connected to a single supply, changing the system frequency in the range 50 Hz $\pm$ 1% has almost no effect on the generated harmonics. Moreover, the studies [51] focused on attenuation and diversity phenomena in PC clusters indicate that diversity effects have a minor impact on the harmonics produced by a cluster of PCs but some kind of harmonic compensation may occur. Also highlighted is that presented factors affecting produced harmonics are related to each other, and if nonlinear devices are supplied by different power systems their harmonics generation and other power quantity levels will be different [52]. This as well as other relevant experimental data can be useful in harmonic characterization studies.

In [47], the energy efficiency of power electronic-based home appliances was analyzed. The authors underline that although most energy-efficient loads based on power electronics consume considerably less power than their older counterparts, they are harmonic polluters and inject a relatively large number of harmonic currents into the power grid relative to their power demand.

2.3. Problem Description

Most of the presented research focuses on the harmonic generated by one type of commonly used nonlinear loads (NLLs), often in sinusoidal conditions of the supply voltage. The main general properties of all NLLs can be distinguished as follows [41,42,53]:

- the current harmonic pollutions generated by a load depend on the voltage harmonics' magnitudes and phase angles. The THD_I and individual harmonic currents for a group of NLLs vary due to the background voltage distortion variations within the recommended standards limits. This means that changes in one voltage harmonic lead to changes in multiple current harmonics;
- there are no even harmonics if the waveform has half-wave symmetry; and
- domination in the amplitude of the lower-order harmonics (sequentially, the 3rd, 5th, and 7th order) is noticeable.

When large amounts of such appliances operate in power distribution systems, the collective effect on the feeder power quality [54,55] has become a large concern to utilities. For predicting the harmonic current injection to the utility grid and implementing compensation technique, several theoretical models and simulations of nonlinear loads were conducted in the past decades [24]. The level of injected harmonic pollutions from a group of nonlinear loads of the same type operating simultaneously can be predicted by scaling the typical harmonic current spectrum of one load in proportion to total load power [52].

However, we seldom must deal with such a situation in real life because many different types of harmonic sources are almost always connected to one point of common coupling

(PCC) [56]. Thus, the simple combined approach for forecasting the level of harmonic current injection to the utility grid is not proper anymore [57]. Therefore, a composite approach is required, which takes into account different nonlinear load types placed in the PCC at the same time [58].

By applying such an approach, not only can the addition effect of many harmonic sources be assessed, but the attenuation outcome of future harmonic distortion in the network can also be better estimated [59]. The main purpose of the experimental tests presented in this paper is to create a first baseline study for office and domestic environments in case of deterioration of PQ by harmonic currents.

3. Case study

3.1. Methodology for Case Study

In Stage 1, the problem was defined and detailed, based on the collected knowledge of the subject, and a description was delivered. The theory of the power quality deterioration phenomenon associated with the operation of nonlinear loads was provided in the Introduction section of this paper.

Stage 2 determined relevant aspects to be included in the study and its scope. For the analysis of the deterioration level of PQ caused by harmonic currents, some sets of various commercial and residential electrical appliances with different technical specifications were tested. For comparison, the research was conducted for selected individual devices as well as for sets of diverse appliances.

Table 1 shows the technical specification of the selected nonlinear loads. The power value included in the table is the value that the manufacturer declares the device consumes. All devices are supplied from a single-phase LV network, in a voltage range of 230–240 V, with a frequency of 50 Hz.

The accuracy of the measurements conducted by PQ Analyser Fluke 435 is as follows:

- THD_I measurement: $\pm 2.5\%$
- $\%f$ measurement: $\pm 0.1\% \pm n \times 0.1\%$
- $\%r$ measurement: $\pm 0.1\% \pm n \times 0.4\%$

Table 1. Technical specifications of the selected equipment.

No.	Type	Brand	Model	I (A)	P (W)
1	Laptop charger (AC Power Adapter)	Samsung	A10–090P1A	1.5	90
2	Smartphone Charger	Samsung	EP–TA20EWE	0.5	0.1
3	Desktop Copier (standby mode)	Canon	IR2016	2.5	3.6
4	LCD monitor–20"	LG	L204WT	1.0	45
5	Vacuum cleaner	Zelmer	321.0.E01E	6.0	1200
6	Coffee machine	Krups	EA81	5.5	1450
7	Microwave oven	FIF	MD 42035	6.5	800
8	Compact fluorescent lamp	OSRAM	20 W/865	0.2	20
9	Fluorescent lamp type T8	Philips	TLD 18 W 830	0.36	18
10	High intensity discharge lamp	Philips	ML 160W E27	0.73	160
11	Dimmer and incandescent lamp	F&F	SCO–812	1.5	60

For each set of devices, the same parameters as those obtained for the appliances individually are measured. One of these sets consists of some typical domestic electrical loads used for a short and long time. For the second test, the office electrical equipment is selected. The third test is conducted for a set of lighting equipment (including an incandescent lamp dimmer) that can be used in both commercial and residential spaces. According to the IEC 61000–3–2 standards [30], incandescent lamp dimmers are not classified into Class C (lighting equipment) in the case of the injection of high-order harmonic currents to the grid, but to Class A, which include difficult-to-classify devices. Thus, the level of harmonic contained in the current drawn from the LV grid by incandescent lamp dimmers is not so strictly limited as other lighting equipment. Table 2 shows the type of set and the appliances that constitute it.

Table 2. Selected sets of equipment.

No.	Type of Loads	Group of Loads
1	Household	LCD monitor–20″ Caffe machine Compact fluorescent lamp Microwave AC Power Adapter (Charger) Vacuum cleaner
2	Office	LCD monitor–20″ Fluorescent lamp type T8 Compact fluorescent lamp Desktop Copier AC Power Adapter (Charger) Adaptive Fast Smartphone Charger
3	Lighting	Compact fluorescent lamp Fluorescent lamp type T8 High–intensity discharge lamp Dimmer and incandescent lamp

After the selection of the appliances, the laboratory test stand to assess the impact of various nonlinear electrical equipment on PQ were constructed. The diagram and photo of the prepared test stand are shown in Figures 4 and 5.

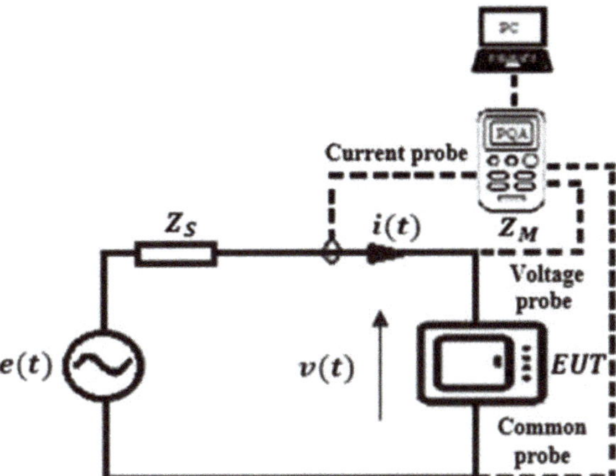

Figure 4. Diagram of the measurement system.

Figure 5. Laboratory measurement stand for testing the impact of nonlinear devices on power quality: (**a**) power supply AC 230 V 50 Hz; (**b**) power quality analyzer Fluke 435; (**c**) adapter for measuring voltage in a single-phase circuit; (**d**) measuring clamp; (**e**) PC with FlukeView software; (**f**) adapter for testing a set of electrical equipment; (**g**) electrical equipment under test.

In Stage 3, the indicators of the impact on power quality associated with their operation were collected. Hence, the current harmonics distortion rate of each of the devices as well three different sets of them are measured. To conduct the measurements, the class A PQ analyzer (FLUKE 435) and the associated equipment were used, including 5 A and 20 A precision current clamps.

In this work, the steady-state operation and respective harmonic pollution of devices are considered. The power quality analysis was performed by the assessment of the current distortion level, which was evaluated via the current total harmonic distortion factor (THD_I) in percentage. The power quality analyzer allows the recording of harmonics up to the 50th order. Thus, the current distortion rate was calculated according to the expression shown in Equation (2), as defined in IEC 61000–2–2 standard [60].

$$THD_I = \frac{\sqrt{\sum_{h=2}^{50} I_h^{2}}}{I_1} \cdot 100\% \tag{2}$$

For organization of the data, Stage 4 in systematic methodology was conducted using the simple mouse-controlled PC software FlukeView (version 3.2), which communicates with the PQ analyzer via the optically isolated OC4USB adapter/cable connected to the USB of the PC. The software enables saving, opening, and printing the data collected by the PQ analyzer, or exporting it to other programs.

The next section of this paper presents the obtained measurement results. It is followed by the discussion section, where the analysis and interpretation of the results are provided (Stage 5).

3.2. Results of Case Study

This section presents the measurement results that were conducted in a laboratory environment with the class A PQ analyzer (FLUKE 435) on the commonly used residential and commercial appliances to assess their impact on PQ.

In Figure 6, the measurement results for the appliances tested individually are presented. The input current waveforms are shown in a time domain, the harmonic spectra are represented up to the 50th harmonic, and the current total harmonic distortion factor (THD_I) is obtained in percentage. Additionally, Table 3 was prepared to sum up the information about appliances and results of THD_I.

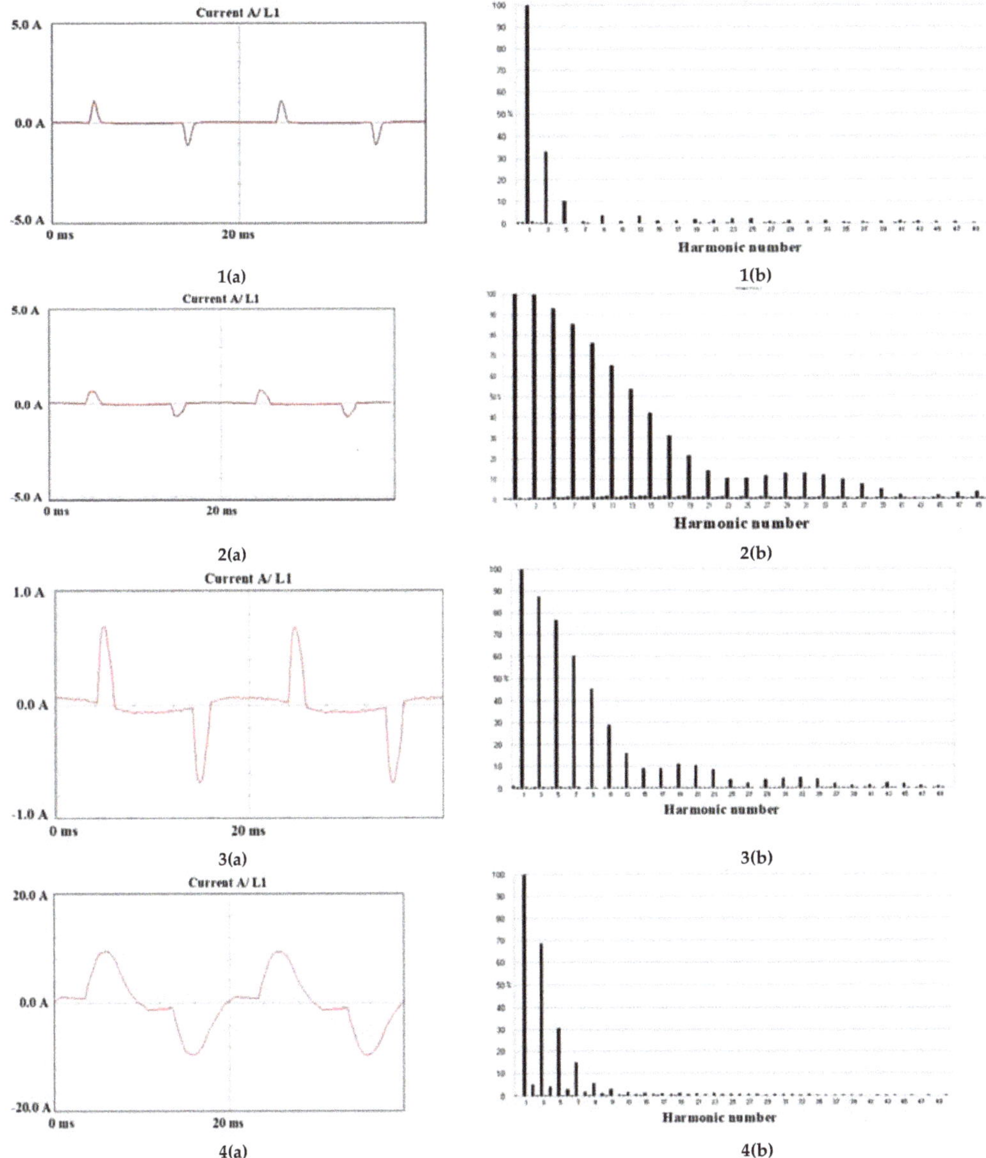

Figure 6. *Cont.*

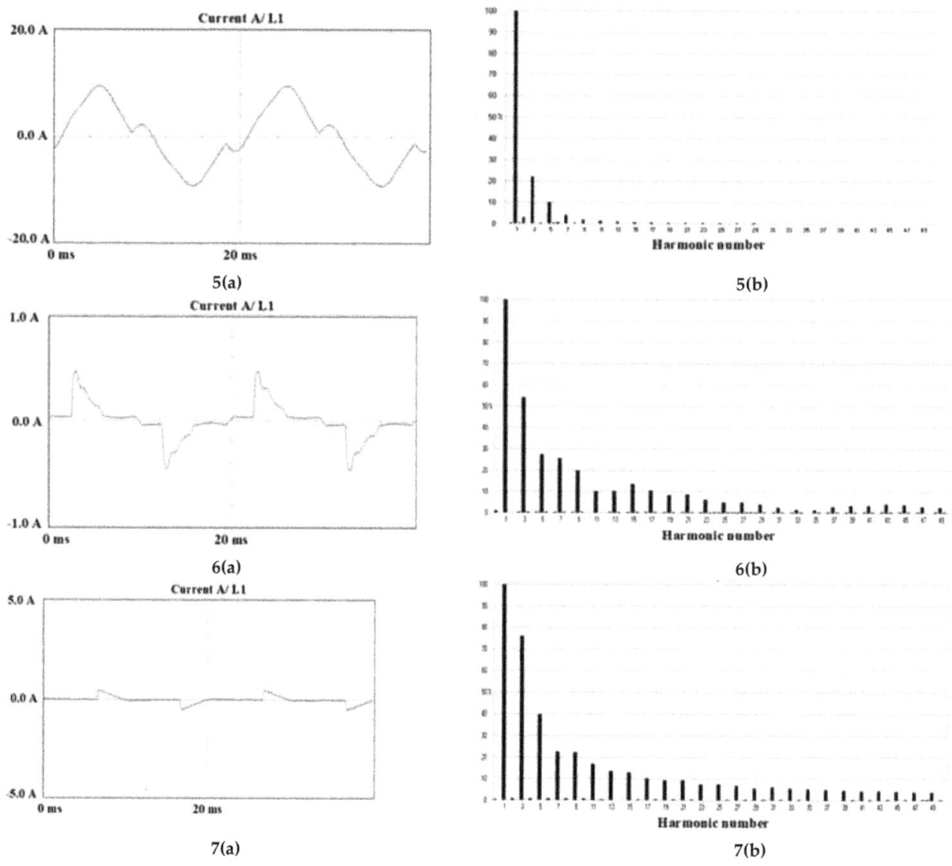

Figure 6. Measurements obtained for individual devices: (**a**) current waveform; (**b**) harmonic spectrum; 1—AC Power Adapter, 2—Adaptive Fast Smartphone Charger, 3—LCD monitor–20″, 4—Vacuum cleaner, 5—Microwave oven, 6—Compact fluorescent lamp, 7—Dimmer and incandescent lamp.

Table 3. Current total harmonic distortion factor (THD_I) measurements obtained for individual devices.

No.	Appliance	THD_I
1	AC Power Adapter	35.3%
2	Smartphone Charger	207.3%
3	LCD monitor–20″	145.3%
4	Vacuum cleaner	75.9%
5	Microwave oven	24.6%
6	Compact fluorescent lamp	73.5%
7	Dimmer and incandescent lamp	97.7%

The input current waveforms in all cases are highly non-sinusoidal in each cycle. They are in different shapes and often flow in a pulsed manner, which indicates a relevant harmonic content in the signals. The level of harmonic distortion is graphically represented by bars that represent the individual harmonic magnitudes (as a percentage of the fundamental component). In all harmonic spectra, there is visible a clear advantage of the odd harmonics (gradually decreasing for larger orders) over even harmonics, which suggests that positive and negative half-cycles of the signals are symmetrical.

Analyzing the numerical indicator of harmonic current distortions (the THD_I) for each device, it can be seen that all of these values are excessively high, which implies that harmonic pollution injected into the network is very significant. The highest measured current distortion rate is equal to 207.3% for the adaptive fast smartphone charger. It is not difficult to imagine the negative impact of a huge number of these devices, which is adequate to the number of smartphones, on the distribution network. The smallest THD_I is noticed for the microwave oven, and it is equal to 24.6%.

As established in the IEC 61000–3–2 standards [30], the equipment with a rated current not greater than or equal to 16 A must not breach specified limits for harmonic current pollution. The literature [28,30,35] studies in detail whether the commercial and residential nonlinear loads supplied from the low-voltage utility grids meet the established criteria or not. This paper presents results that only support the hypothesis that nonlinear electrical loads are one of the main causes of electrical power quality deterioration and require appropriate analysis to evaluate their negative impact on the AC power distribution system.

In subsections below (Sections 3.2.1–3.2.3), the results of experimental tests for sets of devices are presented. The same parameters as those obtained for the appliances individually are measured.

3.2.1. Experimental Test 1: Set of household equipment

The set consists of some typical domestic electrical loads used for a short and long time, such as an LCD monitor–20″, coffee machine, compact fluorescent lamp, microwave oven, AC power adapter (laptop charger), and vacuum cleaner.

Figure 7 shows the waveforms of applied voltage and current drawn by the set of household equipment when connected collectively to the single-phase utility LV network, in the voltage range of 230–240 V. Due to the nonlinear behavior of all connected devices (see Figure 8) the total current drawn from the grid is significantly distorted by high-order harmonic currents, while the supplying voltage waveform is almost sinusoidal in shape.

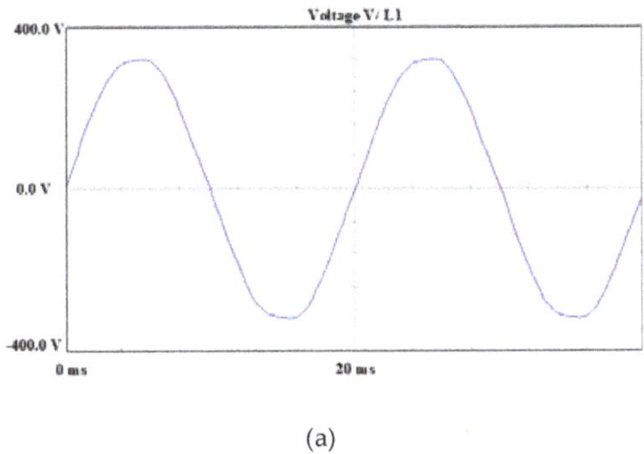

(a)

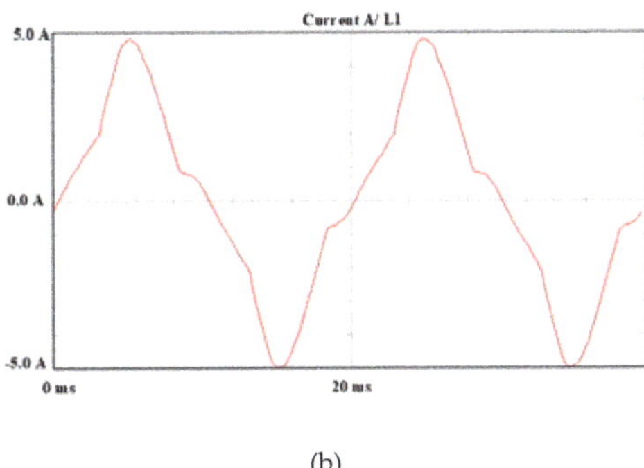

(b)

Figure 7. Set of household equipment: (**a**) input voltage waveform; (**b**) input current waveform.

The obtained harmonic spectrum (Figure 8) consists mainly of third (≈25%), fifth (≈7%), and seventh harmonics (≈4%). The values in brackets indicate the magnitude of the harmonic current components in percentages of the magnitude of the fundamental component. The individual amplitudes decrease for larger orders. The THD_I factor is equal to 25. 7%, which means a relatively low impact on PQ.

As previously mentioned, harmonic cancellation and attenuation phenomena in low-voltage networks occur. The first is due to phase angle diversity between the same-order harmonics produced by different nonlinear loads; the second is due to system impedance and the corresponding voltage distortion.

Thus, the current flowing through the sockets adapter (Figure 6f) has a lower current total harmonic distortion than each piece of previously individually tested household equipment.

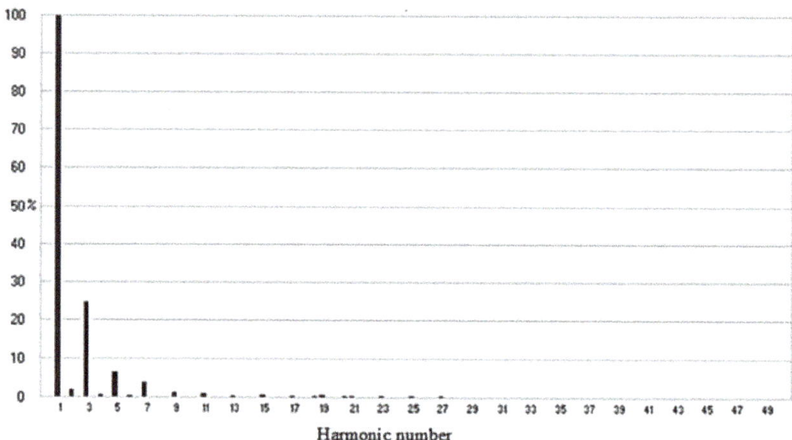

Figure 8. Set of household equipment: harmonic spectrum and current total harmonic distortion factor. $THD_I = 25.7\%$.

3.2.2. Experimental Test 2: Set of Office Equipment

The set consists of some typical office electrical equipment used for a short and long time, such as an LCD monitor–20″, fluorescent lamp type T8, compact fluorescent lamp, desktop copier, AC power adapter (laptop charger), and adaptive fast smartphone charger.

The combination of lighting and office equipment also features the harmonic currents' cancellation and attenuation effects in the low-voltage network. Figure 9 shows that the total current drawn in a pulsed manner by the second set of nonlinear loads is strongly distorted, while supplying voltage is almost free from distortion.

The visible bars in Figure 10 represent the odd harmonic currents' magnitudes (as a percentage of the fundamental component). These results demonstrate that the devices operating simultaneously produce less current distortion ($THD_I = 74.7\%$) than individually, e.g., or one adaptive fast smartphone charger the $THD_I = 207.3\%$ (Figure 6).

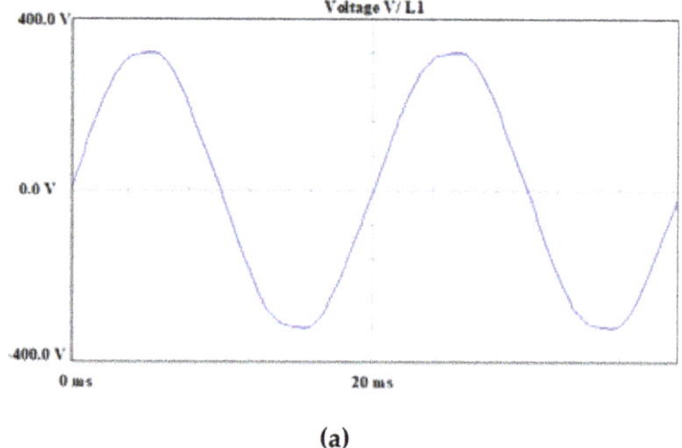

(a)

Figure 9. *Cont.*

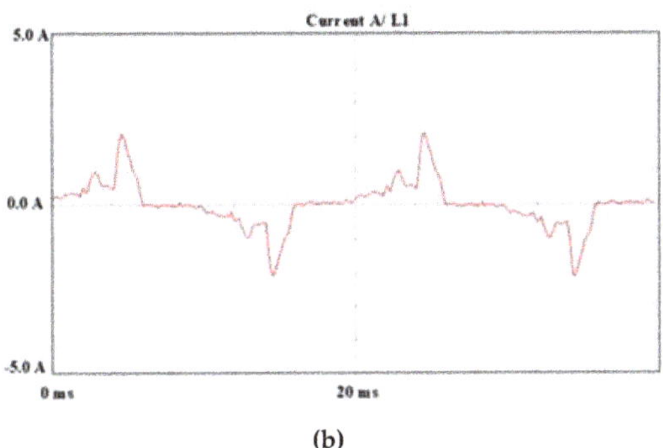

(b)

Figure 9. Set of office equipment: (**a**) input voltage waveform, (**b**) input current waveform.

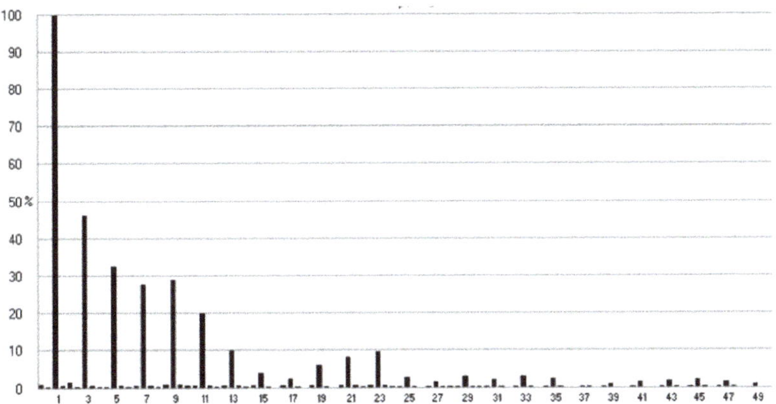

Figure 10. Set of office equipment: harmonic spectrum and current total harmonic distortion factor. $THD_I = 74.7\%$.

3.2.3. Experimental Test 3: Set of Lighting Equipment

The set consists of some lighting equipment that can be used in both commercial and residential spaces, such as a: compact fluorescent lamp, fluorescent lamp type T8, high–intensity discharge lamp, dimmer and incandescent lamp.

Analyzing the harmonics spectrum (Figure 11) for the set of lighting equipment supplied from one socket adapter, it is evident again that the harmonic currents of different nonlinear loads are vectorially superposed. The higher magnitude reaches the third harmonic (c. 35% of the fundamental component). The amplitudes of successive odd harmonics are lower and do not decrease proportionally (the 17th harmonic ≈ 9%, whereas the 19th harmonic ≈ 1%, and then the 21st harmonic ≈ 7%). The current total harmonic distortion factor (THD_I) is equal to 46. 3%, and it is reflected in the oscillatory current waveform shape (Figure 12). This leads to the conclusion that the set of different types of lighting equipment has less impact on PQ than, e.g., the THD_I factor equal to 73.8% for the CFL operating individually (see Figure 6). However, in general, the harmonic currents for CFLs are additive in behavior [51].

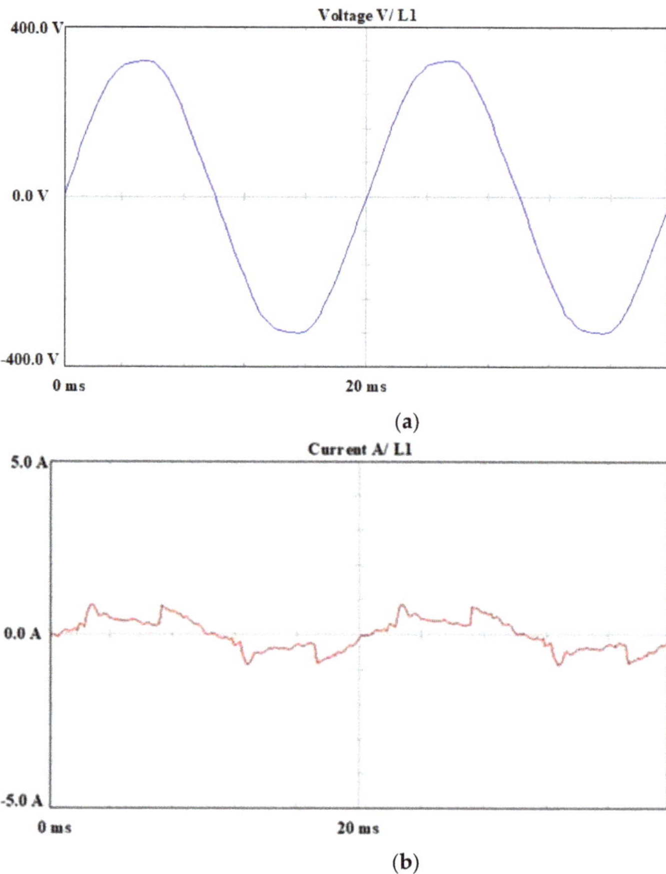

Figure 11. Set of lighting equipment: (**a**) input voltage waveform, (**b**) input current waveform.

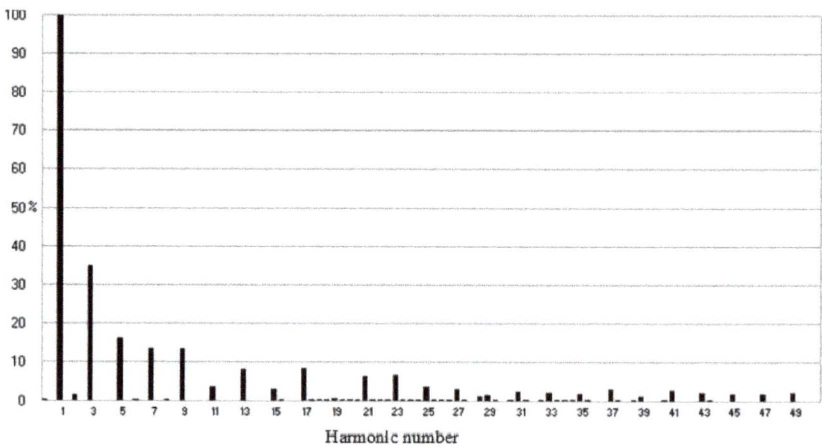

Figure 12. Set of lighting equipment: harmonic spectrum and current total harmonic distortion factor. $THD_I = 46.3\%$.

4. Discussion

The laboratory experiments demonstrated that the harmonic currents produced by multiple types of nonlinear loads were reduced. This effect is often less significant for the 3rd and 5th and 7th harmonics but becomes important for higher-current harmonics. Thus, the current total harmonic distortion factors (THD_I) for the combination of different residential, commercial, or lighting receivers were typically lower than for the individually tested units.

Although the results were based on a relatively small sample of electrical appliances, the obtained results are highly useful. For example, adding from 2 to even 10 CFLs to the set of different types of lighting sources in a typical house is not as harmonically harmful as one may expect.

In the obtained results, there were harmonic current summation and cancellation effects between individual loads working on the same network caused by phase angle diversity as well as a system impedance and the voltage distortions corresponding to them. This statement is in accord with harmonic studies about the impact of electric equipment on the utility grid [26,33,36,41].

Causes of summation and cancellation phenomena are other aspects of the presented analysis that will be investigated in future work, together with considering the interactions between different types of loads typically found in the residential and commercial sectors.

In all the different load groups analyzed, it was found that although there is a slight improvement, the distortion effect on the currents is still high.

There are considerable financial losses associated with poor PQ, and thus associated with harmonics [61,62]. The harmonic currents generated by nonlinear loads decrease the power factor, which can be considered an indicator of the quality of the device's energy conversion and consequently distortion power. The reactive power and distortion power lead to actual energy consumption required for these loads' proper operation that is higher than declared by the manufacturer. The losses associated with harmonics constantly vary with the change of the THD_I factor and thus with connected and disconnected electrical equipment. Even if the amount of wasted energy in one residential area may be negligible, the total amount of energy that is used in the residential and commercial sector is relevant and the losses cannot be neglected anymore.

Working in accordance with the standards often requires some assumptions to be made. In the conducted experimental tests described in this paper, it is simplified that the voltage stays undistorted (sinusoidal, without harmonics); however, in the real low-voltage grid conditions the THD_V changed continuously in respect to time and location. Thus, the same loads put in two different places in the power system may result in various harmonic characteristics because of two different distortion values in supplying voltage [63,64].

Often, the additional difficulty is to measure harmonic characteristics for many non-linear receivers in laboratory conditions due to lack of space and costs. Thus, to simulate harmonic currents generated by sets of large–scale nonlinear loads and to assess their impact on power quality, different software packages such as MATLAB, ETAP, CYME, PSCAD are often used [29]. Some of the research related to the simulation of harmonic character by using this software is presented in [42,49].

There are typical mitigation techniques used in industry today to reduce harmonic loading in a system [65]. They include active filtering, series and shunt passive filtering, multiple pulse rectifiers, and isolation transformers. Each type of these solutions has its advantages and disadvantages, which are described in more detail in [66–68].

The PQ problems and their mitigation techniques are playing a very important role in future electrical engineering. Thus, this area requires a continuous process for research and development.

5. Conclusions

The appliances commonly used in commercial and residential facilities can be significant sources of harmonic currents. This is particularly true when the simple approach for forecasting the level of harmonic current injection to the utility grid is taken into account. The composite approach also contributes to the harmonic-producing characteristics of electrical equipment.

The objective of this paper was to present a comparative study on both approaches. The measurements were conducted for different nonlinear load types served from the same feeder at the same time. The study indicates that harmonics in the current waveforms strongly deteriorate the quality of the electricity. The current harmonic pollution at PCC depends on the summation and/or cancellation effects in total current drawn from the grid by nonlinear loads. This should be a key factor used to determine the guidelines that limit current harmonic pollution in the distribution system.

Another value of this work is the establishment of a framework for comparing the harmonic effect of various appliances consistently. The next effort is to increase the number of appliance types measured so a useful database can be established.

Author Contributions: Conceptualization, Ł.M. and M.J.; methodology, M.J.; software, Ł.M.; validation, T.S., Ł.J., V.S. and Z.L.; formal analysis, Ł.M., M.J. and T.S.; investigation, Ł.M. and M.J.; resources, Ł.M.; data curation, Ł.M. and M.J.; writing—original draft preparation, Ł.M.; writing—review and editing, M.J.; visualization, Ł.M., Ł.J. and V.S.; supervision, M.J., T.S. and Z.L.; project administration, M.J.; funding acquisition, Z.L. All authors have read and agreed to the published version of the manuscript.

Funding: The research was founded by the Chair of Electrical Engineering Fundamentals K38W05D02, Wroclaw University of Science and Technology.

Data Availability Statement: Data will be sent on request by correspondence author.

Acknowledgments: This research was founded by the K38W05D02 of Wroclaw University of Science and Technology.

Conflicts of Interest: The authors declare no conflict of interest.

References

1. Rüstemli, S.; Okuducu, E.; Almalı, M.N.; Efe, S.B. Reducing the effects of harmonics on the electrical power systems with passive filters. *Bitlis Eren Univ. J. Sci. Technol.* **2015**, *5*. [CrossRef]
2. Singh, B.; Chandra, A.; Al-Haddad, K. *Power Quality Problems and Mitigation Techniques*; Wiley: Hoboken, NJ, US, 2014; Volume 9781118922.
3. Milankov, R.; Radić, M. Harmonics: Examples of negative impacts. In Proceedings of the International Conference on Harmonics and Quality of Power, ICHQP, Bucharest, Romania, 25–28 May 2014.
4. Da Silva, R.P.B.; Quadros, R.; Shaker, H.R.; Da Silva, L.C.P. Analysis of the electrical quantities measured by revenue meters under different voltage distortions and the influences on the electrical energy billing. *Energies* **2019**, *12*, 4757. [CrossRef]
5. Nassar, S.R.; Eisa, A.A.; Saleh, A.A.; Farahat, M.A.; Abdel-Gawad, A.F. Evaluating the Impact of Connected Non Linear Loads on Power Quality- a Nuclear Reactor case study. *J. Radiat. Res. Appl. Sci.* **2020**, *13*. [CrossRef]
6. Jasiński, M.; Sikorski, T.; Kostyła, P.; Kaczorowska, D.; Leonowicz, Z.; Rezmer, J.; Szymańda, J.; Janik, P.; Bejmert, D.; Rybiański, M.; et al. Influence of Measurement Aggregation Algorithms on Power Quality Assessment and Correlation Analysis in Electrical Power Network with PV Power Plant. *Energies* **2019**, *12*, 3547. [CrossRef]
7. Song, J.; Xie, Z.; Zhou, J.; Yang, X.; Pan, A. Power Quality Indexes Prediction Based on Cluster Analysis and Support Vector Machine. In Proceedings of the 24th International Conference on Electricity Distribution, Glasgow, UK, 12–15 June 2017; pp. 12–15.
8. Liu, Y.; Xu, L.; Zhou, S.; Yang, L.; Li, Y.; Feng, D. Identification of Major Power Quality Disturbance Sources in Regional Grid based on Monitoring Data Correlation Analysis. In Proceedings of the 2018 International Conference on Power System Technology (POWERCON), Guangzhou, China, 6–9 November 2018; pp. 4257–4263.
9. Vinogradov, A.; Vinogradova, A.; Bolshev, V. Analysis of the quantity and causes of outages in LV/MV electrical grids. *CSEE J. Power Energy Syst.* **2020**. [CrossRef]
10. Zhang, J.; Cheng, Y.; Du, J.; Zhou, F.; Xiao, J.; Chang, S. Non-intrusive load monitoring and power quality optimization technology of major power customer basing on RFID. In *Proceedings of the IOP Conference Series: Earth and Environmental Science*; IOP Publishing LTD: Bristol, UK, 2020; Volume 440.
11. Nazirov, K.B.; Ganiev, Z.S.; Dzhuraev, S.D.; Ismoilov, S.T.; Rahimov, R.A. Elaboration of the Method for Providing the Level of Power Quality in the Nodes of the Energy System while Power Quality Monitoring. In Proceedings of the 2020 IEEE Conference of Russian Young Researchers in Electrical and Electronic Engineering, EIConRus 2020, St. Petersburg and Moscow, Russia, 27–30 January 2020.

12. Mahaddalkar, S.L.; Shet, V.N. Real Time Monitoring of LED Lighting Loads—A study assessing Power Quality. In Proceedings of the International Conference on Emerging Trends in Information Technology and Engineering, ic-ETITE 2020, Vellore, India, 24–25 February 2020.

13. Sinvula, R.; Abo-Al-Ez, K.M.; Kahn, M.T. Harmonic Source Detection Methods: A Systematic Literature Review. *IEEE Access* **2019**, *7*, 74283–74299. [CrossRef]

14. Olivares-Galván, J.C.; Georgilakis, P.S.; Ocon-Valdez, R. A review of transformer losses. *Electr. Power Compon. Syst.* **2009**, *37*, 1046–1062. [CrossRef]

15. Khan, S.; Maximov, S.; Escarela-Perez, R.; Olivares-Galvan, J.C.; Melgoza-Vazquez, E.; Lopez-Garcia, I. Computation of stray losses in transformer bushing regions considering harmonics in the load current. *Appl. Sci.* **2020**, *10*, 3527. [CrossRef]

16. Song, W.; Fang, J.; Jiang, Z. Numerical AC Loss Analysis in HTS Stack Carrying Nonsinusoidal Transport Current. *IEEE Trans. Appl. Supercond.* **2019**, *29*. [CrossRef]

17. 2020 19th InternaOonal Conference on Harmonics and Quality of Power (ICHQP). In Proceedings of the International Conference on Harmonics and Quality of Power, ICHQP, Dubai, United Arab Emirates, 6–7 July 2020.

18. Ojo, A.; Awodele, K.; Sebitosi, A. Power Quality Monitoring and Assessment of a Typical Commercial Building. In Proceedings of the IEEE Africon Conference, Accra, Ghana, 25–27 September 2019; Volume 2019-Septe.

19. Song, W.; Fang, J.; Jiang, Z.; Staines, M.; Badcock, R. AC Loss Effect of High-Order Harmonic Currents in a Single-Phase 6.5 MVA HTS Traction Transformer. *IEEE Trans. Appl. Supercond.* **2019**, *29*. [CrossRef]

20. Khelifa, R.F.; Jelassi, K. An energy and power quality monitoring system of a power distribution. In Proceedings of the 2016 International Conference on Electrical Sciences and Technologies in Maghreb, CISTEM 2016, Marrakech & Bengrir, Morocco, 26–28 October 2017.

21. De La Rosa, F.C. *Harmonics, Power Systems and Smart Grids*, 2nd ed.; CRC Press: Boca Raton, FL, USA, 2015.

22. Ghorbani, M.J.; Mokhtari, H. Impact of harmonics on power quality and losses in power distribution systems. *Int. J. Electr. Comput. Eng.* **2015**, *5*, 166–174. [CrossRef]

23. Vinogradov, A.; Vasiliev, A.; Bolshev, V.; Semenov, A.; Borodin, M. Time Factor for Determination of Power Supply System Efficiency of Rural Consumers. In *Handbook of Research on Renewable Energy and Electric Resources for Sustainable Rural Development*; IGI Global: Hershey, PA, USA, 2018; pp. 394–420. ISBN 9781522538677.

24. Constantinescu, F.; Gheorghe, A.G.; Marin, M.E.; Taus, O.S. Harmonic balance analysis of home appliances power networks. In Proceedings of the 2017 14th International Conference on Engineering of Modern Electric Systems, EMES 2017, Oradea, Românía, 1–2 June 2017.

25. Sivaraman, P.; Sharmeela, C. Power system harmonics. In *Power Quality in Modern Power Systems*; Academic Press: London, UK, 2021.

26. Committee, D.; Power, I.; Society, E. IEEE Std 519-2014 (Revision IEEE Std 519-1992). In Proceedings of the 2017 IEEE Power & Energy Society General Meeting, Chicago, IL, USA, 16–20 July 2014; Volume 2014.

27. International Electrotechnical Commission. *Electromagnetic Compatibility (EMC) Part 3-6: Limits—Assessment of Emission Limits for the Connection of Distorting Installations to MV, HV and EHV Power Systems*; International Electrotechnical Commission: Geneva, Switzerland, 2008.

28. Liang, X.; Andalib -Bin- Karim, C. Harmonics and Mitigation Techniques Through Advanced Control in Grid-Connected Renewable Energy Sources: A Review. *IEEE Trans. Ind. Appl.* **2018**, *54*, 3100–3111. [CrossRef]

29. Collocott, C.L.; Awodele, K.O.; Adebayo, A.V. Harmonic emission of non-linear loads in distribution systems—A computer laboratory case study. In Proceedings of the 2020 International SAUPEC/RobMech/PRASA Conference, SAUPEC/RobMech/PRASA 2020, Cape Town, South Africa, 29–31 January 2020.

30. Abidin, M.N.Z. IEC. In *61000-3-2 Harmonics Standards Overview*; Schaffner EMC Inc.: Edsion, NJ, USA, 2014.

31. Wang, Z.; Li, Q.; Tang, Y.; Liu, S.; Dai, S. Comparison of harmonic limits and evaluation of the international standards. *MATEC Web Conf.* **2019**, *277*, 03009. [CrossRef]

32. Cho, N.; Lee, H.; Bhat, R.; Heo, K. Analysis of Harmonic Hosting Capacity of IEEE Std. 519 with IEC 61000-3-6 in Distribution Systems. In Proceedings of the 2019 IEEE PES GTD Grand International Conference and Exposition Asia (GTD Asia), Bangkok, Thailand, 21–23 March 2019; IEEE: Piscataway, NJ, USA, 2019; pp. 730–734.

33. Dartawan, K.; Najafabadi, A.M. Case study: Applying IEEE Std. 519-2014 for harmonic distortion analysis of a 180 MW solar farm. In Proceedings of the 2017 IEEE Power & Energy Society General Meeting, Chicago, IL, USA, 16–20 July 2017; pp. 1–5.

34. Mishra, A.; Tripathi, P.M.; Chatterjee, K. A review of harmonic elimination techniques in grid connected doubly fed induction generator based wind energy system. *Renew. Sustain. Energy Rev.* **2018**, *89*, 1–15. [CrossRef]

35. Vinayagam, A.; Aziz, A.; PM, B.; Chandran, J.; Veerasamy, V.; Gargoom, A. Harmonics assessment and mitigation in a photovoltaic integrated network. *Sustain. Energy Grids Netw.* **2019**, *20*, 100264. [CrossRef]

36. Gong, J.; Li, D.; Wang, T.; Pan, W.; Ding, X. A comprehensive review of improving power quality using active power filters. *Electr. Power Syst. Res.* **2021**, *199*, 107389. [CrossRef]

37. Montoya, F.G.; Baños, R.; Alcayde, A.; Arrabal-Campos, F.M. A new approach to single-phase systems under sinusoidal and non-sinusoidal supply using geometric algebra. *Electr. Power Syst. Res.* **2020**, *189*, 106605. [CrossRef]

38. Chiradeja, P.; Ngaopitakkul, A.; Jettanasen, C. Energy savings analysis and harmonics reduction for the electronic ballast of T5 fluorescent lamp in a building's lighting system. *Energy Build.* **2015**, *97*. [CrossRef]

39. Alammari, R.; Islam, M.S.; Chowdhury, N.A.; Sakil, A.K.; Iqbal, A.; Khandakar, A. Impact on power quality due to large-scale adoption of compact fluorescent lamps–a review. *Int. J. Ambient Energy* **2017**, *38*. [CrossRef]

40. Henao-Muñoz, A.C.; Herrera-Murcia, J.G.; Saavedra-Montes, A.J. Experimental characterization of compact fluorescent lamps for harmonic analysis of power distribution systems. *TecnoLógicas* **2018**, *21*. [CrossRef]

41. Mathwai, T.; Awodele, K.; Ojo, A. Power quality evaluation of electrical loads in a typical commercial building. In Proceedings of the 2020 International SAUPEC/RobMech/PRASA Conference, SAUPEC/RobMech/PRASA 2020, Cape Town, South Africa, 29–31 January 2020.

42. Blanco, A.M.; Yanchenko, S.; Meyer, J.; Schegner, P. Impact of supply voltage distortion on the current harmonic emission of non-linear loads. *DYNA* **2015**, *82*. [CrossRef]

43. Graña-López, M.Á.; Filgueira-Vizoso, A.; Castro-Santos, L.; García-Diez, A.I. Analysis of the real energy consumption of energy saving lamps. *Appl. Sci.* **2020**, *10*, 8446. [CrossRef]

44. Montoya, F.G.; Castillo, J. Power quality in modern lighting: Comparison of LED, microLED and CFL lamps. *Renew. Energy Power Qual. J.* **2016**, *1*. [CrossRef]

45. Gil-De-Castro, A.; Medina-Gracia, R.; Ronnberg, S.K.; Blanco, A.M.; Meyer, J. Differences in the performance between CFL and LED lamps under different voltage distortions. In Proceedings of the International Conference on Harmonics and Quality of Power, ICHQP, Ljubljana, Slovenia, 13–16 May 2018.

46. Nikum, K.; Saxena, R.; Wagh, A. Power Quality Issues in Commercial Load—Impact and Mitigation Difficulties in Present Scenario. In *Proceedings of the Lecture Notes in Electrical Engineering*; Springer: Singapore, 2021; Volume 698.

47. Ahir, J.; Upadhyay, C. Harmonic Analysis and Mitigation for Modern Home Appliances. In *Proceedings of the 4th International Conference on Electrical Energy Systems, ICEES 2018*; Institute of Electrical and Electronics Engineers Inc.: Piscataway, NJ, USA, 2018; pp. 218–223.

48. Manias, S.N. *Power Electronics and Motor Drive Systems*; Academic Press: London, UK, 2016.

49. Djokic, S.Z.; Collin, A.J. Cancellation and attenuation of harmonics in low voltage networks. In Proceedings of the International Conference on Harmonics and Quality of Power, ICHQP, Bucharest, Romania, 25–28 May 2014.

50. Blanco, A.M.; Gupta, M.; Gil de Castro, A.; Ronnberg, S.; Meyer, J. Impact of flat-top voltage waveform distortion on harmonic current emission and summation of electronic household appliances. *Renew. Energy Power Qual. J.* **2018**, *1*, 698–703. [CrossRef]

51. Rawa, M.J.H.; Thomas, D.W.P.; Sumner, M. Factors affecting the harmonics generated by a cluster of personal computers. In Proceedings of the International Conference on Harmonics and Quality of Power, ICHQP, Bucharest, Romania, 25–28 May 2014.

52. Mesas, J.J.; Sainz, L.; Sala, P. Statistical study of personal computer cluster harmonic currents from experimental measurements. *Electr. Power Components Syst.* **2015**, *43*. [CrossRef]

53. Rawa, J.H.; Factors, M. Affecting the Harmonics Generated by a Group of CFLs: Experimental Measurements. *Am. J. Electr. Power Energy Syst.* **2015**, *4*. [CrossRef]

54. Ghanbari, T.; Farjah, E.; Naseri, F. Power quality improvement of radial feeders using an efficient method. *Electr. Power Syst. Res.* **2018**, *163*, 140–153. [CrossRef]

55. Santha Kumar, C.; Ramesh, P.; Kasilingam, G.; Ragul, D.; Bharatiraja, C. The power quality measurements and real time monitoring in distribution feeders. *Mater. Today Proc.* **2021**, *45*, 2987–2992. [CrossRef]

56. Ujile, A.; Ding, Z. A dynamic approach to identification of multiple harmonic sources in power distribution systems. *Int. J. Electr. Power Energy Syst.* **2016**, *81*, 175–183. [CrossRef]

57. Çiçek, A.; Erenoğlu, A.K.; Erdinç, O.; Bozkurt, A.; Taşcıkaraoğlu, A.; Catalão, J.P.S. Implementing a demand side management strategy for harmonics mitigation in a smart home using real measurements of household appliances. *Int. J. Electr. Power Energy Syst.* **2021**, *125*, 106528. [CrossRef]

58. Pérez Vallés, A.; Salmerón Revuelta, P. A new distributed measurement index for the identification of harmonic distortion and/or unbalance sources based on the IEEE Std. 1459 framework. *Electr. Power Syst. Res.* **2019**, *172*, 96–104. [CrossRef]

59. Kalair, A.R.; Stojcevski, A.; Seyedmahmoudian, M.; Abas, N.; Kalair, A.; Khan, N.; Saleem, M.S. Steady-state and time-varying harmonics in distribution system. In *Uncertainties in Modern Power Systems*; Elsevier: Amsterdam, The Netherlands, 2021; pp. 485–539.

60. IEC. *IEC 61000-2-2: Electromagnetic Compatibility (EMC)—Part 2-2: Environment—Compatibility Levels for Low-Frequency Conducted Disturbances and Signalling in Public Low-Voltage Power Supply Systems*; IEC Comission: New York, NY, USA, 2002.

61. Yanchenko, S.; Kulikov, A.; Tsyruk, S. Modeling harmonic amplification effects of modern household devices. *Electr. Power Syst. Res.* **2018**, *163*. [CrossRef]

62. Qingliang, W.; Shuaiqi, T.; Zhengdong, H.; Qian, N. Calculation and Analysis of Transformer Loss Cost under Harmonic Environment. In Proceedings of the 2019 IEEE 2nd International Conference on Electronics and Communication Engineering (ICECE), Xi'an, China, 9–11 December 2019; pp. 392–396.

63. Moradi, A.; Yaghoobi, J.; Alduraibi, A.; Zare, F.; Kumar, D.; Sharma, R. Modelling and prediction of current harmonics generated by power converters in distribution networks. *IET Gener. Transm. Distrib.* **2021**, gtd2.12166. [CrossRef]

64. Almutairi, M.S.; Hadjiloucas, S. Harmonics Mitigation Based on the Minimization of Non Linearity Current in a Power System. *Designs* **2019**, *3*, 29. [CrossRef]

65. Yazdani-Asrami, M.; Sadati, S.M.B.; Samadaei, E. Harmonic study for MDF industries: A case study. In Proceedings of the 2011 IEEE Applied Power Electronics Colloquium (IAPEC), Johor Bharu, Malaysia, 18–19 April 2011; pp. 149–154.

66. Senthil Kumar, R.; Surya Prakash, R.; Yokesh Kiran, B.; Sahana, A. Reduction and elimination of harmonics using power active harmonic filter. *Int. J. Recent Technol. Eng.* **2019**, *8*. [CrossRef]
67. Park, B.; Lee, J.; Yoo, H.; Jang, G. Harmonic Mitigation Using Passive Harmonic Filters: Case Study in a Steel Mill Power System. *Energies* **2021**, *14*, 2278. [CrossRef]
68. Lumbreras, D.; Gálvez, E.; Collado, A.; Zaragoza, J. Trends in Power Quality, Harmonic Mitigation and Standards for Light and Heavy Industries: A Review. *Energies* **2020**, *13*, 5792. [CrossRef]

Article

Off-Grid Rural Electrification in India Using Renewable Energy Resources and Different Battery Technologies with a Dynamic Differential Annealed Optimization

Polamarasetty P Kumar [1,*], Vishnu Suresh [2,3,*], Michal Jasinski [2] and Zbigniew Leonowicz [2]

[1] Department of Electrical and Electronics Engineering, GMR Institute of Technology, Rajam 532127, India
[2] Faculty of Electrical Engineering, Wroclaw University of Science and Technology, 50-370 Wroclaw, Poland; jasinski@ieee.org (M.J.); zbigniew.leonowicz@pwr.edu.pl (Z.L.)
[3] Hitachi ABB Power Grids Research, ul. Pawia 7, 31-154 Kraków, Poland
* Correspondence: praveenkumar.p@gmrit.edu.in (P.P.K.); vishnu.suresh@pwr.edu.pl (V.S.)

Citation: Kumar, P.P.; Suresh, V.; Jasinski, M.; Leonowicz, Z. Off-Grid Rural Electrification in India Using Renewable Energy Resources and Different Battery Technologies with a Dynamic Differential Annealed Optimization. *Energies* **2021**, *14*, 5866. https://doi.org/10.3390/en14185866

Academic Editor: Nicu Bizon

Received: 15 August 2021
Accepted: 10 September 2021
Published: 16 September 2021

Publisher's Note: MDPI stays neutral with regard to jurisdictional claims in published maps and institutional affiliations.

Abstract: Several families in India live in remote places with no access to grid-connected power supply due to their remoteness. The study area chosen from the Indian state of Odisha does not have an electrical power supply due to its distant location. As a result, this study analyzed the electrification process using Renewable Energy (RE) resources available in the locality. However, these RE resources are limited by their dependency on weather conditions and time. So, a robust battery storage system is needed for a continuous power supply. Hence, the Nickel Iron (Ni-Fe), Lithium-Ion (Li-Ion) and Lead Acid (LA) battery technologies have been analyzed to identify a battery technology that is both technologically and economically viable. Using the available RE resources in the study area, such as photovoltaic and biomass energy resources, as well as the various battery technologies, three configurations have been modelled, such as Photovoltaic Panels (PVP)/Biomass Generator(BIOMG)/BATTERY$_{(Ni\text{-}Fe)}$, PV/BIOMG/BATTERY$_{(Li\text{-}Ion)}$ and PVP/BMG/BATTERY$_{(LA)}$. These three configurations have been examined using nine prominent metaheuristic algorithms, in which the PVP/BIOMG/BATTERY$_{(Ni\text{-}Fe)}$ configuration provided the optimal Life Cycle Cost value of 367,586 USD. Among the all metaheuristic algorithms, the dynamic differential annealed optimization algorithm was given the best Life Cycle Cost values for all of the three configurations.

Keywords: optimization techniques; different batteries; off-grid microgrid; integrated renewable energy system

1. Introduction

Despite the fact that freshwater and energy are essential for human survival, existing and future energy and freshwater demands are having a significant impact on the earth. Renewable Energy (RE) resources play a critical role in addressing these concerns. This is especially true in rural and distant areas where the grid expansion is neither feasible nor cost-effective. For a multitude of reasons, planning and constructing an off-grid microgrid is difficult from a both technological and economic perspective. One of them is the reliance of RE resources on meteorological conditions. A suitable battery storage system is needed for power supply continuity. In order to fulfil the energy requirements, most off-grid systems are either over or under-sized. While the larger system is more expensive and produces more energy, the smaller system is incapable of meeting the required load demands, resulting in a power supply shortage. To overcome these challenges and fully utilize the benefits of an off-grid microgrid built with RE sources, an Energy Management Strategy (EMS) is necessary [1]. Furthermore, renewable energy resources, particularly solar energy resource can be used in a wide range of applications, including solar dryers, solar home systems, solar cookers, thermal power generation and water heaters. The biomass generator can produce both heat and electricity. The hydro energy resource can be used for irrigation, drinking water supply and hydropower generation. Wind energy has many applications, including power generation, water pumps and windmills. Some commercial and residential applications of battery energy storage systems are explained as follows. From a commercial point of view, battery energy storage systems are useful in the following applications such as peak shaving, emergency backup, microgrids, load shifting, grid services and renewable energy resource integration. From a residential standpoint, battery energy storage systems are useful in the following applications: storing excess energy produced by the solar PV panels, emergency backup and off-grid applications.

Numerous studies on the sizing of microgrids have been published in reputed journals. These approaches can be classified into three broad categories, which are categorized as follows: (i) software tools such as IHOGA, RETScreen, HOMER and HOGA, [2] (ii) deterministic methodologies such as linear programming, numerical, iterative, graphical design, analytical and probabilistic methods [3] and (iii) metaheuristic algorithms such as Grasshopper Optimization Algorithm (GOA), Moth Flame Optimization (MFO), Genetic Algorithm (GA) and Dragonfly Algorithm (DA). Nevertheless, while software tools are simple to use the relevant components inside them cannot be picked. Furthermore, its algorithms and calculations are hidden from users. When it comes to estimating the size of a microgrid, software tools are constrained by certain assumptions. Deterministic approaches, on the other hand, outperform the software tools but the solutions are trapped at the local optima due to its complex design. They are unable to discover the global best optimal values under such conditions. It should be run several times with random beginning conditions to prevent the local optima trapping. Hence, the solution is unlikely to be the global best optimum value. Obtaining the global best optimal solution may necessitate multiple efforts. Finally, metaheuristic algorithms were chosen as the most promising methods, and they are widely used for optimization problems [4].

Last decade, metaheuristic algorithms were built to manage microgrid sizing issues. Surprisingly, certain algorithms, such as Differential Evolutionary Algorithm (DE), Particle Swarm Optimization (PSO) [5] and GA, are well-known among most software engineers as well as other researchers from other disciplines. They are adaptable techniques and offer better results than deterministic methods by preventing the local optima entanglement. These algorithms offer a range of properties that allow them to tackle any optimization problem and they solve optimization problems by imitating natural processes. On the other hand, the No-Free-Lunch theorem states that while a specific metaheuristic algorithm can produce optimal results for a specific objective function, the same algorithm produces lower performance in other objective functions. For such reasons, researchers working on microgrid sizing issues have looked at robust metaheuristic algorithms [6].

A review of the literature on off-grid microgrid size issues revealed that evaluating the techno-economic feasibility analysis with various battery technologies to deliver an uninterrupted electrical supply is quite limited. Moreover, researchers can determine a specific algorithm's robustness and convergence efficiency using various algorithms. However, such studies are limited.

The limitations stated above should be addressed while analyzing perfect off-grid rural electrification from a techno-economic perspective. So, taking these limitations into account, some research has been undertaken to supply power to four un-electrified villages in the Indian state of Odisha. After conducting some research in the study area, it was discovered that the accessible RE resources in the study area are solar and biomass. Solar energy is the most volatile of the two RE resources due to its time-dependent nature and weather conditions. So, the power supply is not continuous. Hence, a robust battery system is essential to ensure a continuous electrical supply. Therefore, in this study, a detailed analysis with three v kinds of batteries, such as Ni-Fe, Li-Ion and LA, has been carried out. Three types of configurations are modelled using available RE resources and battery technologies to identify the most viable configuration for electrifying the study region, such as Photovoltaic Panels (PVP)/ Biomass Generator (BIOMG)//BATTERY$_{(Ni-Fe)}$, PVP/BIOMG/BATTERY$_{(Li-Ion)}$ and PVP/BIOMG/BATTERY$_{(LA)}$. In order to find out a feasible configuration to electrify the study area, nine proven metaheuristic algorithms have been used in the study, such as Differential Evolutionary Algorithm (DE) [5], Genetic Algorithm (GA) [6], Moth Flame Optimization (MFO) [7], Dragonfly Algorithm (DA) [8], Particle Swarm Optimization (PSO) [9], Ant Lion Optimization (ALO) [10], Grey Wolf Optimization (GWO) [11], Grasshopper Optimization Algorithm (GOA) [12] and Dynamic Differential Annealed Optimization Algorithm [13]. Finally, after identifying the optimal configuration, it is evaluated with various loss of power supply probability values.

The novelty in this work can be summarized as:

- A renewable energy-based plan for electrification of an un-electrified village in a remote location.
- Identification of a suitable optimization algorithm amongst nine prominent ones in the literature for energy management.
- Exploration of the most cost-effective and technically suitable battery storage technology for the given setup.
- Identify the best configuration cost wise amongst three different possibilities.

2. The Study Area Identification

The study examined the off-grid rural electrification procedure using an off-grid remote rural area located more than 50 km away. Power supply to this area via grid connection is not possible due to the area's hilly terrain and its location in the middle of a dense forest. Four villages are included in the study area, which are located in the Chandrapur block of the Rayagada district of Odisha state in India. Figure 1 depicts the study area's location on a map of India. The study area's latitude and longitude are 19.5927° N and 83.8712° E, respectively. The whole population of the villages is accommodated in 153 houses. People in this area continue to rely on candles, kerosene lanterns and solar lamps for illumination.

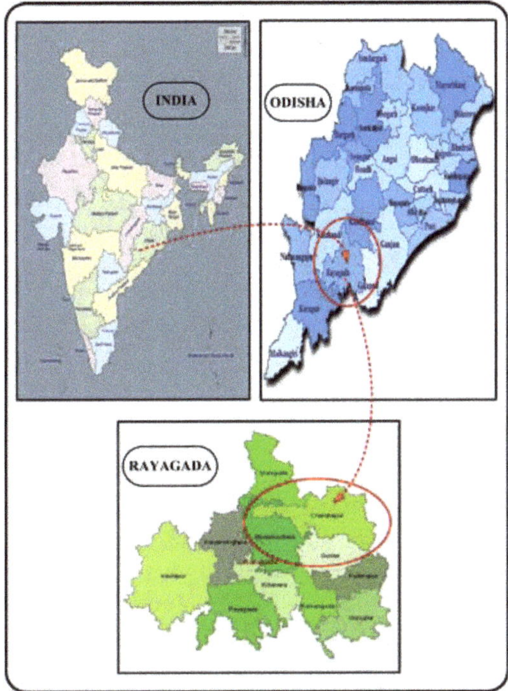

Figure 1. The position of the research region on the Indian map.

3. Renewable Energy Resources Identification in the Study Region

Natural RE resources can be exploited to electrify off-grid areas. The study area has been completely covered by solar and biomass energy resources. The study area's annual ambient temperature is 26 °C. The simulation used a ten-year average, i.e., from 2005 to 2015, of ambient temperature and solar radiation values obtained from the National Renewable Energy Laboratory, which are provided, respectively, in Figures 2 and 3. The study area is located in the middle of a high-density forest, which is around 196 hectares. With a foliage production assumption of 16 tons per hectare. Each year, the foliage for the 196 hectares of forest will be 31.36 tons. If the yearly foliage collecting rate is 60%, the annual foliage collection will be around 19 tons.

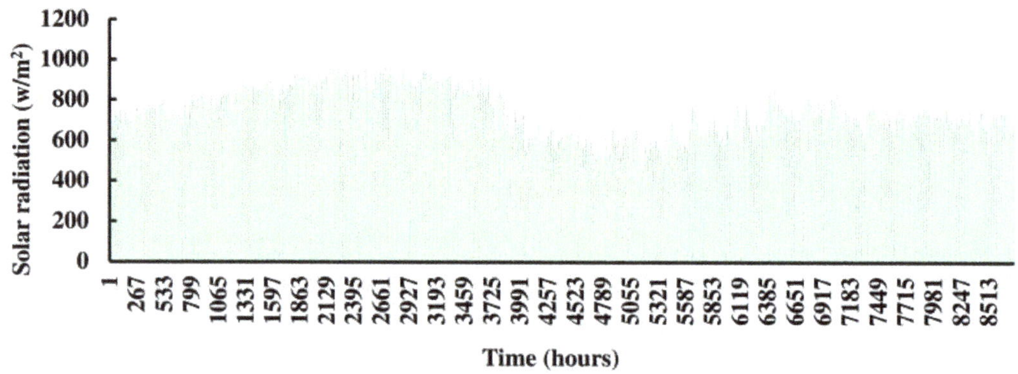

Figure 2. Study area's annual solar radiation.

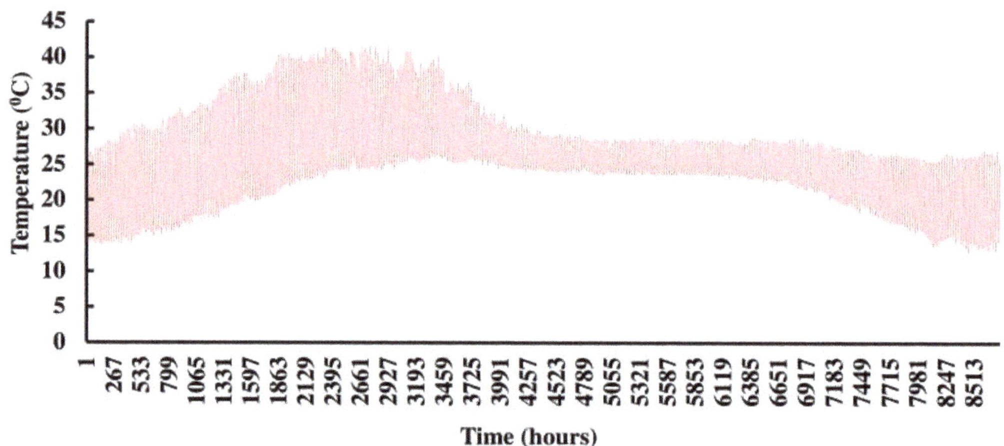

Figure 3. Study areas ambient temperature.

4. Study Area's Load Estimation

The load demands have been estimated based on the villagers' load requirements. According to villagers' needs, the total number of load sectors is anticipated to be domestic, community, agricultural, small scale industrial and commercial. Table 1 clearly explains the complete load estimation details of the rated power, quantity and time of usage of the appliances for the study areas winter (November–February) and summer (March–October) seasons of the study area. Figure 4 depicts the corresponding load demand graphs.

Energies 2021, 14, 5866

Table 1. Complete information on the study area's load demands.

Load Sector →	Domestic Load				Community Load										Agricultural Load		Commercial Load				SIL	Hourly Energy Demand (kWh)
					School			Hospital			Community Hall		SL	PW	MCTM	Shops		MDP	Flour Mill	Saw Mills	
Appliance	LED Lamp	Fan	T.V+Dish	MC	LED Lamp	Fan	Computer	LED Lamp	Fan	Refrigerator	LED Lamp	Fan	LED Lights	Motor (2 hp)	Motor (5 hp)	LED Lamp	Fan	Motor (4 hp)	Motor (5 hp)	Saw Machine	
Rated Power →	20 W	75 W	100 W	5 W	20 W	75 W	250 W	20 W	75 W	200 W	20 W	75 W	40 W	1.5 kW	3.73 kW	20 W	75 W	2.983 kW	3.73 kW	1.8 kW	
Quantity →	2	1	1	1	12	12	12	5	5	1	3	3	27	6	1	5	5	3	1	1	
Time (h) ↓		S/W				S/W			S/W			S/W						S/W				S/W
0:00–1:00		11.48/0						0.1	0.38/0	0.2			0.6								12.76	12.76/0.9
1:00–2:00		11.48/0						0.1	0.38/0	0.2			0.6								12.76	12.76/0.9
2:00–3:00		11.48/0						0.1	0.38/0	0.2			0.6								12.76	12.76/0.9
3:00–4:00		11.48/0						0.1	0.38/0	0.2			0.6								12.76	12.76/0.9
4:00–5:00	3.06	11.48/0						0.1	0.38/0	0.2			0.6								15.82	15.82/3.96
5:00–6:00	3.06	11.48/0	3.83					0.1	0.38/0	0.2			0.6								19.65	19.65/7.79
6:00–7:00		11.48/0	3.83	0.77					0.38/0	0.2				4.5							21.16	21.16/9.30
7:00–8:00		11.48/0	3.83	0.77					0.38/0	0.2				4.5							21.16	21.16/9.30
8:00–9:00		11.48/0	3.83	0.38					0.38/0	0.2				4.5							20.77	20.77/8.91
9:00–10:00		5.74/0	3.83		0.24	0.9/0	3		0.38/0	0.2					3.73	0.1	0.38/0		1.8		20.30	20.30/12.90
10:00–11:00		5.74/0	3.83		0.24	0.9/0	3		0.38/0	0.2					3.73	0.1	0.38/0		1.8		20.30	20.30/12.90
11:00–12:00		5.74/0	7.65		0.24	0.9/0	3		0.38/0	0.2	0.06	0.23/0			3.73	0.1	0.38/0		1.8		24.41	24.41/16.78
12:00–13:00		5.74/0	7.65		0.24	0.9/0	3		0.38/0	0.2	0.06	0.23/0			3.73	0.1	0.38/0		1.8		24.41	24.41/16.78
13:00–14:00		5.74/0	7.65		0.24	0.9/0	3		0.38/0	0.2	0.06	0.23/0			3.73	0.1	0.38/0	3.73			26.34	26.34/18.71
14:00–15:00		5.74/0	7.65		0.24	0.9/0	3		0.38/0	0.2	0.06	0.23/0				0.1	0.38/0	3.73			22.61	22.61/14.98
15:00–16:00		5.74/0	7.65		0.24	0.9/0	3		0.38/0	0.2	0.06	0.23/0				0.1	0.38/0	3.73			22.61	22.61/14.98
16:00–17:00		5.74/0	7.65		0.24	0.9/0	3		0.38/0	0.2	0.06	0.23/0				0.1	0.38/0	3.73			22.61	22.61/14.98
17:00–18:00		5.74/0	7.65						0.38/0	0.2	0.06	0.23/0				0.1	0.38/0	3.73			18.47	18.47/11.74

Table 1. *Cont.*

Load Sector →	Domestic Load				Community Load									Agricultural Load			Commercial Load				SIL	Hourly Energy Demand (kWh)
					School			Hospital			Community Hall		SL	PW	MCTM	Shops		MDP	Flour Mill	Saw Mills		
Appliance	LED Lamp	Fan	T.V+ Dish	MC	LED Lamp	Fan	Computer	LED Lamp	Fan	Refrigerator	LED Lamp	Fan	LED Lights	Motor (2 hp)	Motor (5 hp)	LED Lamp	Fan	Motor (4 hp)	Motor (5 hp)	Saw Machine		
Rated Power →	20 W	75 W	100 W	5 W	20 W	75 W	250 W	20 W	75 W	200 W	20 W	75 W	40 W	1.5 kW	3.73 kW	20 W	75 W	2.983 kW	3.73 kW	1.8 kW		
Quantity →	2	1	1	1	12	12	12	5	5	1	3	3	27	6	1	5	5	3	1	1		
Time (h) ↓	S/W				S/W			S/W			S/W					S/W				S/W		
18:00–19:00	6.12	11.48/0	15.3	0.77				0.1	0.38/0	0.2	0.06	0.23/0	0.6			0.1	0.38/0			35.72		35.72/23.25
19:00–20:00	6.12	11.48/0	15.3	0.77				0.1	0.38/0	0.2	0.06	0.23/0	0.6			0.1	0.38/0			35.72		35.72/23.25
20:00–21:00	6.12	11.48/0	15.3	0.38				0.1	0.38/0	0.2			0.6							34.56		34.56/22.7
21:00–22:00	6.12	11.48/0	15.3	0.38				0.1	0.38/0	0.2			0.6							34.56		34.56/22.7
22:00–23:00	3.06	11.48/0	7.65	0.38				0.1	0.38/0	0.2			0.6							23.85		23.85/11.99
23:00–24:00		11.48/0						0.1	0.38/0	0.2			0.6							12.76		12.76/0.9
Total Load Demand																						528.83/282.4

SIL = Small Industrial Load; MCTM = Multi Crop Threshing Machine; PW = Pumping Water; S/W = Summer/Winter; SL = Street Lighting; MDP = Mini Diary Plant; MC = Mobile Charger".

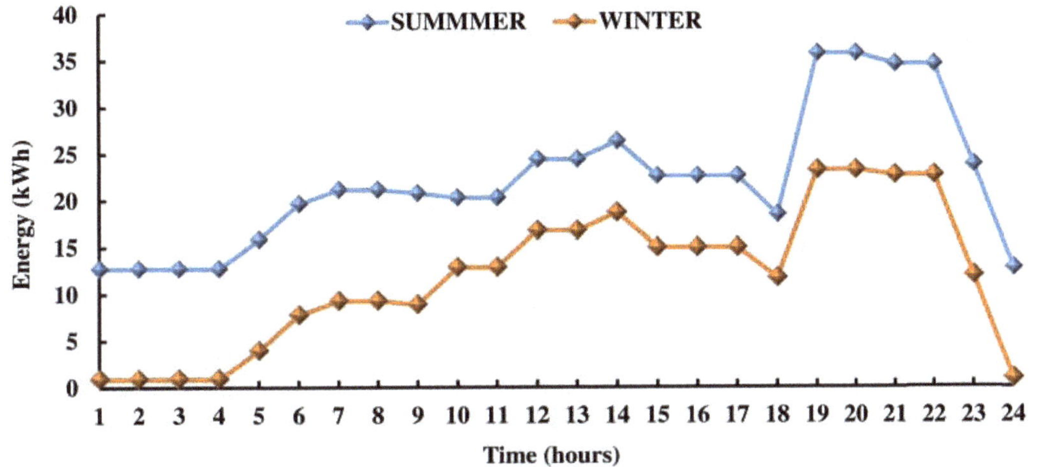

Figure 4. The daily load demand of the study area.

5. The Components of Mathematical Modelling

The optimal sizing of an off-grid microgrid is primarily determined by the components utilized in the analysis, hence accurate component analysis is essential before optimal sizing of the off-grid microgrid. The current study includes PV panels, a biomass generator, a battery bank and a bi-directional converter with a charge controller. The schematic diagram is shown in Figure 5.

5.1. Solar Energy System

A variety of approaches can be used to compute the PV output power from the PV panels. A normal framework was employed in this study to compute the output power of the PV panels, which was determined by taking into consideration of hourly ambient temperature and solar radiation values [14]:

$$P_{PV}(t) = PV_{rated} \times (G(t)/G_{ref}) \times \left[1 + K_T \times \left(T_C - T_{ref}\right)\right] \tag{1}$$

where, T_{ref} = the standard test condition temperature value i.e., 25° degrees Celsius for PV cells. PV_{rated} = the PV panel's rated power, K_T = at maximum power, the temperature coefficient, its value is 3.7×10^{-3} (1/°C) and G_{ref} = the solar radiation value at reference condition, its value is 1000 W/m².

The PV panel cell temperature (T_C) can be determined as:

$$T_C = T_{amb}(t) + (0.0256 \times G(t)) \tag{2}$$

The amount of energy produced by the PV panels (E_{PV}) can be determined as follows:

$$E_{PV}(t) = N_{PV} \times P_{PV}(t) \times \Delta t \tag{3}$$

where, Δt = time period.

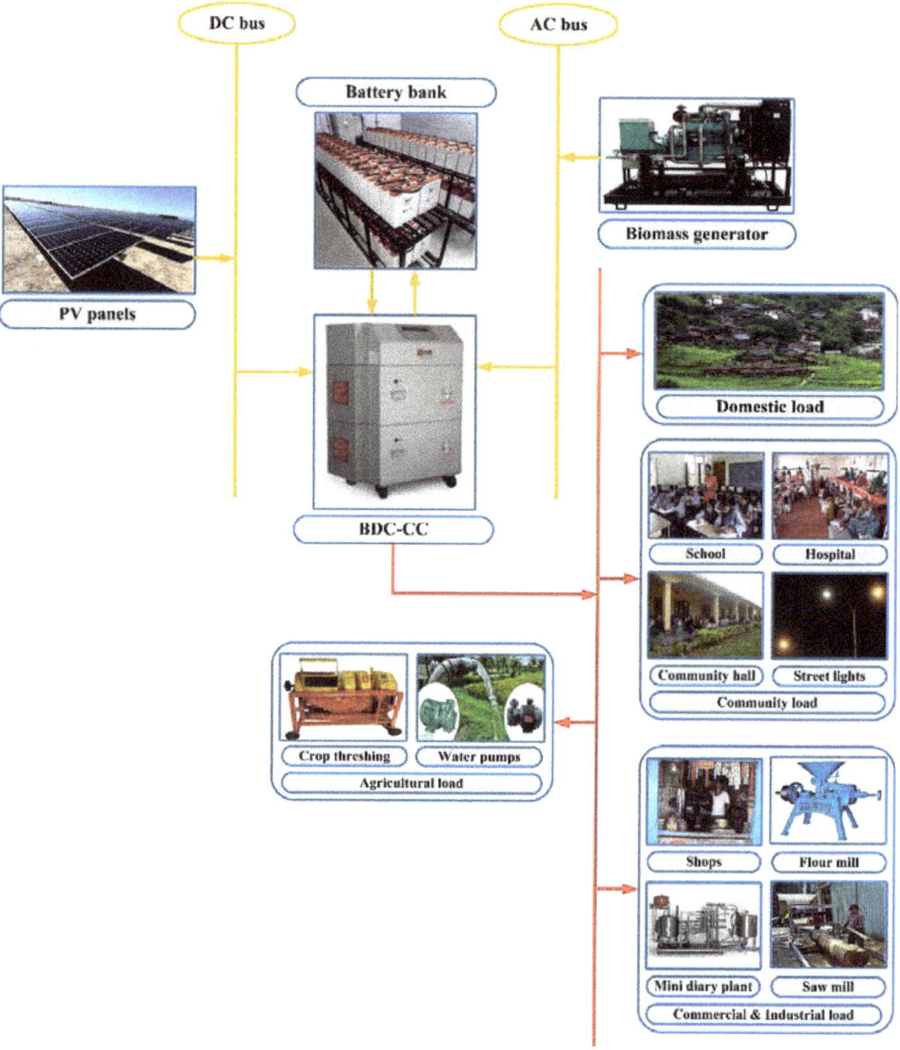

Figure 5. Schematic diagram of the study area electrification model.

5.2. Biomass Generator

In this study, we are advised to use the down draught gasifier model of biomass generator. The following equation can be used to calculate the power generated by the Biomass Generator [1]:

$$P_{BMG}(t) = \frac{Q_{BM} \times \eta_{BMG} \times CV_{BM} \times 1000}{DOH_{BMG} \times 365 \times 860} \tag{4}$$

where, the conversion factor from kcal to kWh is 860, Q_{BM} = biomass quantity (tons/year), η_{BMG} = efficiency of the biomass generator, DOH_{BMG} = per day operative hours of the biomass generator and CV_{BM} = the biomass has a calorific value of 4015 kcal/kg.

The biomass generator's hourly energy output is determined as follows:

$$E_{BMG}(t) = P_{BMG}(t) \times \Delta t \tag{5}$$

5.3. Battery Bank

When RE sources are not available to deliver power, the battery bank will supply power in off-grid systems. In general, the battery bank stores the extra energy generated by the RE sources. This phenomenon is known as a battery bank charging process, and it is expressed by the equation below [14,15]:

$$E_{Bat}(t) = (1 - \sigma) \times E_{Bat}(t - 1) + (E_G(t) - E_L(t)/\eta_{Conv}) \times \eta_{CC} \times \eta_{rbat} \qquad (6)$$

where, $E_{Bat}(t)$, and $E_{Bat}(t - 1)$ are the energy levels of a battery bank, respectively, at time "$t - 1$" and "t", σ = hourly self-discharge rate of the battery, η_{CC} = efficiency of the charge controller, E_L = electrical energy demand, E_G = generated electrical energy, η_{Conv} = the bidirectional converter's efficiency and η_{rbat} = the battery's round-trip efficiency.

The total electrical energy generation is calculated as:

$$E_G(t) = [E_{DC}(t) + E_{AC}(t)] \times \eta_{Conv} \qquad (7)$$

where, (E_{DC}) denotes the DC energy produced by the RE resources:

$$E_{DC}(t) = E_{PV}(t) \qquad (8)$$

The AC energy (E_{AC}) produced by RE sources is calculated as follows:

$$E_{AC}(t) = E_{BMG}(t) \qquad (9)$$

Whenever the energy produced by the RE sources is inadequate to meet the load demands or when RE resources completely fail to meet the load demand. The energy in the battery bank will supply the load demand; this occurrence is known as the battery bank discharge process, and it is described in the following equation:

$$E_{Bat}(t) = (1 - \sigma) \times E_{Bat}(t - 1) - (E_L(t)/\eta_{Conv} - E_G(t))/\eta_{rbat} \qquad (10)$$

5.4. Bi-Directional Converter with a Charge Controller

The bi-directional converter is essential in off-grid systems because it collects energy from both DC and AC sources. Its primary function is to convert direct current to alternating current and vice versa. The charge controller inside the converter is useful for controlling the battery bank's overcharging and discharging. The following equation can be used to compute the converter's rated power for the off-grid microgrid [2]:

$$P_{BDC-CC} = E_{T,max} \times 1.1 \qquad (11)$$

The multiplication factor 1.1 specifies the converter's overloading capacity, which means that the converter power rating should be 10% greater than the system's peak load demand.

6. Economic Analysis of the Off-Grid Microgrid

The Life Cycle Cost (LCC) technique is widely used for economic analysis. The LCC has been estimated in this study by adding the Initial Capital Costs (ICC), Erection Costs (EREC), the Present Value of O&M Costs ($P_{V,O\&M}$), the Present Value of Replacement Costs ($P_{V,REP}$) and the Present Value of Fuel Costs ($P_{V,FUEL}$) using all system components, which is expressed as follows [2]:

$$LCC = ICC + EREC + P_{V,O\&M} + P_{V,REP} + P_{V,FUEL} \qquad (12)$$

The initial capital cost (ICC) of the IRES components are calculated as follows [2]:

$$ICC = \begin{bmatrix} (C_{BMG,cap}) + (N_{PV} \times C_{PV,cap}) + \\ (N_{BAT} \times C_{BAT,cap}) + (C_{BDC-CC,cap}) \end{bmatrix} \qquad (13)$$

where, $C_{BMG,cap}$, $C_{PV,cap}$, $C_{BAT,cap}$ and $C_{BDC-CC,cap}$ are the initial capital costs of the biomass generator, PV panels, batteries and bi-directional converter with a charge controller respectively.

The erection costs ($EREC$) of the IRES components are calculated as follows [2]:

$$EREC = \begin{bmatrix} (N_{PV} \times C_{PV,erect}) + \left((N_{BAT} \times C_{BAT,erect}) \times \sum_{b=1}^{N_r} \frac{(1+x)^{bN_c-1}}{(1+y)^{bN_c}} \right) + \\ \left(C_{BDC-CC,erect} \times \sum_{d=1}^{N_r} \frac{(1+x)^{dN_c-1}}{(1+y)^{dN_c}} \right) + \left(C_{BMG,erect} \times \sum_{g=1}^{N_r} \frac{(1+x)^{gN_c-1}}{(1+y)^{gN_c}} \right) \end{bmatrix} \quad (14)$$

where, $C_{PV,erect}$, $C_{BAT,erect}$, $C_{BDC-CC,erect}$ and $C_{BMG,erect}$ are erection costs of the PV panels, batteries and bi-directional converter with a charge controller and biomass generator, respectively.

The present value of annual O&M ($P_{V,O\&M}$) costs of the IRES components are calculated as follows [2]:

$$P_{V,O\&M} = \begin{bmatrix} (N_{PV} \times C_{PV,o\&m}) + (C_{BMG,o\&m}) \\ (N_{BAT} \times C_{BAT,o\&m}) + (C_{BDC-CC,o\&m}) \end{bmatrix} \times \sum_{i=1}^{N} \frac{(1+x)^{i-1}}{(1+y)^i} \quad (15)$$

where, $C_{PV,o\&m}$, $C_{BMG,o\&m}$, $C_{BAT,o\&m}$ and $C_{BDC-CC,o\&m}$ are O&M costs of the PV panels, biomass generator, batteries and bi-directional converter with a charge controller, respectively, and y is defined as follows [2]:

$$y = \frac{I_{nom} - x}{1 + x}. \quad (16)$$

where, I_{nom} = nominal interest rate, y = discount rate, N = life span of the project and x = inflation rate of the project.

The IRES components, such as a bi-directional converter with a charge controller, biomass generator are needed to replace because they have a shortened lifespan than the project lifetime. The present value of annual replacement cost ($P_{V,REP}$) of the microgrid is calculated as follows [2]:

$$P_{V,REP} = \begin{bmatrix} \left(N_{BAT} \times C_{BAT,rep} \times \sum_{b=1}^{N_r} \frac{(1+x)^{bN_c-1}}{(1+y)^{bN_c}} \right) + \left(C_{BMG,rep} \times \sum_{g=1}^{N_r} \frac{(1+x)^{gN_c-1}}{(1+y)^{gN_c}} \right) + \\ + \left(C_{BDC-CC,rep} \times \sum_{d=1}^{N_r} \frac{(1+x)^{dN_c-1}}{(1+y)^{dN_c}} \right) \end{bmatrix} \quad (17)$$

where, $C_{BAT,rep}$, $C_{BMG,rep}$ and $C_{BDC-CC,rep}$ are the replacement costs of the batteries, biomass generator and bi-directional converter with a charge controller, respectively, and the N_r is defined as follows [2]:

$$N_r = int\left(\frac{N - N_c}{N_c} \right) \quad (18)$$

where, N_r = the number of replacements needed for the system components and N_c = life span of each system component.

The present value of annual fuel cost ($P_{V,FUEL}$) of the microgrid is calculated as follows [2]:

$$P_{V,FUEL} = [(C_{BM} \times Q_{BM})] \times \sum_{i=1}^{N} \frac{(1+x)^{i-1}}{(1+y)^i} \quad (19)$$

where, Q_{BM} and C_{BM} are the quantity and cost of the biomass.

7. The Objective Function and Its Constraints

The objective function of the system is finding the optimal Life Cycle Cost (LCC) of the system. The system cost mainly depends on two variable parameters such as the number of PV panels and batteries. The objective function is defined as follows:

$$min \ LCC(N_{PV}, N_{BAT}) = \sum_{C=PV,BMG,BAT,BDC-CC}^{min} (LCC)_C \tag{20}$$

7.1. Upper and Lower Bounds

In this study, the biomass generator is considered as a fixed energy resource. Which is operational every day for 5 h, i.e., from 6 p.m. to 10 p.m. The biomass generator's rated power is considered to be 10 kW. The biomass generator produces around 9 kWh of energy each hour. As a result, the biomass generator is not bound by any constraints. PV energy is the remaining energy resource, which is bound by the following constraint.

$$1 \leq N_{PV} \leq N_{PV-max} \tag{21}$$

where, N_{PV} is the number of PV panels.

The battery bank is bound by the following constraint.

$$1 \leq N_{BAT} \leq N_{BAT-max} \tag{22}$$

where, N_{BAT} is the number of batteries.

7.2. Battery Bank Energy Storage Limits

During the charging and discharging processes, some energy will be stored or discharged from the battery bank, which will be limited by the following constraints [3]:

$$E_{Bat\_min} \leq E_{Bat}(t) \leq E_{Bat\_max} \tag{23}$$

The battery bank's min-max energy storage levels are computed as follows:

$$E_{Bat\_max} = \left(\frac{N_{BAT} \times V_{BAT} \times S_{BAT}}{1000} \right) \times SOC_{max-bat}. \tag{24}$$

$$E_{Bat\_min} = \left(\frac{N_{BAT} \times V_{BAT} \times S_{BAT}}{1000} \right) \times SOC_{min-bat} \tag{25}$$

where, S_{BAT} = rated capacity of the battery (Ah) and V_{BAT} = battery's voltage.

The battery's min-max state of charges are computed as follows:

$$SOC_{min-bat} = 1 - DOD \tag{26}$$

$$SOC_{max-bat} = SOC_{min-bat} + DOD \tag{27}$$

where, DOD is the depth of discharge of the battery.

7.3. Power Reliability Index

A microgrid's power reliability can be described as its capacity to supply power on a continuity basis. The *LPSP* is a critical microgrid criterion that will define the microgrid's power supply continuity. Which is calculated as follows [4]:

$$LPSP = \frac{\sum_{t=1}^{T} LPS(t)}{\sum_{t=1}^{T} E_L(t)} \tag{28}$$

where, $LPS(t)$ is the loss of power supply at any hour "t"

$$LPS(t) = \frac{E_L(t)}{\eta_{Conv}} - E_G(t) - [(1-\sigma) \times E_{Bat}(t-1) - E_{Bat\_min}] \times \eta_{rbat} \tag{29}$$

8. Methodology

The methodology of the power flow among the various components used in the study are explained in the following modes of operation [5]:

Mode 1: In this operating mode, the power produced by the RE sources is equal to the load demand, hence there is no deficit or excess energy in the system.

Mode 2: In this operating mode, whenever the RE sources generate excess energy, it will be stored in the battery bank through the charging process of the battery bank.

Mode 3: In this operating mode, both the energy generated by the RE resources and the energy stored in the bank are at their maximum capacity. At this time, the extra energy generated will be delivered to the dump load.

Mode 4: This mode of operation demonstrates that whenever the energy provided by the RE resources is insufficient to supply the load demand, the energy stored in the battery bank will be delivered to the loads through the discharging process of the battery bank.

Mode 5: In this operating mode, the energy provided by the RE resources does not meet the load demand and the battery bank is also at its minimum energy storage levels, resulting in a loss of power supply to the system.

The above-mentioned modes of operation are depicted pictorially in the form of a flowchart as shown in Figures 6 and 7.

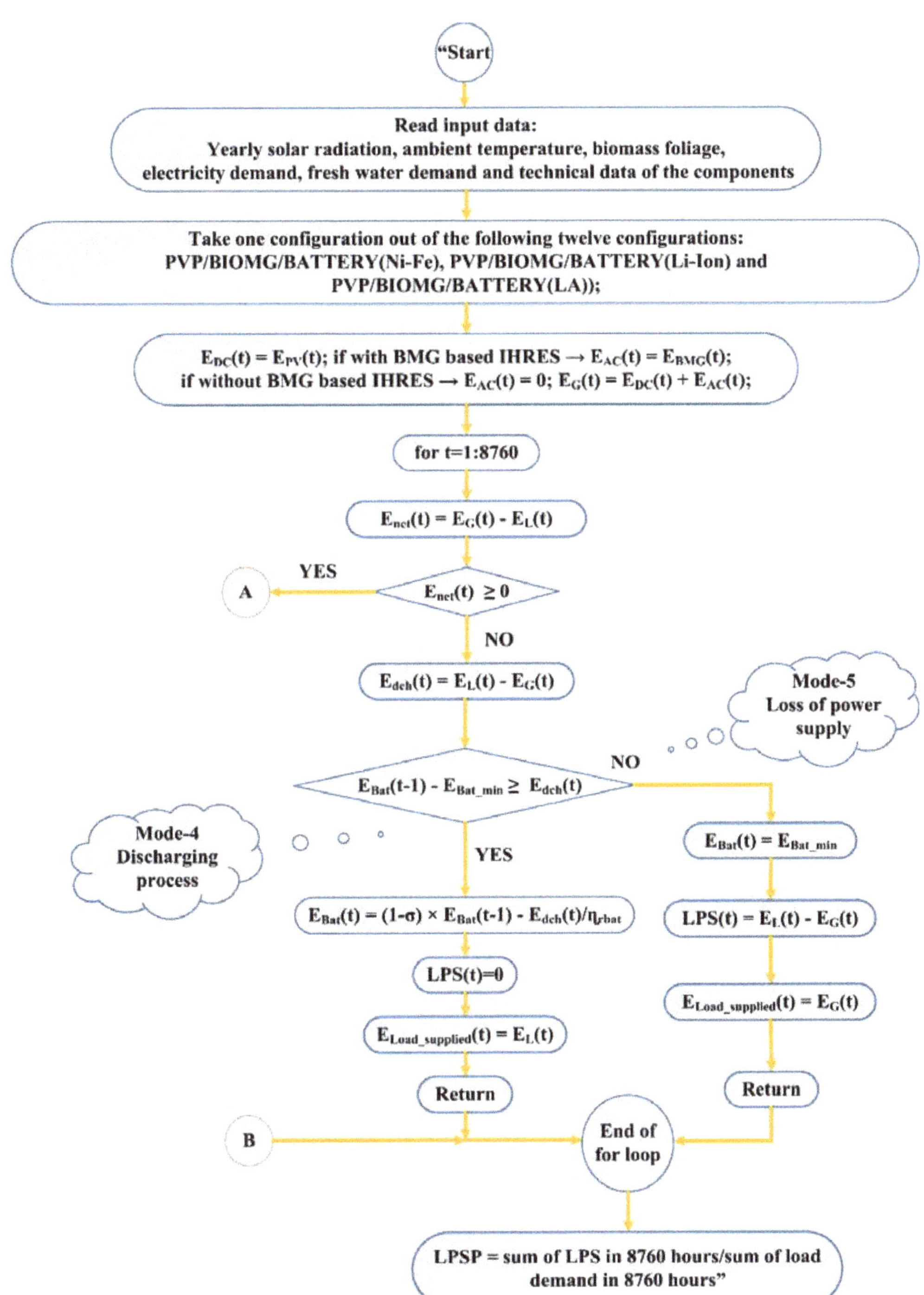

Figure 6. Modes of operation for the off-grid microgrid.

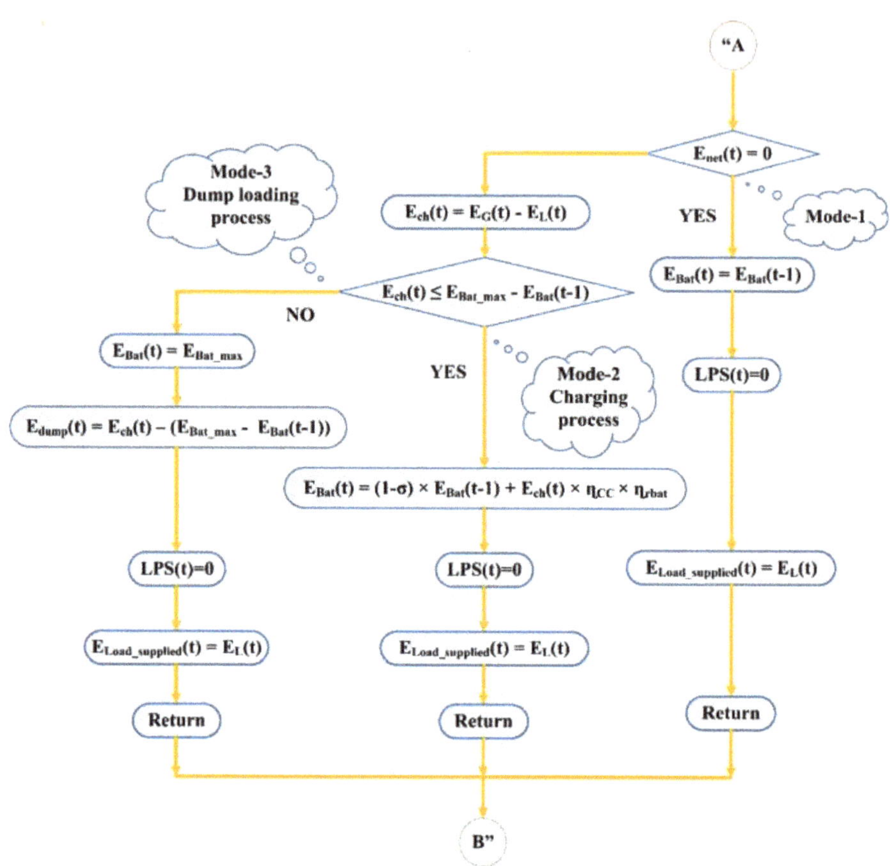

Figure 7. Modes of operation for the off-grid microgrid.

9. Results and Discussion

In general, using the existing RE resources, rural distant communities can be electrified. However, the fundamental disadvantage of RE resources is their inherent nature, as well as their energy production depends on the metrological conditions in the chosen location. As a result of the RE resources' time and metrological dependence, the power supply to the study area is not continuous.

As a result, while RE resources were used to electrify the study area, one of the key problems with these resources is that they cannot provide a continuous power supply. Hence, a reliable battery storage system is required to provide a continuous power supply. So, three different types of battery technologies have been proposed in this study to provide a continuous power supply in order to select a battery technology that is technologically and economically feasible for the study area.

Finally, we have been modelled three different types of configurations employing available RE resources and battery technologies, such as PVP/BIOMG/BATTERY$_{(Ni-Fe)}$, PVP/BIOMG/BATTERY$_{(Li-Ion)}$ and PVP/BIOMG/BATTERY$_{(LA)}$ to Identify a configuration that is technologically and economically feasible to electrify the study area. PV panels and batteries are the variable parameters in all of the three configurations and their quantities can be determined using optimization algorithms based on the required load demand and constraints. However, the biomass generator is regarded as a fixed energy resource, with a rated power of 10 kW. Hence, no optimization algorithm is required to estimate its size because it is already fixed as a 10 kW. The converter size is also considered as a fixed parameter because it is already predesigned to meet the peak load demands of the study area. Because the study area's peak load demand is 35.72 kW, the converter power rating is set to 40 kW. For safety reasons, the converter power rating is set to be 10% higher than the peak load demand. As a result, there is no need to use optimization algorithms to optimize the converter size.

The three above-mentioned configurations have been optimized using well-known algorithms such as Differential Evolutionary Algorithm (DE) [10], Genetic Algorithm (GA) [11], Moth Flame Optimization (MFO) [12], Dragonfly Algorithm (DA) [13], Particle Swarm Optimization (PSO) [14], Ant Lion Optimization (ALO) [15], Grey Wolf Optimization (GWO) [16], Grasshopper Optimization Algorithm (GOA) [17] and Dynamic Differential Annealed Optimization Algorithm [18]. These algorithms that were run with their default control parameter values with the population and iterations are considered to be 100. The technical and cost values used in the study are listed in Tables 2 and 3.

Table 2. The study's technical and financial values.

Parameters	Value	Parameters	Value
Project lifetime	25 years	Erection cost of BIOMG [2]	450.5 USD
Nominal interest rate [2]	13%	Rated Power of BIOMG [6]	10 kW
Inflation rate [7]	5%	BIOMG Capital cost [6]	9010 USD
Manufacturer of PV Panel [19]	Vikram solar	AO&M cost of BIOMG [2]	$271
Model No. of PV Panel [19]	Somera 385	Quantity of biomass	19 tons/year
Rated power of PV Panel [19]	385 Wp	Cost of biomass [2]	15 USD/ton
Lifetime of PV Panel [8]	25 years	Efficiency of BIOMG [2]	20%
Capital cost of PV Panel [8]	128 USD	Rated power of BDC-CC	40 kW
AO&M cost of PV Panel [2]	3.2 USD	Lifetime of converter [2]	10 years
Mechanical structure cost of PV Panel [20]	41 USD	C&R of converter [2]	4320 USD
Life time of mechanical structure of PV panel [20]	25 years	AO&M cost of converter [2]	108 USD
Erection cost of PV panel [2]	25.6 USD	Efficiency of converter [2]	95%
Erection cost of battery [2]	3% of CC of battery	Erection cost of converter [2]	129.6 USD

Table 3. The study's technical and financial values of the batteries.

Battery Type	Lead Acid (PbSO$_4$)	Lithium Iron Phosphate (LiFePO$_4$)	Nickel Iron (Ni-Fe)
Manufacturer	Trojan [21]	Victron [22]	Iron Edison [23]
Model	SSIG 06 490	LFP-12.8/200-a	TN 1000
Nominal capacity (S_{BAT})	490 Ah	300 Ah	1000 Ah
Nominal voltage (V_{BAT})	6 V	12.8 V	1.2 V
Round trip efficiency (η_{rbat})	85%	92%	80%
Lifespan in years	2.5 years	15 years	30 years+
Self-discharge rate (%/day) (σ)	0.3%	0.2%	1%
Capital cost (CC) in USD	410	3317	1057
Annual O&M cost in USD	2.5% of CC	No maintenance	2% of CC
Operating temperature	$-20\,°C$ to $+45\,°C$	$-20\,°C$ to $+50\,°C$	$-30\,°C$ to $+60\,°C$
Cycle life of the batteries	750 cycles	5000 cycles	11,000+ cycles

The results shown in Table 4 were obtained with an LPSP value of 0% for the above-mentioned three configurations using the algorithms shown above. For each configuration, this table shows the optimal LCC values as well as the optimal number of PV panels and batteries.

Table 4. Optimization results of all configurations at an LPSP value of 0%.

Configuration	Q&C	GA	PSO	DE	GWO	ALO	DA	MFO	GOA	DDAO
PVP/ BIOMG/ BATTERY$_{(Ni\text{-}Fe)}$	N_{PV}	511	511	513	511	511	513	511	516	511
	N_{BAT}	351	351	351	351	352	352	351	352	351
	LCC in USD	367,586	367,586	368,088	367,586	368,153	368,655	367,586	369,407	367,586
	STIS	8687	8906	20538	8938	8980	9080	8806	8972	8152
PVP/ BIOMG/ BATTERY$_{(LA)}$	N_{PV}	475	474	474	478	474	476	477	474	474
	N_{BAT}	136	136	136	136	136	137	136	136	136
	LCC in USD	607,281	607,030	607,030	608,032	607,030	610,821	607,781	607,030	607,030
	STIS	8871	8669	20820	9007	8980	9120	8889	8992	8083
PVP/ BIOMG/ BATTERY$_{(Li\text{-}Ion)}$	N_{PV}	436	438	439	436	437	436	436	439	436
	N_{BAT}	155	155	157	155	155	155	155	156	155
	LCC in USD	1,019,683	1,020,185	1,031,657	1,019,683	1,019,934	1,019,683	1,019,683	1,026,046	1,019,683
	STIS	8814	8845	20542	8874	8834	8953	8792	8811	8092

9.1. The Impact of Various Battery Technologies on System Performance

According to the results given in Table 4, the configuration made up of Ni-Fe battery technology provided the lowest LCC value when compared to the configurations made up of LA and Li-Ion battery technologies. The Ni-Fe battery configuration offered the optimal LCC value as 367,586 USD and its required number of PV panels and Ni-Fe batteries to electrify the study area are 511 and 351, respectively.

The next optimal configuration to electrify the study area is the configuration made up of LA battery technology, which has provided the optimal LCC value as 607,030 USD, which is about 65% higher than the LCC value obtained with the Ni-Fe battery technology and about 41% lower than the LCC obtained with the Li-Ion battery technology. To electrify the study area using the configuration made up of LA battery technology, 474 PV panels and 136 LA batteries are required.

The final optimal configuration to electrify the study area is the configuration made up of Li-Ion battery technology, which produced the optimal LCC value as 1,019,683 USD. Which is about 177% and 68% higher than the LCCs obtained with the configurations made up of Ni-Fe and LA battery technologies, respectively. Finally, the number of PV panels and Li-Ion batteries required to electrify the study area with this configuration are 436 and 155, respectively.

Figure 8 depicts the annual energy production of the PV panels and biomass generator, as well as the charging and discharging process of the battery bank based on load demand.

9.2. Robustness of the Algorithms

The robustness of the algorithm can be determined by its ability to provide the global best optimal values for the proposed configurations. Three configurations were modelled in this study, and these configurations were tested using well-known algorithms as well as a new proposed metaheuristic algorithm Dynamic Differential Annealed Optimization (DDAO). For each of these three configurations, the proposed algorithm, DDAO, provided the optimal LCC values. The number of times each algorithm provided the global best optimal values for three configurations are shown in Table 5.

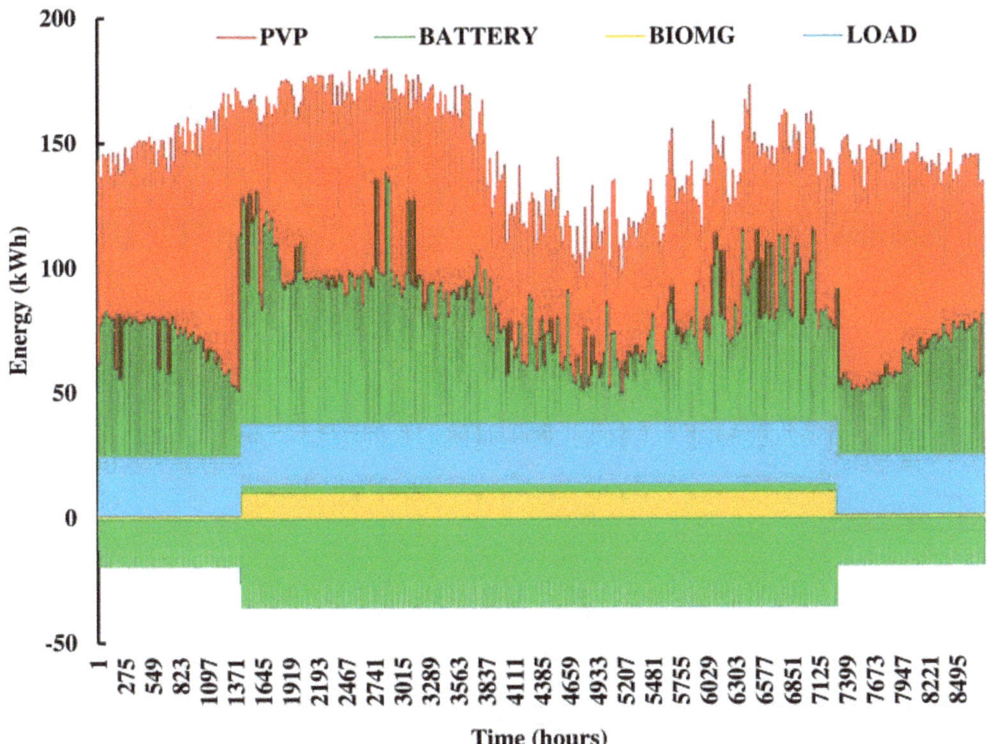

Figure 8. Energy output graph for different components of a Ni-Fe battery configuration over a year.

Table 5. Rankings of algorithms in providing the optimal values.

S.No.	Algorithms	Optimal Values Given Times	Ranking
1	DDAO	3	1
2	GA, PSO, GWO and MFO	2	2
3	DE, DA and GOA	1	3

According to Table 4, when compared to the remaining algorithms in terms of Simulation Time in Seconds (STIS), the DDAO algorithms performed the best in both finding global best optimal values and simulation completion process.

The convergence curves for the three configurations proposed in the study are depicted in Figures 9–11.

9.3. Loss of Power Supply Probability

The configuration made up of Ni-Fe battery technology is identified as the optimal configuration for electrifying the study area. This configuration was examined with various probabilities of power supply failure. As a result, the optimal configuration was examined with different LPSP values ranging from 1% to 5%. Table 6 shows the optimal values for this configuration with various LPSP values. Table 6 and Figure 12 shows that at the LPSP value of 1%, there is a significant change in all values when compared to the values obtained at the LPSP value of 0%. When we compare the difference between the 1% LPSP and the remaining LPSP values ranging from 2% to 5% optimal values, the difference in LCC and component values is not that much of significant. At an LPSP value of 1%, the number of power supply hours lost is also 123 only out of 8760 h. As a result, electrification of the study area at an LPSP value of 1% is also recommendable.

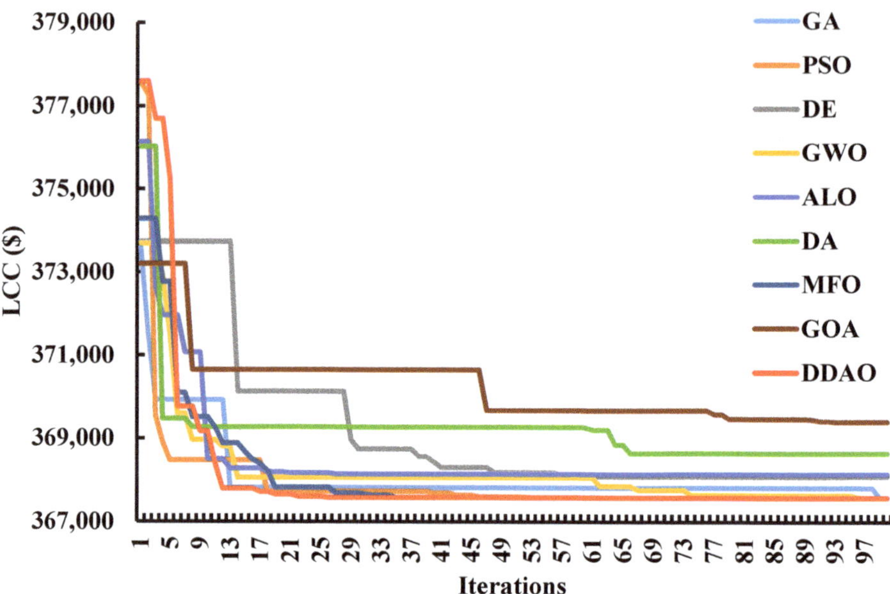

Figure 9. Convergence curves of the configuration made up of Ni-Fe battery.

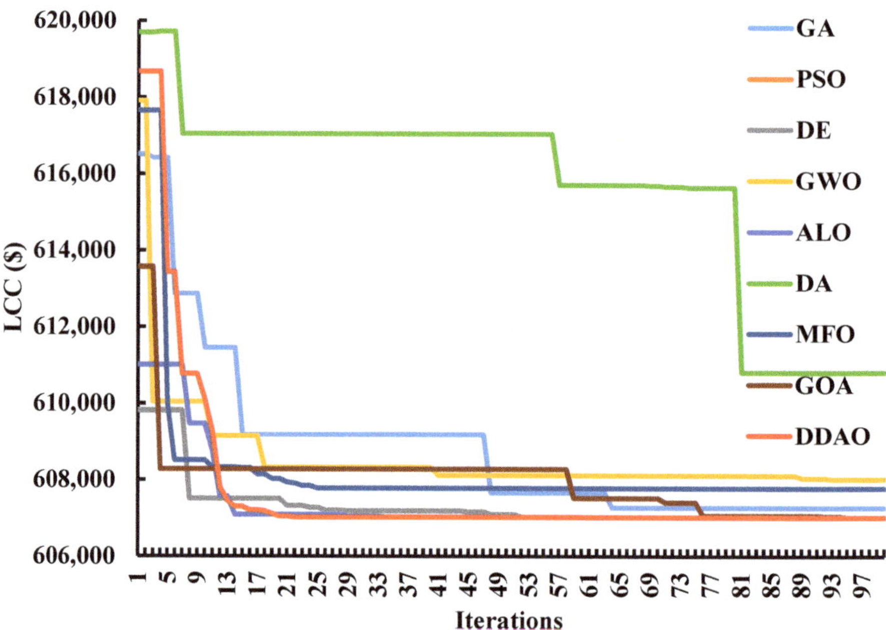

Figure 10. Convergence curves of the configuration made up of LA battery.

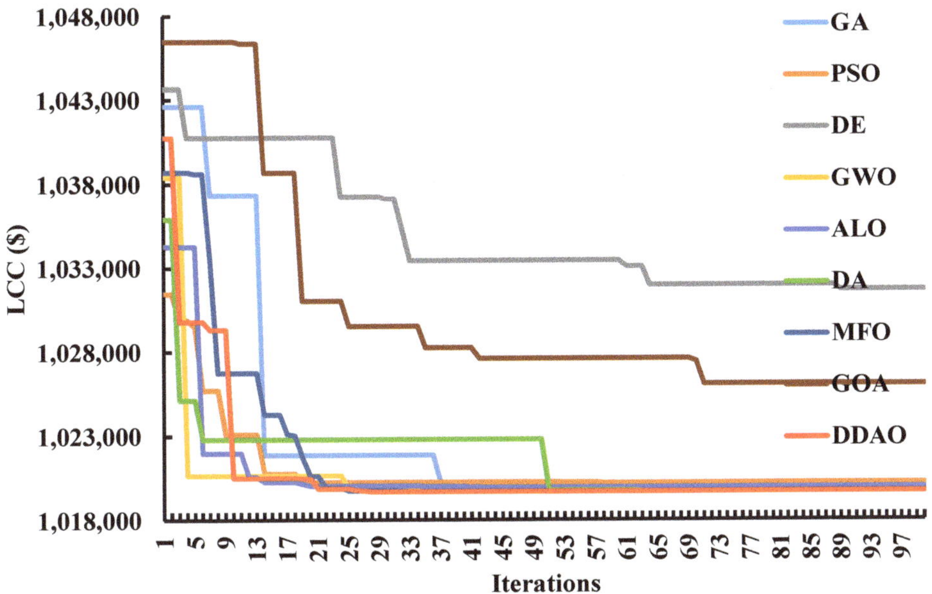

Figure 11. Convergence curves of the configuration made up of Li-Ion battery.

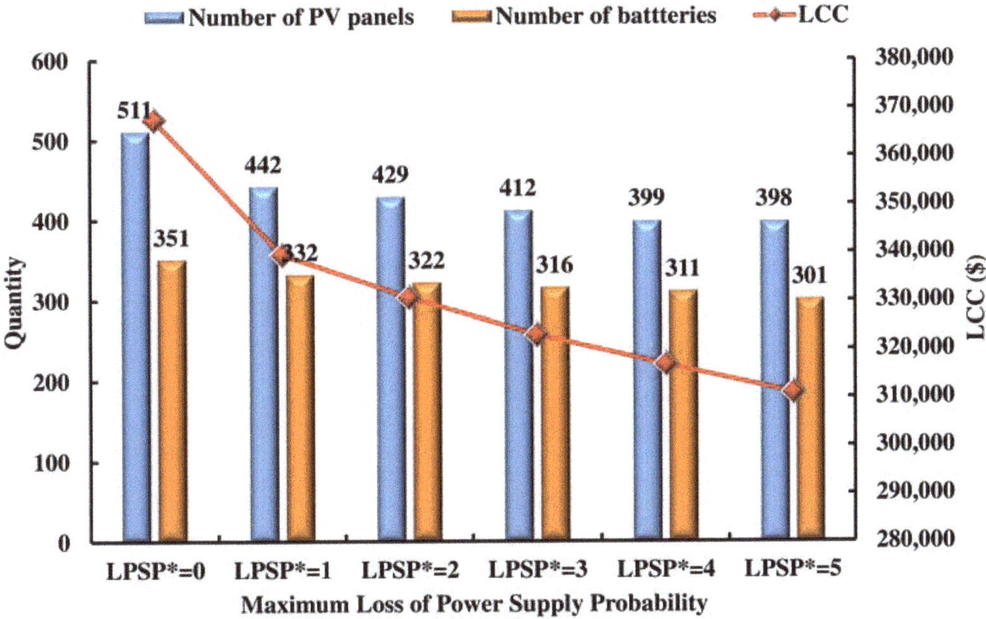

Figure 12. The Ni-Fe battery configuration analysis with different LPSP values.

Table 6. Optimum configuration results for various LPSP values.

Configuration	Quantity & Cost	LPSP (0%)	LPSP (1%)	LPSP (2%)	LPSP (3%)	LPSP (4%)	LPSP (5%)
PVP/ BIOMG/ BATTERY (Ni-Fe)	N_{PV}	511	442	429	412	399	398
	N_{BAT}	351	332	322	316	311	301
	LCC in USD	367,586	339,523	330,600	322,939	316,847	310,933
	LPSH	0	182	317	430	564	632

10. Conclusions

The current study has focused on off-grid rural electrification in a remote rural area of the Indian state of Odisha. The location is still not electrified due to its remote location because grid extension to that area is not feasible for two reasons: first, extending the grid to that area is more expensive because it is more than 50 km away. The second reason is that grid extension is not a viable option for the location since it is located in a densely forested area surrounded by hilly mountains. As a result, electrification of the study area is planned, utilizing the study area's readily available RE resources, such as solar and biomass. However, due to the time dependence and metrological weather conditions dependence of these RE resources, the power supply is not continuous. As a consequence, three different types of battery technologies have been proposed for the current study. To identify a battery technology that is technologically and economically suitable for the study area.

The current study used available RE resources and battery technologies to model three configurations: PVP/BIOMG/BATTERY$_{(Ni\text{-}Fe)}$, PVP/BIOMG/BATTERY$_{(Li\text{-}Ion)}$ and PVP/-BIOMG/BATTERY$_{(LA)}$. The main goal here is to identify the configuration that is appropriate for electrifying the study area. The identification process incorporated with nine well-known and established algorithms, such as Differential Evolutionary Algorithm, Genetic Algorithm, Moth Flame Optimization, Dragonfly Algorithm, Particle Swarm Optimization, Grey Wolf Optimization, Ant Lion Optimization, Grasshopper Optimization Algorithm and Dynamic Differential Annealed Optimization (DDAO) Algorithm. When DDAO was compared to the remaining algorithms, it produced the global best optimal results. Based on the results, the PVP/BIOMG/BATTERY$_{(Ni\text{-}Fe)}$ configuration was identified as the feasible configuration to electrify the study area.

When compared to the configurations made up of Li-Ion and LA battery, the configuration of Ni-Fe battery provided the lowest LCC value as 367,586 USD, which is 39% and 64% lower in percentage terms. The optimal configuration has been further analyzed with different loss of power supply probability values ranging from 1% to 5%. According to the findings, at a loss of power supply probability value of 1%, the optimal configuration may be recommendable for electrifying the study area.

Author Contributions: Conceptualization P.P.K.; methodology, P.P.K.; software, P.P.K.; validation, P.P.K. and M.J.; formal analysis, P.P.K.; investigation, P.P.K. and V.S.; resources, M.J. and V.S.; data curation, M.J. and P.P.K.; writing—original draft preparation, P.P.K.; writing—review and editing, M.J. and V.S.; visualization, P.P.K., V.S. and M.J.; supervision, Z.L.; project administration, Z.L.; funding acquisition, Z.L, V.S. and M.J. All authors have read and agreed to the published version of the manuscript.

Funding: This research was funded by the Chair of Electrical Engineering Fundamentals, Wroclaw University of Science and Technology K38W05D02.

Data Availability Statement: Not Applicable.

Conflicts of Interest: No conflict of Interest.

References

1. Patel, A.M.; Singal, S.K. Optimal component selection of integrated renewable energy system for power generation in stand-alone applications. *Energy* **2019**, *175*, 481–504. [CrossRef]
2. Man, K.F.; Tang, K.S.; Kwong, S. Genetic algorithms: Concepts and applications. *IEEE Trans. Ind. Electron.* **1996**, *43*, 519–534. [CrossRef]
3. Kumar, P.; Saini, R.P. Optimization of an off-grid integrated hybrid renewable energy system with various energy storage technologies using different dispatch strategies. *Energy Sources Part A Recovery Util. Eff.* **2020**, *32*, 1–30. [CrossRef]
4. Kumar, P.P.; Saini, R.P. Optimization of an off-grid integrated hybrid renewable energy system with different battery technologies for rural electrification in India. *J. Energy Storage* **2020**, *32*, 101912. [CrossRef]
5. Salameh, T.; Kumar, P.P.; Sayed, E.T.; Abdelkareem, M.A.; Rezk, H.; Olabi, A.G. Fuzzy modeling and particle swarm optimization of Al_2O_3/SiO_2 nanofluid. *Int. J. Thermofluids* **2021**, *10*, 100084. [CrossRef]
6. Storn, R. Differential Evolution—A Simple and Efficient Heuristic for Global Optimization over Continuous Spaces. *J. Glob. Optim.* **1997**, *11*, 341–359. [CrossRef]
7. Mirjalili, S. Moth-flame optimization algorithm: A novel nature-inspired heuristic paradigm. *Knowl. Based Syst.* **2015**, *89*, 228–249. [CrossRef]
8. Mirjalili, S. Dragonfly algorithm: A new meta-heuristic optimization technique for solving single-objective, discrete, and multi-objective problems. *Neural Comput. Appl.* **2016**, *27*, 1053–1073. [CrossRef]
9. Kennedy, J.; Eberhart, R. Particle Swarm Optimization. In Proceedings of the ICNN'95-International Conference on Neural Networks, Perth, WA, Australia, 27 November–1 December 1995; pp. 1942–1948.
10. Mirjalili, S. The Ant Lion Optimizer. *Adv. Eng. Softw.* **2015**, *83*, 80–98. [CrossRef]
11. Mirjalili, S.; Mohammad, S.; Lewis, A. Grey Wolf Optimizer. *Adv. Eng. Softw.* **2014**, *69*, 46–61. [CrossRef]
12. Saremi, S.; Mirjalili, S.; Lewis, A. Advances in Engineering Software Grasshopper Optimisation Algorithm: Theory and Application. *Adv. Eng. Softw.* **2017**, *105*, 30–47. [CrossRef]
13. Ghafil, H.N.; Jármai, K. Dynamic differential annealed optimization: New metaheuristic optimization algorithm for engineering applications. *Appl. Soft Comput. J.* **2020**, *93*, 106392. [CrossRef]
14. Bukar, A.L. Optimal sizing of an autonomous photovoltaic/wind/battery/diesel generator microgrid using grasshopper optimization algorithm. *Solar Energy* **2019**, *188*, 685–696. [CrossRef]
15. Chauhan, A.; Saini, R.P. Discrete harmony search based size optimization of Integrated Renewable Energy System for remote rural areas of Uttarakhand state in India. *Renew. Energy* **2016**, *94*, 587–604. [CrossRef]
16. Maleki, A. Design and optimization of autonomous solar-wind-reverse osmosis desalination systems coupling battery and hydrogen energy storage by an improved bee algorithm. *Desalination* **2018**, *435*, 221–234. [CrossRef]
17. Enersol Biopower. Available online: http://enersolbiopower.com/ (accessed on 5 May 2020).
18. Reserve Bank of India. Available online: https://www.rbi.org.in/home.aspx (accessed on 12 February 2020).
19. Vikram Solar. Available online: https://www.vikramsolar.com/ (accessed on 25 May 2020).
20. Mechanical Structure Cost of PV Panel. Available online: https://www.loomsolar.com/products/loom-solar-2-row-design-6-panel-stand-375-watt?gclid=CjwKCAjwsan5BRAOEiwALzomXxjAoKPGLLYkGRoWRgkR7mAros-FbPGeWeNhzvOqId-TRQESPrWq3xoC3CAQAvD_BwE (accessed on 12 May 2020).
21. Trojan Battery Company. Available online: https://www.trojanbattery.com/ (accessed on 12 January 2020).
22. Victron Energy Blue Power. Available online: https://www.victronenergy.com/batteries (accessed on 15 January 2020).
23. Iron Edison. Available online: https://ironedison.com/store (accessed on 10 January 2020).

Article

Clustering Methods for Power Quality Measurements in Virtual Power Plant

Fachrizal Aksan, Michał Jasiński *, Tomasz Sikorski, Dominika Kaczorowska, Jacek Rezmer, Vishnu Suresh, Zbigniew Leonowicz, Paweł Kostyła, Jarosław Szymańda and Przemysław Janik

Faculty of Electrical Engineering, Wroclaw University of Science and Technology, 50-370 Wroclaw, Poland; 254212@student.pwr.edu.pl (F.A.); tomasz.sikorski@pwr.edu.pl (T.S.); dominika.kaczorowska@pwr.edu.pl (D.K.); jacek.rezmer@pwr.edu.pl (J.R.); vishnu.suresh@pwr.edu.pl (V.S.); zbigniew.leonowicz@pwr.edu.pl (Z.L.); pawel.kostyla@pwr.edu.pl (P.K.); jaroslaw.szymanda@pwr.edu.pl (J.S.); przemyslaw.janik@pwr.edu.pl (P.J.)
* Correspondence: michal.jasinski@pwr.edu.pl; Tel.: +48-713202022

Abstract: In this article, a case study is presented on applying cluster analysis techniques to evaluate the level of power quality (PQ) parameters of a virtual power plant. The conducted research concerns the application of the K-means algorithm in comparison with the agglomerative algorithm for PQ data, which have different sizes of features. The object of the study deals with the standardized datasets containing classical PQ parameters from two sub-studies. Moreover, the optimal number of clusters for both algorithms is discussed using the elbow method and a dendrogram. The experimental results show that the dendrogram method requires a long processing time but gives a consistent result of the optimal number of clusters when there are additional parameters. In comparison, the elbow method is easy to compute but gives inconsistent results. According to the Calinski–Harabasz index and silhouette coefficient, the K-means algorithm performs better than the agglomerative algorithm in clustering the data points when there are no additional features of PQ data. Finally, based on the standard EN 50160, the result of the cluster analysis from both algorithms shows that all PQ parameters for each cluster in the two study objects are still below the limit level and work under normal operating conditions.

Keywords: power quality; cluster analysis; K-means; agglomerative; virtual power plant

Citation: Aksan, F.; Jasiński, M.; Sikorski, T.; Kaczorowska, D.; Rezmer, J.; Suresh, V.; Leonowicz, Z.; Kostyła, P.; Szymańda, J.; Janik, P. Clustering Methods for Power Quality Measurements in Virtual Power Plant. *Energies* **2021**, *14*, 5902. https://doi.org/10.3390/en14185902

Academic Editor: Abu-Siada Ahmed

Received: 25 June 2021
Accepted: 23 July 2021
Published: 17 September 2021

Publisher's Note: MDPI stays neutral with regard to jurisdictional claims in published maps and institutional affiliations.

1. Introduction

The electrical power system aims to generate electrical power and deliver it through the transmission and distribution system to customers' devices in a stable, secure, reliable, and sustainable manner [1]. However, nowadays, various electronic devices such as AC/DC converters, switching power supplies, and industrial non-linear load are becoming the factors responsible for the increasing PQ disturbances{XE "PQ"\t "Power Quality"} [2,3]. These devices tend to significantly distort the waveform of the supply and voltage [4,5]. The term power quality refers to a wide range of electromagnetic phenomena that characterize the voltage and current quality at a given time and location in the power system [6]. Generally, PQ disturbances are defined into two types based on the characteristics: voltage variations [7,8] and voltage events [9,10]. The analysis of PQ can be used to monitor the characteristic disturbances to capture PQ events that potentially detect faults associated with power quality problems in electrical power systems [11–13].

Power quality assessment has become a critical issue since the increase in sensitive electronic equipment in industrial and household applications [14,15]. The original lack of standardization of the measurement method has led to significant differences in each of the main parameters calculated by different devices. Thus, the consequences of power quality disturbances can lead to equipment malfunctions and process shutdowns [16,17]. The IEC 61000-4-30 standard [18] satisfies the standardization problem related to power quality by specifying a precise procedure, mathematical relationships, and required measurement accuracy for power quality analyzers [19,20]. During power quality monitoring, a short averaging time may be sufficient to evaluate the performance and disturbances related to PQ problems [21,22]. The ideal analysis time for PQ surveying is usually over one week, which is typically an integration period set at 10 min [23,24]. A network survey over a long period means that a large amount of data needs to be collected, which can be difficult and time-consuming to process. It is necessary to use appropriate tools or techniques to analyze the power quality measurement data to obtain valuable information or patterns within the huge amount of data. The data mining technique is one of the proposed solutions to process the huge amount of power quality measurement data to discover useful information related to power quality disturbances [25,26]. However, when working with data mining techniques, a basic understanding of the workflow and processing steps is required to achieve optimal results [27,28].

This research concerns the application of data mining techniques, especially on the clustering analysis method with the non-hierarchical and hierarchical approach for power quality problems in a virtual power plant (VPP). In general, a VPP consists of three effective components. The components used are the conventional dispatchable power plants, energy storage systems, and responsive or flexible loads [29]. Since cluster analysis is a part of unsupervised learning that has no labeling [30], it is used for exploratory data analysis of a VPP to group a collection of data items that are similar to each other and dissimilar from data items in other clusters [31].

In terms of a non-hierarchical approach, reference [32] introduces cluster analysis with the K-means algorithm on the long-term measurement of power quality data in the real virtual power plant that operates in Poland. The research aimed to identify the different working conditions of the VPP based on data features. Then, the different input power quality databases that consist of global index values as PQ parameters were used to identify how the algorithm works with different input features. In contrast, identifying the optimal number of clusters was defined by using v-fold cross-validation and a cost sequence chart. Reference [33] proposed the K-means algorithm to identify the energy features of the prosumers. The algorithm allows one to obtain the categories of prosumers from two specific indicators, which can assist the distribution network operators in optimizing networks' operation.

On the other hand, with the hierarchical approach, reference [34] proposed the application of the Ward algorithm to detect short-term working conditions of a VPP by analyzing the classical PQ parameters as the input data, and the qualitative assessment of the clustering process was realized by using the cubic clustering criterion. In comparison, reference [35] involves the same approach but using the dendrogram to select the final number of clusters, which was the unquestionable disadvantage of the hierarchical approach. Reference [36] uses the Ward algorithm for a prosumer fair load sharing and surplus trading approach based on the micro-grid concept of the transactive energy concept. This article presented a dendrogram to separate the daily energy consumption and maximum power from each micro-grid bus.

The exploration of PQ data in mining electrical power networks has been introduced in reference [37]. The proposed solution is using both the K-means algorithm and the Ward algorithm for cluster analysis. The K-means algorithm was used to classify the non-flagged data and flagged data, while the Ward algorithm was proposed to obtain a dendrogram for determining the optimal number of clusters in non-flagged data to identify the different working conditions of the electrical network. Since PQ monitoring collects a huge amount of measurement data, it is necessary to explore the pattern of data to obtain useful information that can inform upgrades or changes. Reference [26] proposed the methodology known as mixture modeling or intrinsic classification to recognize network problems at medium voltage (MV) electrical distribution systems, while reference [38] presented the use of the K-means algorithm and a hierarchical method to identify the reference nodes in distribution generation (DG).

Due to working with clustering analysis, it is very important to define the optimal number of clusters of the dataset for further analysis [39] because the main problem in applying many of the existing clustering techniques is that the optimal number of clusters needs to be pre-specified before the clustering is carried out [40]. Various methods have been proposed in the literature for determining the optimal number of clusters. Previous literature presents the use of a dendrogram and v-fold cross-validation, while reference [41] proposed the elbow method by calculating a sum of squares at each number of clusters and graphing. The optimal number of clusters is the one that looks like an arm in the graph. Reference [11] also proposed applying the elbow method to determine the optimal number of clusters on the K-means method. The result shows that the algorithm is sensitive to selecting the initial centroid of the cluster, and the elbow method allows the achievement of the same number of clusters on different data.

On the other hand, reference [42] reveals that the K-means algorithm is widely used for processing quantitative data with numeric attributes. However, this algorithm has drawbacks because it needs to determine the number of clusters before being applied. This research proposed applying the elbow method to assume the optimal number of clusters for the initial selection of centroids to fill the basic requirement of the K-means algorithm. The result shows that the elbow method can reduce 25% of iterations of processing using the K-means algorithm.

Based on the literature review, cluster analysis is suitable to analyze the power quality disturbance in a VPP. This research focuses on comparing two cluster analysis algorithms on a hierarchical approach with a non-hierarchical approach to the power quality measurement data of a VPP. The analysis concerns the medium voltage distribution network in a VPP, consisting of a 1.25 MW hydropower plant (HPP) and 0.5 MW energy storage system (ESS). The research is based on PQ measurements, which were realized synchronically in measurement points on the HPP, ESS, and associated with medium voltage lines. The measurement values consist of classic PQ parameters and an application of the global index, which is taken from the duration 1 May to 29 October 2020. Due to the huge amount of data collection varying on different feature units and scales, the dataset should be standardized to achieve a compatible input for the cluster analysis algorithms. The K-means algorithm has been selected from a non-hierarchical approach, while the agglomerative algorithm is a representative of a hierarchical approach, and both of the algorithms were performed by using a Euclidean distance metric. Additionally, the method for determining the optimal number of clusters for the K-means algorithm using the elbow method and determining the optimal number of clusters of the agglomerative algorithm was conducted by observing the dendrogram. Finally, the qualitative assessment of clustering results from both algorithms was presented to identify the power quality characteristic of a VPP based on the standard EN 50160 [43]. The analysis research and calculation method was performed using a scikit-learn machine learning library [44] for python [45] users. The contribution of this article is described below:

- The power quality dataset based on the long-term measurement in a VPP was standardized to cluster analysis.
- The proposed K-means algorithm and agglomerative algorithm were performed to compare the hierarchical and non-hierarchical approach for PQ data, which have different sizes of input data.
- The elbow method and dendrogram were performed to obtain the optimal number of clusters for PQ data.
- The global index was used for the comparative assessment of PQ parameters between clustering results of the K-means algorithm and agglomerative algorithm.
- The cluster algorithm evaluation and comparison are used to determine which algorithm is suitable for the investigation object.

This article is organized into five sections to achieve these contributions. Section 2 contains the description of the VPP, power quality data parameters, and the cluster analysis methodology. Section 3 presents the optimal number of clusters for PQ data of the VPP and the qualitative assessment of PQ for each clustering result. Section 4 reveals the discussion of the results. Finally, Section 5 covers the conclusion.

2. Research Object Description and Methodology

2.1. Object Investigated

The research object of this article is to study the power quality problem in a virtual power plant operating in Lower Silesia, Poland. The studied subject refers to reference research [32], which concerns a case study of a VPP that analyzed medium voltage (MV) distribution networks. The real VPP consists of a 1.25 MW hydroelectric power plant (HPP) connected to a 0.5 MW battery energy storage system (ESS), which relies on a 20 kV distribution network connected to the HV/MV substation via an MV line (1-MV). The study was conducted by analyzing PQ measurement data from power quality recorders marked "R" in Figure 1. For further analysis, the research object in this article is divided into two subsections: the first investigation object and the second investigation object.

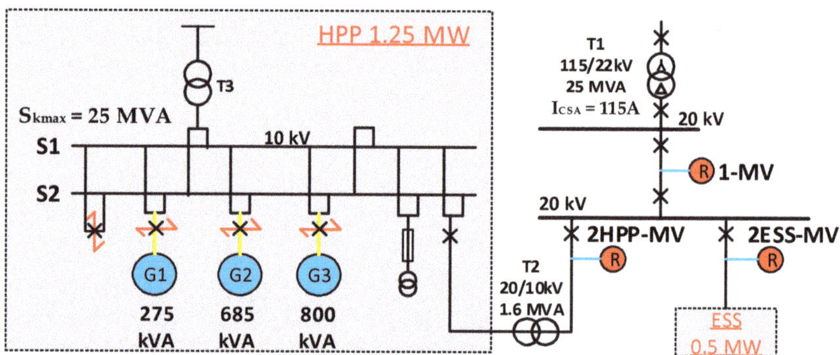

Figure 1. Investigated object of VPP with the location of PQ recorders, where 2HPP-MV: Hydro power plant with medium voltage, 2ESS-MV: Medium voltage energy storage system, and 1-MV: medium voltage line.

2.1.1. First Investigation Object

The first object of study includes the power quality measurement data from PQ recorders at the measurement points 2HPP-MV and 2EES-MV, presented in Figure 1. They are treated as one point but reported as two for PQ issues because the hydropower plant and the energy storage system are connected to one node, and their PQ recorders are connected to the same voltage transformer [32]. However, the active power level of 2ESS-MV and 2HPP-MV is measured separately.

2.1.2. Second Investigation Object

The second object of study has a wider scope than the previous object. It includes the PQ data of the first investigation object and additional PQ data from the PQ recorders of 1-MV, shown in Figure 1. For further analysis, the extension of the dataset into two sub-datasets is due to observation on how the different sizes and features of the input dataset will influence the performance of cluster analysis algorithms and on how the optimal number of clusters will be determined.

2.2. Parameter of Dataset Description

The most common power quality standards used nowadays [46] are IEC 61000-4-30 [18] and EN 50160 [43]. IEC 61000-4-30 [18] is an IEC standard that specifies procedures for measuring and interpreting power quality parameters in 50/60 Hz AC networks. This standard also describes measurement procedures to obtain reliable, repeatable, and comparable results with any compliant measuring instrument.

In this section, the PQ measurement parameters from all PQ recorders of the VPP were used as the input dataset corresponding to the demand of the standard IEC 61000-4-30 [18]. The global power quality index consists of a classical power quality measurement, and the active power level was included in the proposed dataset. The measurement from PQ recorders was taken for a 26-week period, which already went through the preprocessing stage to exclude the voltage events. Thus, the dataset of the VPP that was used has parameters from a 10-min interval of about 24,612 data points. Since there are two study objects in this article, the first study object and the second study object, the dataset is independently divided into two different datasets, as shown in Figure 2.

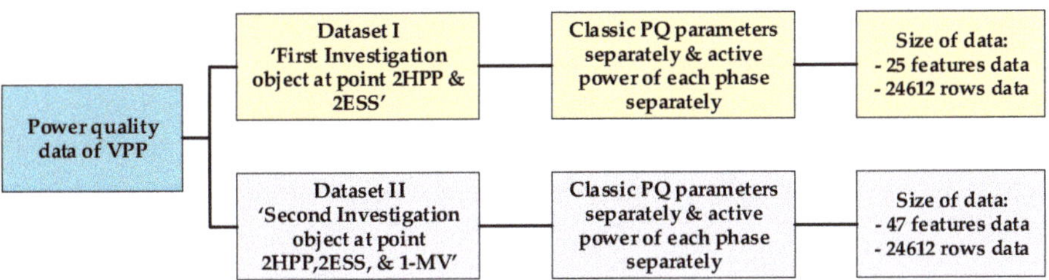

Figure 2. Proposed dataset schema: description of dataset separation, parameters, and size of data.

Dataset I was used for the first study object and contains classical 10-min PQ parameters and active power levels of PQ recorders at points 2HPP-MV and 2ESS-MV. The dataset has 25 features as input parameters and 24,612-row data, and it contains about 615,300 single cells. On the other hand, dataset II is larger than dataset I, and it contains classical 10-min PQ parameters and active power levels of PQ recorders at points 2HPP-MV, 2ESS-MV and 1-MV. This dataset is in the form of an array with 47 features as input parameters and 24,612-row data, and the size of the second dataset is 1,156,764 individual cells. Both the first and second datasets consist of classical power quality parameters, as shown below:

- Three phases of voltage;
- Three phases of 200 ms minimal voltage;
- Three phases of 200 ms maximal voltage;
- Voltage unbalance;
- Three phases of active power;
- Three phases of total harmonic distortion in voltage;
- Three phases of 200 ms maximum of total harmonic distortion in voltage.

2.3. Proposed Methodology

This section proposes cluster analysis as a data mining technique to group the data points based on their similarity [47]. The different algorithms were used to compare the grouping process and clustering result between hierarchical and non-hierarchical methods in this research approach. K-means is one of the algorithms from the non-hierarchical family, while the agglomerative technique is representative of the hierarchical method. The summary of processing steps to explore the dataset by using clustering techniques can be seen in the following schema in Figure 3.

The proposed methodology schema consists of five major steps: load dataset, feature engineering, applying clustering algorithm, qualitative assessment of PQ parameters, and the last is cluster algorithm evaluation and comparison. These five stages are described below:

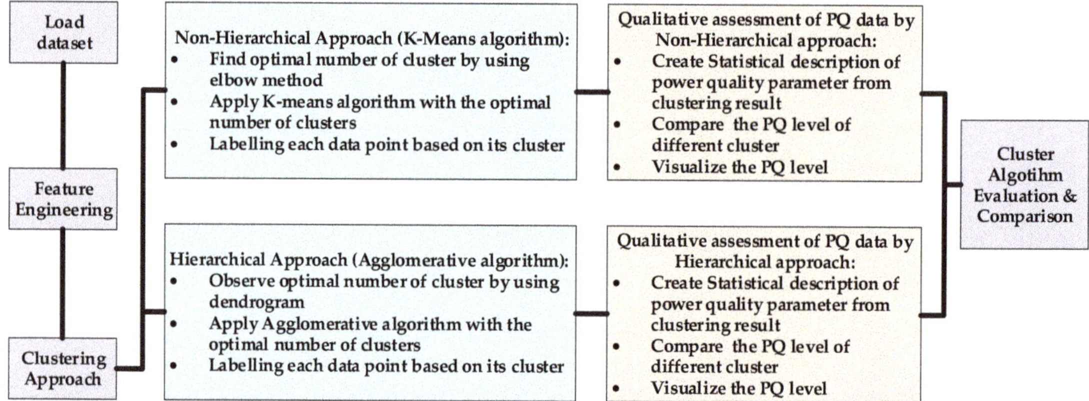

Figure 3. Proposed methodology schema.

2.3.1. Load Dataset

The data are obtained from reliable single or multiple sources. This step is crucial for clustering since the quantity and quality of the dataset affect the accuracy of the output [41]. The dataset used in this article contains power quality parameters without flagged data. In this work, two different datasets are loaded to conduct the research: dataset I was used for the first research object, and dataset II was used for the second research object.

2.3.2. Feature Engineering

Feature engineering is a process of preparing the right input dataset that is compatible with the requirements of the clustering analysis algorithm with the distance calculation [48], and it can improve the performance of the clustering analysis model [49]. In this stage, standardization is needed because the dataset contains features with different measurement unit scales, where these differences in the ranges of initial features cause trouble for the clustering algorithm. To prevent this problem, the standardization scaling technique by the Z-score method was used to transform the features to comparable scales. The authors used the Standard Scaler library from scikit-learn [50] and Pandas library [51] as tools to standardize the features of the dataset.

2.3.3. Clustering Approach

In this step, the K-means [52] and the agglomeration [53] methods are used to determine the cluster of the data points. To meet this requirement, the elbow method is proposed to obtain the optimal number of clusters in the dataset by measuring the value of the sum of squares within the clusters when using the K-means algorithm. When using the agglomeration algorithm, the optimal number of clusters does not need to be defined specifically, but the dendrogram graph was proposed to observe the optimal number of clusters for this method. After finding the optimal number of clusters for each algorithm, the authors applied the clustering algorithms to the dataset to assign each data point to belong to their clusters. The summary of the cluster analysis stage can be seen in Figure 3.

2.3.4. Qualitative Assessment

Qualitative assessment was required to analyze the PQ problems that belong to the cluster. To perform the power quality assessment, the data points were grouped based on their cluster, and the statistical value of each cluster was measured to observe the qualitative data analysis in each cluster and compare the average value of PQ parameters from each cluster. In this article, the standard limit based on the European standard EN

50160 [43] was used as an indicator to observe the visualization of the PQ level. The acceptable values based on the standard used for the global index are presented in Table 1.

Table 1. Voltage characteristic of standard EN 50160 [43] used for the global index.

Parameter	Voltage Characteristic
Voltage	10% of declared voltage
Short-term flicker severity	1.0
Total harmonic distortion in voltage	8%
Voltage unbalance	2%

2.3.5. Cluster Algorithm Evaluation and Comparison

Cluster evaluation determines how well the clustering algorithm can separate the dataset with the optimal number of clusters [54]. Since cluster analysis is not part of supervised learning and is performed as a part of exploratory data analysis, it is challenging to assess and evaluate the algorithm. However, there are several methods to define the validation of the clustering result. In this research, the Calinski–Harabasz index and silhouette score were used to evaluate the clustering algorithms. In this article, the metric was performed by using the scikit-learn library [44]. The silhouette function computes the mean silhouette coefficient of all samples based on the mean intra-cluster distance and the mean nearest-cluster distance for each sample. The formula is described mathematically [55,56]:

$$s = \frac{b - a}{max\,(a, b)} \tag{1}$$

where:

- s: the silhouette coefficient score;
- a: the mean distance between all other points and a sample in the same class;
- b: the mean distance between all other points and a sample in the next closest cluster.

The Calinski–Harabasz index, also known as the Variance Ratio Criterion, is the ratio of the sum of the dispersion between clusters and the dispersion between clusters for all clusters—the higher the value, the better the performance [57,58]. The math formula for this method is shown below [59]:

$$CH = \frac{SS_B}{SS_w}\,x\,\frac{N - k}{k - 1} \tag{2}$$

where:

- CH: the Calinski–Harabasz score;
- k: the number of clusters;
- N: the total number of observations (data points);
- SS_w: the overall within-cluster variance;
- SS_B: the overall between-cluster variance.

3. Result

This section presents the observation of the selected optimal number of clusters using the elbow method and a dendrogram for the first and second investigated object. Then, after applying the selected optimal number of clusters, the qualitative assessment was performed to compare the PQ level for different clusters. Finally, the clustering algorithm evaluation and comparison were calculated to determine which algorithm is suitable as a cluster analysis algorithm approach for the investigation objects.

3.1. Optimal Number of Clusters

The optimal number of clusters is a fundamental problem in partitioning clustering such as K-means clustering or hierarchical clustering [60]. It is necessary to know the number of clusters to achieve an optimal result. In this section, the optimal number of clusters of the first investigation object and second investigation object was performed using the elbow method provided by the Yellowbrick library [61] for machine learning visualization and the dendrogram provided by using SciPy library [62] for scientific computing.

The first investigation object, the elbow method associated with the K-means algorithm, was performed for the minimal number of clusters equal to 1, and the maximal is equal to 9. However, the analysis presents that the optimal number of clusters is equal to 3, shown in the elbow curve in Figure 4a, while on a hierarchical approach by using the agglomerative algorithm with a dendrogram, the optimal number of clusters is determined equal to 3, as presented in Figure 4b, after truncating the threshold distance at height 300 in the dendrogram.

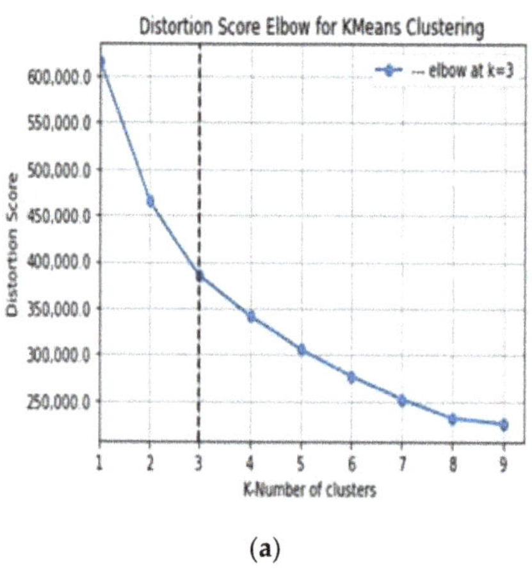

(**a**)

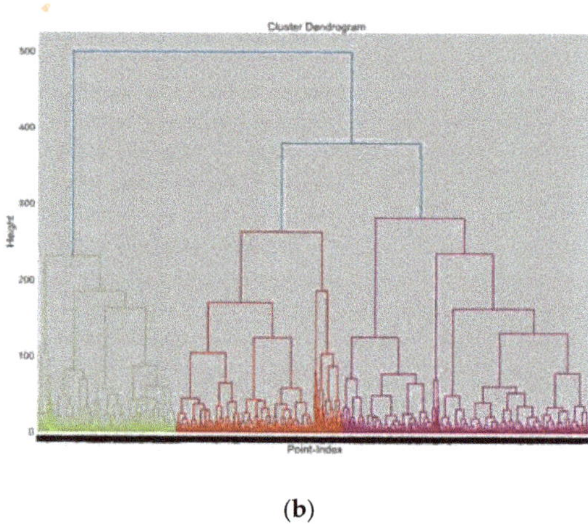

(**b**)

Figure 4. The optimal number of clusters for the first investigation object: the elbow method (**a**) shows that the optimal number of clusters is equal to 3, while on the dendrogram (**b**), the possibility of the optimal number of clusters is also equal to 3.

On the other hand, dataset II belongs to the second investigation object with a bigger input size than the previous object. By using the elbow method associated with the K-means algorithm as a non-hierarchical approach, the minimal number of clusters is equal to 1, and the maximal is equal to 9. However, the analysis shows that the optimal number of clusters is equal to 4, which can be seen in the elbow curve presented in Figure 5a. On the contrary, with the hierarchical approach using a dendrogram, which is shown in Figure 5b, the optimal number of clusters is equal to 3 after determining to truncate the threshold distance at height 400 in the dendrogram.

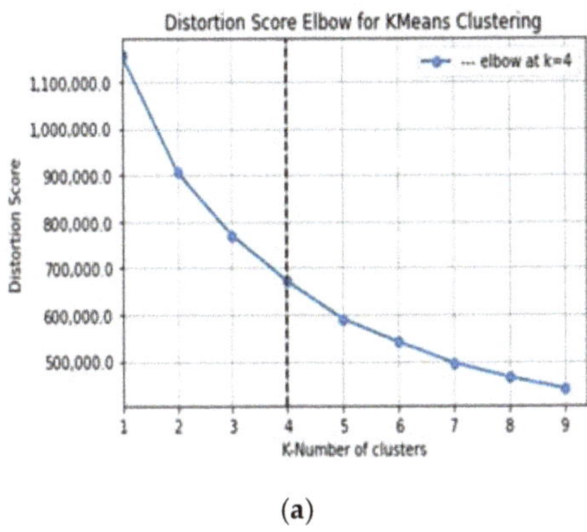

(a)

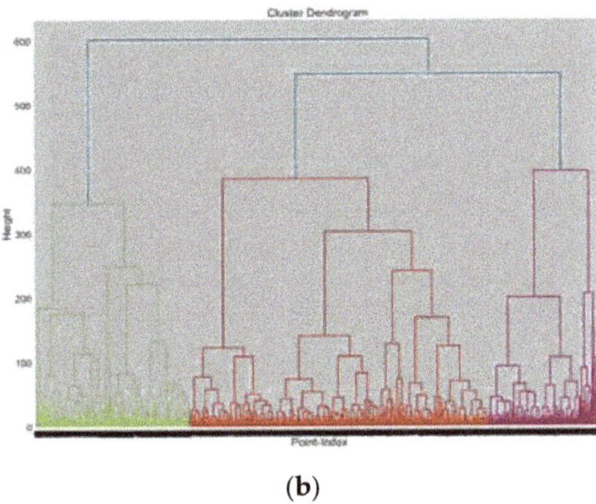

(b)

Figure 5. The optimal number of clusters for the second investigation object: the elbow method (**a**) shows that the optimal number of clusters is equal to 4, while on the dendrogram (**b**), the possibility of the optimal number of clusters is also equal to 3.

3.2. Qualitative Assessment of Clusters

As defined in the previous subsection, the optimal number of clusters has been determined. Thus, the qualitative assessment from the clustering result of the K-means algorithm approach and agglomerative algorithm approach is performed in this section. The authors deal with power quality parameters that are only considered in the elements of the power quality report [63]: short-term flicker magnitude and total harmonic distortion in voltage.

The assessment leads to comparing the average value of the selected parameters for different clusters based on the K-means clustering result and agglomerative clustering result to achieve knowledge about the PQ parameter level of the investigation object. The visualization of the PQ parameter was graphed on each cluster to observe the limit level of the PQ parameter through the standard EN 50160 [43,64].

3.2.1. Qualitative Assessment of the First Investigation Object

The qualitative assessment of the first investigation object, Table 2, shows the comparison of the statistical value of the PQ parameter level for different clusters by the K-means approach and agglomerative approach at measurement points 2HPP-MV and 2ESS-MV.

The clustering result by the K-means algorithm shows that the average value of cluster "1" is characterized by a medium level of Pst and a high level of THDu compared to other clusters. Cluster "2" is characterized by a high level of Pst and a low level of THDu compared to the other clusters. However, cluster "3" has a low level of Pst and a medium level of THDu.

On the other approach, by using an agglomerative algorithm, the comparison of the average value of the PQ level for different clusters presented in Table 2 indicates that cluster "1" represents a medium level of Pst and a high percentage level of THDu compared to others. Cluster "2" represents a medium level of THDu with a low level of Pst. Only cluster "3" has a high value for Pst and a low level of THDu.

Table 2. Comparison of PQ level for different clusters on the first investigation object.

| The PQ Parameter Level at Measurement Points 2HPP-MV and 2ESS-MV | | | | | | | | |
| Clustering Algorithm | Cluster | Value | Pst | | | THDu (%) | | |
			L1	L2	L3	L1	L2	L3
K-means Approach	Cluster 1	Mean	0.111	0.115	0.125	1.3	1.3	1.3
		Min	0.032	0.028	0.03	0.6	0.5	0.6
		Max	1.358	1.704	1.584	2.2	2.4	2.3
	Cluster 2	Mean	0.138	0.143	0.147	0.8	0.8	0.8
		Min	0.034	0.032	0.04	0.4	0.3	0.4
		Max	1.586	1.63	2.788	1.5	1.6	1.6
	Cluster 3	Mean	0.101	0.107	0.111	1	1	1
		Min	0.028	0.024	0.03	0.3	0.3	0.4
		Max	2.386	2.18	2.864	1.7	1.8	1.8
Agglomerative Approach	Cluster 1	Mean	0.104	0.111	0.119	1.158	1.108	1.142
		Min	0.028	0.028	0.03	0.36	0.375	0.403
		Max	0.36	0.352	0.412	2.234	2.384	2.319
	Cluster 2	Mean	0.096	0.1	0.104	0.994	1.007	1.031
		Min	0.028	0.024	0.03	0.317	0.308	0.378
		Max	0.814	0.658	0.82	1.907	1.938	1.99
	Cluster 3	Mean	0.159	0.163	0.167	0.905	0.886	0.894
		Min	0.04	0.038	0.036	0.366	0.345	0.351
		Max	2.386	2.18	2.864	1.846	2.011	1.914

The level of PQ parameters by the K-means algorithm approach. Cluster "2" represents the average of short-term flicker severity (Pst) in each phase at a higher level than the other clusters, while cluster "1" represents the highest average value of THDu in each phase compared to the other clusters. This can be seen clearly in Figure 6.

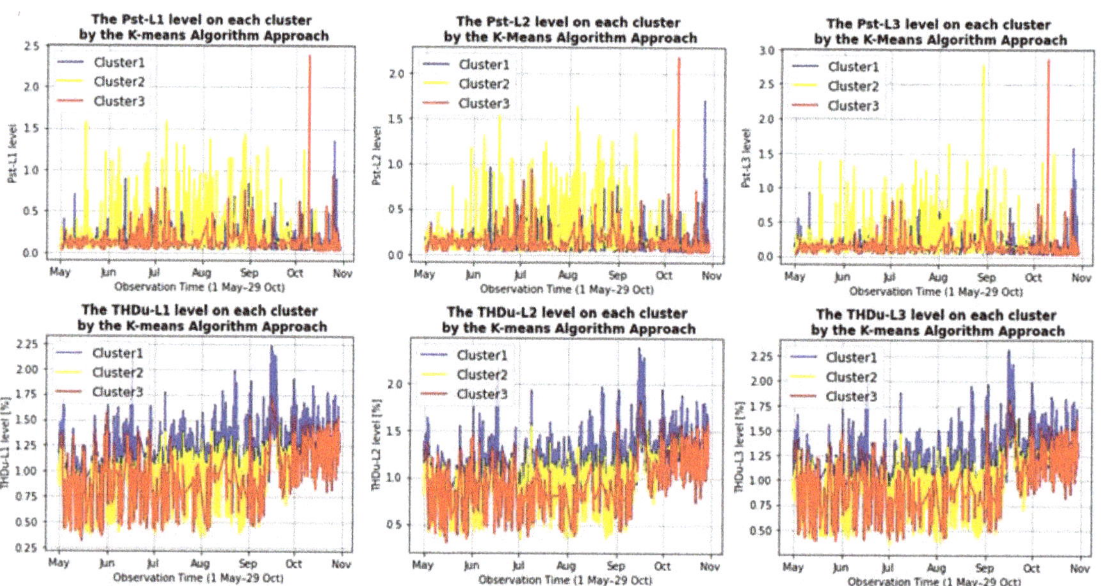

Figure 6. Visualization of PQ parameters' level of the first investigation object at measurement points 2HPP-MV and 2ESS-MV by the K-means algorithm approach.

Based on the standard EN 50160 [43], the visualization of the recorded set of PQ parameters of the first investigation object that is graphed in Figure 6 indicates that short-term flicker severity and total harmonic distortion in voltage on each cluster are under normal operating conditions due to the fact that all the levels of PQ parameters are below compatibility for 95% of the time.

The agglomerative algorithm approach shows that cluster "3" is indicated as the group with the highest average value of Pst in each phase compared to the other clusters, but in terms of THDu, cluster "1" has the highest average value over 1.1% among other clusters. This can be proven by the visualization of PQ parameters' level in Figure 7. According to the standard EN 50160 [43], the recorded set of THDu and Pst, which are shown in Figure 7, is below the limit level and working under normal operating conditions during any period.

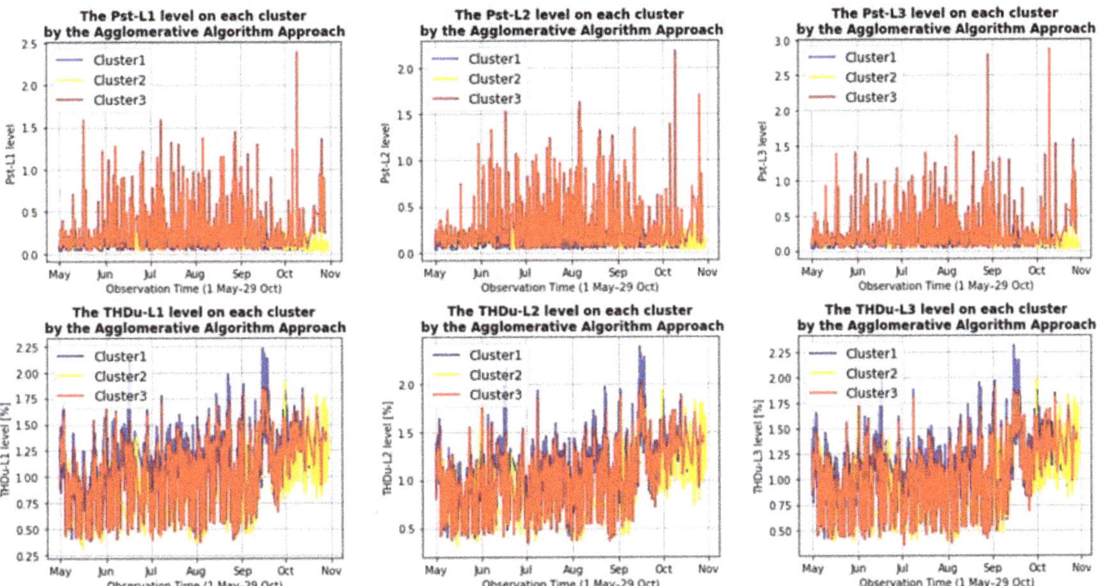

Figure 7. Visualization of PQ parameters' level of the first investigation object at measurement points 2HPP-MV and 2ESS-MV by the agglomerative algorithm approach.

3.2.2. Qualitative Assessment of the Second Investigation Object

Table 3 shows the comparison of the statistical value of the PQ level for different clusters by the K-means algorithm approach and agglomerative algorithm approach at the measurement points 2HPP-MV and 2ESS-MV and additionally at the measurement point 1-MV for the second investigation object.

With a non-hierarchical approach using the K-means algorithm, when examining the average values of the PQ level in certain clusters at the measurement point 1-MV, cluster "1" represents a low level of Pst but an intermediate level of THDu compared to other clusters. Cluster "2" represents a high level of Pst with a low level of THDu. Cluster "3" represents a low level of Pst with a high level of THDu. Cluster "4" indicates a medium level of Pst and THDu compared to other clusters. While the average comparison value of the PQ level at measurement points 2HPP-MV and 2ESS-MV shows that cluster "1" has a medium level of Pst and low level of THDu among other clusters, cluster "2" has a high level of Pst but a low level of THDu, and cluster "3" has a high level of THDu but a medium level of Pst. Cluster "4" has a low level of Pst and a high level of THDu compared to the other clusters.

Through a hierarchical approach, the agglomerative algorithm represents the comparative evaluation of the average value of the PQ level in specific clusters. At the measurement point 1-MV, it reveals that cluster "1" has a high value of Pst and a low value of THDu compared to the others. Cluster "2" shows a medium level of both the THDu value and

Pst value. Cluster "3" shows a low level for Pst and a high level for THDu compared to the other clusters. Regarding the comparative assessment of the average value of the PQ level in specific clusters at the measurement points 2HPP-MV and 2ESS-MV, it is found that only cluster "1" has a high Pst value with a low THDu value compared to the others, while cluster "2" is characterized by a medium value at the THDu level but still has a low Pst value. Cluster "3" is the opposite of cluster "1" as it has a high THDu value with a medium Pst value compared to the other clusters.

Table 3. Comparison of PQ level for different clusters on the second investigation object.

| The PQ Parameter Level at Measurement Point 1-MV | | | | | | | | |
| Clustering Algorithm | Cluster | Value | Pst | | | THDu (%) | | |
			L1	L2	L3	L1	L2	L3
K-means Approach	Cluster 1	Mean	0.064	0.06	0.058	0.933	0.878	0.913
		Min	0.006	0.008	0.006	0.308	0.357	0.4
		Max	1.074	1.012	1.154	1.553	1.498	1.422
	Cluster 2	Mean	0.121	0.113	0.114	0.778	0.785	0.803
		Min	0.012	0.014	0.014	0.293	0.354	0.409
		Max	3.47	2.864	3.47	1.587	1.584	1.56
	Cluster 3	Mean	0.065	0.064	0.061	1.382	1.338	1.33
		Min	0.004	0	0.006	0.61	0.519	0.671
		Max	0.87	0.968	0.844	2.218	2.252	2.191
	Cluster 4	Mean	0.07	0.068	0.066	1.121	1.082	1.112
		Min	0.006	0.002	0.004	0.351	0.397	0.446
		Max	1.07	1.226	1.11	1.755	1.743	1.727
Agglomerative Approach	Cluster 1	Mean	0.148	0.138	0.14	0.758	0.764	0.784
		Min	0.014	0.014	0.008	0.293	0.354	0.409
		Max	3.47	2.864	3.47	1.953	1.929	1.944
	Cluster 2	Mean	0.067	0.064	0.063	1.068	1.04	1.071
		Min	0.006	0.002	0.004	0.335	0.378	0.433
		Max	0.506	0.444	0.654	2.057	1.999	2.014
	Cluster 3	Mean	0.061	0.06	0.057	1.147	1.096	1.108
		Min	0.004	0	0.006	0.308	0.357	0.4
		Max	0.57	0.744	0.528	2.218	2.252	2.191
The PQ parameter level at measurement points 2HPP-MV and 2ESS-MV								
K-means Approach	Cluster 1	Mean	0.111	0.119	0.124	0.934	0.874	0.919
		Min	0.028	0.028	0.032	0.384	0.378	0.415
		Max	0.846	0.766	1.078	1.569	1.526	1.526
	Cluster 2	Mean	0.152	0.156	0.16	0.81	0.802	0.807
		Min	0.04	0.038	0.04	0.36	0.345	0.351
		Max	2.386	2.18	2.864	1.471	1.556	1.608
	Cluster 3	Mean	0.111	0.116	0.125	1.392	1.366	1.383
		Min	0.032	0.028	0.03	0.552	0.537	0.613
		Max	0.736	0.982	1.254	2.234	2.384	2.319
	Cluster 4	Mean	0.097	0.1	0.105	1.027	1.033	1.057
		Min	0.028	0.024	0.03	0.317	0.308	0.385
		Max	0.944	0.942	0.996	1.648	1.727	1.718
Agglomerative Approach	Cluster 1	Mean	0.171	0.174	0.179	0.792	0.779	0.786
		Min	0.04	0.038	0.04	0.366	0.345	0.351
		Max	2.386	2.18	2.864	1.801	1.865	1.914
	Cluster 2	Mean	0.097	0.101	0.105	0.991	1.004	1.026
		Min	0.028	0.024	0.03	0.317	0.308	0.378
		Max	0.452	0.398	0.602	1.907	1.938	1.99
	Cluster 3	Mean	0.11	0.117	0.124	1.159	1.11	1.14
		Min	0.028	0.028	0.03	0.388	0.391	0.418
		Max	0.54	0.742	0.552	2.234	2.384	2.319

Table 3 contains two sets of the statistical value of the PQ parameters at different measurement point locations. Therefore, the qualitative analysis is performed separately. Using the K-means algorithm approach, the group of average values at the measurement point 1-MV shows that cluster "2" has the highest average value of three-phase short-term flicker severity with a value above 0.11 for each phase, and cluster "3" is a group with a high value of THDu (these parameters are visualized in Figure 8a), while the analysis at the measurement points 2HPP-MV and 2ESS-MV shows that cluster "2" has the highest average value at the Pst level and cluster "3" has a high average value of THDu, as presented in Figure 8b. According to the standard EN 50160 [43], the visualization of the recorded set of PQ parameters on the second investigation object at measurement points 2HPP-MV and 2ESS-MV (a) and 1-MV (b) by the K-means approach indicates that short-term flicker severity and total harmonic distortion in voltage on each cluster are under normal operating conditions, due to the fact that all the levels of PQ parameters are below compatibility for 95% of the time.

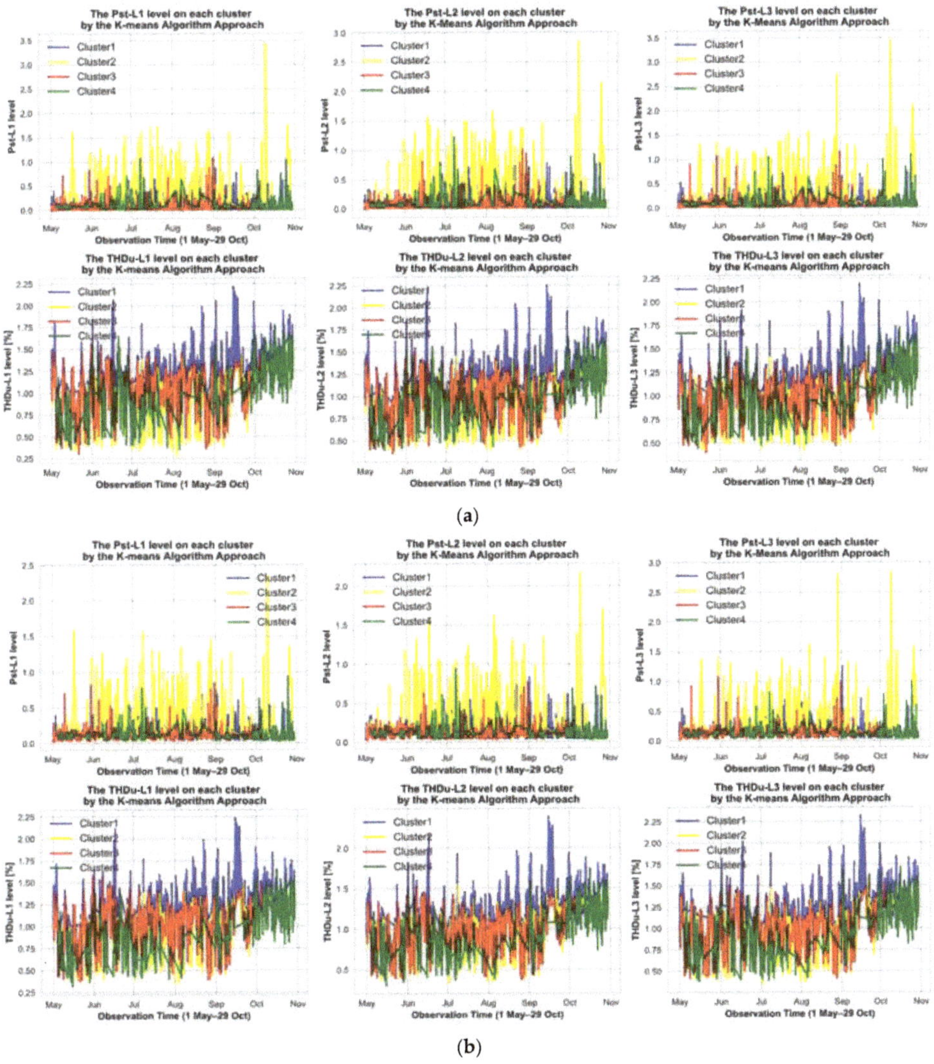

Figure 8. Visualization of PQ parameters' level of the second investigation object by the K-means algorithm approach: Part (**a**) is a PQ level's graph at measurement point 1-MV and part (**b**) is a PQ level's graph at measurement points 2HPP-MV and 2ESS-MV.

In terms of the agglomerative algorithm approach, the average value of PQ phenomena at the measurement point 1-MV shows that cluster "1" has the highest mean value of three-phase short-term flicker severity with an average value on each phase around 0.14, followed by cluster "2" and cluster "3". For THDu, cluster "3" represents the highest value among the other clusters. The PQ parameters' level at the measurement point 1-MV (a) can be seen in Figure 9. Therefore, the qualitative evaluation of the PQ data of the measurement points 2HPP-MV and 2ESS-MV shows that cluster "1" has the highest value for Pst compared to the other clusters, and cluster "3" is the group with a high value for THDu. This can be observed in the graph of measurement points 2HPP-MV and 2ESS-MV (b), as presented in Figure 9.

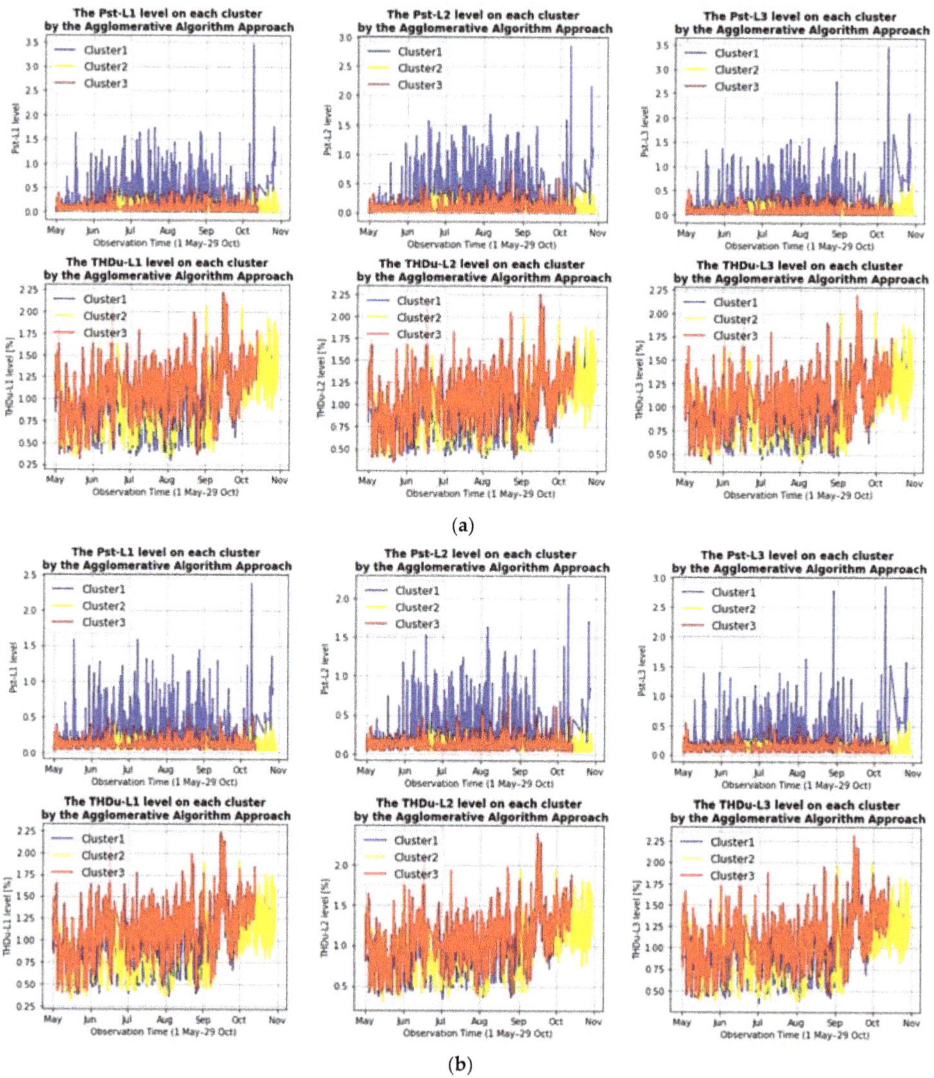

Figure 9. Visualization of PQ parameters' level of the second investigation object by the agglomerative algorithm approach: Part (**a**) is a PQ level's graph at measurement point 1-MV and part (**b**) is a PQ level's graph at measurement points 2HPP-MV and 2ESS-MV.

When referring to the standard EN 50160 [43], the visualization of the recorded set of PQ parameters' level at measurement point 2HPP-MV (b) and the measurement point 1-MV (a), presented in Figure 9, indicates that short-term flicker severity and total harmonic distortion in voltage on each cluster are under normal operating conditions, due to the fact that all the levels of PQ parameters are below compatibility for 95% of the time.

3.3. Cluster Algorithm Evaluation and Comparison

In the first study object, dataset I was used to observe the power quality parameter issues only at measurement points 2HPP-MV and 2ESS-MV. The proposed cluster analysis algorithms, K-means and the agglomerative method, are considered to conduct qualitative data analysis on power quality measurement data. In this section, the evaluation of the algorithms is performed to investigate how well the algorithms group the data points to belong to their clusters. The silhouette coefficient and Calinski–Harabasz index were used as the clustering performance evaluation metric.

The silhouette coefficient validates clustering performance by computing the pairwise difference between and within clusters. Here, the silhouette has a value that expresses how similar an object is to its cluster compared to other clusters. The coefficient varies between -1 and 1, where a low value or close to -1 indicates that the object is assigned to the wrong cluster. A high value or close to 1 means that the object is close to its cluster and matches well with its cluster [54,65]. At the same time, the Calinski–Harabasz index is an evaluation index based on the degree of dispersion between clusters. A higher value indicates that the clusters are dense and well-separated, which is the standard concept of a cluster [66].

Table 4 contains the optimal number of clusters determined by the elbow method and the dendrogram and the evaluation of clustering performance scores by the K-means and agglomeration algorithms performed in dataset I and dataset II.

Table 4. Clustering algorithm performance evaluation on the investigation object.

Object Dataset	Cluster Algorithm	Find Optimal Number Method	Optimal Number of Clusters	Clustering Performance Evaluation Metric	
				Silhouette Coefficient	Calinski–Harabasz Index
Dataset I	K-Means	Elbow method	3	0.235	7336.44
	Agglomerative	Dendrogram	3	0.201	5811.29
Dataset II	K-Means	Elbow method	4	0.213	5923.020
	Agglomerative	Dendrogram	3	0.219	4954.945

The silhouette coefficient score and Calinski–Harabasz index score indicate that the K-means algorithm performs better than the agglomerative algorithm to cluster the data points of dataset I, where the optimal number of clusters is equal to 3.

In the second object of study, dataset II was used to investigate the PQ disturbances at measurement points 2HPP-MV and 2ESS-MV with additional measurement point 1-MV. By the elbow method associated with the K-means algorithm, the optimal number of clusters is equal to 4, while the agglomerative algorithm determines that the optimal number of clusters is equal to 3, based on the consideration of the dendrogram. Due to the different values for the optimal number of clusters, it is challenging to compare these algorithms as they lead to a different view. However, according to the silhouette coefficient, the agglomerative algorithm performs well in separating the data points belonging to their cluster compared to the K-means algorithm. However, for the Calinski–Harabasz index, the K-means algorithm performs better than the agglomerative algorithm.

4. Discussion

This research aimed to apply clustering analysis as a data mining technique to study the qualitative assessment of the power quality level from a virtual power plant operating at medium voltage levels of the distribution network in Lower Silesia in Poland. The object of this study focuses on the analysis of the measurement data of three PQ recorders of the VPP. The measurement dataset consists of 10 min of classical power quality parameters with a measurement period of 26 weeks, recorded from 1 May to 29 October 2020. In previous research [32], the application of the global index was proposed to reduce the size of the input dataset from all PQ parameters to maintain the data features for cluster analysis. However, this article is a further development of the reference study [32]. This study only deals with analyzing a part of the measured data from PQ recorders to obtain comparative results between two clustering algorithms.

The object of study is divided into two sub-studies for further analysis. The first study object includes only the measurement data of PQ recorders at the measurement points 2HPP-MV and 2ESS-MV, which have 25 parameters, while the second study object for further analysis includes the first study object's data with the additional dataset of measurement point 1-MV. Hence, it has 45 features. The reason to extend the dataset into two sub-datasets is to observe how different sizes and features of the input of the dataset will influence the algorithm of cluster analysis to separate the data points and how the optimal number of clusters will be determined.

The measurement data coming from PQ recorders are not standardized. In the first attempt to perform cluster analysis without scaling the dataset, the optimal number of clusters from both algorithms was inconsistent. For this reason, the authors are concerned with the scaling of measurement data parameters to maintain the consistency and regularity of the value between features, and the standardization of the dataset can also improve the performance of the clustering analysis model.

Determining the optimal number of clusters is difficult because there is no specific answer to this question [67,68]. Prior domain knowledge can help choose the optimal number of clusters [40], but when working with an unknown dataset, it is necessary to study the dataset to determine the number of clusters [69]. In this article, the elbow method and the dendrogram are proposed to observe the relationship between the data. The elbow method is related to the K-means algorithm, and the dendrogram refers to the application of the agglomerative hierarchical algorithm. All clustering algorithms were built on Python [45] using Visual Studio Code IDE [70]. The machine learning library was used to run the K-means algorithm and the agglomerative algorithm is from scikit-learn [44]. The computer used for this experiment has a processor with a specification of intel® Core™ i5-4300U CPU@1.90 GHz~2.49 GHz, with 8 GB RAM and a total memory 2160 MB graphics card from intel® HD Graphics Family installed. The operating system of the computer is Windows 10 Pro 64-bit.

The experimental results of this study show that the execution of the dendrogram method provided by the SciPy library [62] required a long processing time of about 1350.28 s for dataset I, which belongs to the first study object, and for dataset II, which belongs to the second study object, the execution time was about 1421.64 s. This method is difficult to compute, and a crowded graph also makes it challenging to obtain the specific number of clusters.

On the other hand, the elbow method provided by Yellowbrick [61] requires the fast computation of the algorithm to calculate the sum of the squared distance between each point and the centroid in a cluster (WCSS). The execution time of the elbow method for dataset I, which belongs to the first study object, was about 13.33 s, and for dataset II, which belongs to the second study object, it was 18.08 s.

A comparison cluster analysis between the two algorithms reveals that in the first object of study, using the K-means algorithm associated with the elbow method, the optimal number of clusters is determined to equal 3, while the hierarchical relationship between the objects in the dendrogram shows that the optimal number of clusters is also equal to

3, which is suitable for the agglomerative algorithm. On the other hand, for the second study item, there was a shift in the value of the optimal number of clusters by the elbow method from 3 to 4. The current study shows that the additional features of the dataset can change the interpretation of the K-means algorithm for grouping data points based on their centroid. However, by using the dendrogram, the authors were able to decide on the optimal number of clusters, which was still set equal to 3 for the dataset of the second study object.

There are some advantages and disadvantages between the two methods presented. The elbow method with K-means provides a fast calculation, but it gives an inconsistent value of the optimal number when there are additional parameters or features, while the dendrogram for the agglomerative algorithm is hard to calculate but still produces a consistent result even if there are additional parameters or features.

5. Conclusions

In this article, a case study of the cluster analysis technique for analyzing the power quality level of a virtual power plant was presented. The algorithm was developed to compare the qualitative assessment of the dataset. The input dataset was standardized to have the same scale between the different feature units to achieve a compatible input for the cluster analysis algorithm. The optimal number of clusters was still difficult to choose because the additional dataset parameters may affect the performance of the clustering algorithm. The result shows that the K-means algorithm is a simple algorithm that allows fast computation, while the agglomerative algorithm is the opposite. In order to run the K-means algorithm, the specific number of clusters must be determined first, which is a disadvantage of this algorithm. On the other hand, there is no specific number of clusters required for the agglomerative algorithm, which must be determined first.

The study at the first object investigation shows that there is a possibility that the optimal number of clusters is equal to 3, which determines both the elbow method and the dendrogram. The comparison result of cluster algorithm evaluation in this object study shows that the K-means algorithm works better than the agglomerative algorithm to separate the data points of dataset I, with a higher score of silhouette coefficient and Calinski–Harabasz index of 0.235 and 7336.49, respectively.

When examining the second object, the optimal number of clusters for the two algorithms is different. The elbow method defines 4 as the optimal number of clusters for the K-means algorithm, while by the dendrogram, it is found that 3 is the optimal number of clusters for the agglomerative algorithm. Moreover, due to the different values of the optimal number, it is not fair to compare the cluster evaluation results between the two algorithms since the score is also generated differently. For the silhouette coefficient, the agglomerative algorithm performs better than the K-means algorithm. However, based on the Calinski–Harabasz index, the K-means algorithm performs better than the agglomerative algorithm.

A qualitative assessment was performed to define the comparison of statistical data analysis between clusters to determine the characteristics of the power quality parameters. It is necessary to use the standard EN 50160 [43] to consider the work of the electrical distribution network in normal or abnormal operating conditions. The cluster analysis results show that all PQ parameters on each cluster for both the first study object and the second study object indicate that all PQ levels are still below the limit value and work under normal operating conditions.

Author Contributions: Conceptualization, M.J. and T.S.; methodology, F.A., M.J. and T.S.; software, F.A., M.J. and T.S.; validation, D.K., J.R. and V.S.; formal analysis, F.A., M.J., T.S., D.K. and J.R.; investigation, F.A., M.J. and T.S.; resources, P.K. and P.J.; data curation, F.A., V.S. and J.S.; writing—original draft preparation, F.A. and M.J.; writing—review and editing, T.S., D.K. and J.R.; visualization, F.A., D.K. and J.R.; supervision, M.J., T.S., J.R. and Z.L.; project administration, T.S. and P.J.; funding acquisition, T.S. and P.J. All authors have read and agreed to the published version of the manuscript.

Funding: This research was funded by the National Center of Research and Development in Poland, the project "Developing a platform for aggregating generation and regulatory potential of dispersed renewable energy sources, power retention devices and selected categories of controllable load" supported by European Union Operational Programme Smart Growth 2014–2020, Priority Axis I: Supporting R&D carried out by enterprises, Measure 1.2: Sectoral R&D Programmes, POIR.01.02.00-00-0221/16, performed by TAURON Ekoenergia Ltd.

Institutional Review Board Statement: Not applicable.

Informed Consent Statement: Not applicable.

Data Availability Statement: The data are not publicly available due to the policy of the associate company.

Acknowledgments: The authors would like to acknowledge the support of Tauron Ekoenergia Ltd.

Conflicts of Interest: The authors declare no conflict of interest.

References

1. Hong, Y.-Y. Electric Power Systems Research. *Energies* **2016**, *9*, 824. [CrossRef]
2. Atputharajah, A.; Ramachandaramurthy, V.K.; Pasupuleti, J. Power Quality Problems and Solutions. In Proceedings of the IOP Conference Series Earth and Environmental Science, Putrajaya, Malaysia, 5–6 March 2013; Volume 16, p. 012153. [CrossRef]
3. Naik, C.A.; Kundu, P. Identification of Short Duration Power Quality Disturbances Employing S-Transform. In Proceedings of the 2011 International Conference on Power and Energy Systems, Chennai, India, 22–24 December 2011; pp. 1–5.
4. Mekhamer, S.F.; Abdelaziz, A.Y.; Ismael, S.M. Design Practices in Harmonic Analysis Studies Applied to Industrial Electrical Power Systems. *Eng. Technol. Appl. Sci. Res.* **2013**, *3*, 467–472. [CrossRef]
5. More, T.G.; Asabe, P.R.; Chawda, S. Power Quality Issues and It's Mitigation Techniques. *Int. J. Eng. Res. Appl.* **2014**, *4*, 8.
6. Yadav, J.R.; Vasudevan, K.; Kumar, D.; Shanmugam, P. Power Quality Assessment for Industrial Plants: A Comparative Study. In Proceedings of the 2019 IEEE 13th International Conference on Compatibility, Power Electronics and Power Engineering (CPE-POWERENG), IEEE, Sonderborg, Denmark, 23–25 April 2019; pp. 1–6.
7. Jena, R. Electrical Power Quality. *Dep. Electr. Eng. CET BBSR 66.* Available online: https://www.cet.edu.in/noticefiles/227_Electrical_Power_Quality-PEEL5403-8th_Sem-Electrical.pdf (accessed on 10 June 2021).
8. Crotti, G.; Giordano, D.; D'Avanzo, G.; Femine, A.D.; Gallo, D.; Landi, C.; Luiso, M.; Letizia, P.S.; Barbieri, L.; Mazza, P.; et al. Measurement of Dynamic Voltage Variation Effect on Instrument Transformers for Power Grid Applications. In Proceedings of the 2020 IEEE International Instrumentation and Measurement Technology Conference (I2MTC), Dubrovnik, Croatia, 25–28 May 2020; pp. 1–6.
9. Barros, J.; Pérez, E.; Diego, R.I. Measurement and Analysis of Voltage Events. In *Power Quality*; Moreno-Muñoz, A., Ed.; Power Systems; Springer: London, UK, 2007; pp. 73–102. ISBN 978-1-84628-771-8.
10. Tur, M.R.; Bayindir, R. Comparison of Power Quality Distortion Types and Methods Used in Classification. In Proceedings of the 2020 International Conference on Computational Intelligence for Smart Power System and Sustainable Energy (CISPSSE), Odisha, India, 29–31 July 2020; pp. 1–7.
11. Syakur, M.A.; Khotimah, B.K.; Rochman, E.M.S.; Satoto, B.D. Integration K-Means Clustering Method and Elbow Method for Identification of the Best Customer Profile Cluster. In Proceedings of the IOP Conference Series Earth and Environmental Science, Banda Aceh, Indonesia, 26–27 September 2018; Volume 336, p. 012017. [CrossRef]
12. Lin, S.; Xie, C.; Tang, B.; Liu, R.; Pan, A. The Data Mining Application in the Power Quality Monitoring Data Analysis. In Proceedings of the 2016 IEEE 11th Conference on Industrial Electronics and Applications (ICIEA), Hefei, China, 5–7 June 2016; pp. 338–342.
13. Sangepu, R. Effect of Power Quality Issues in Power System and Its Mitigation by Power Electronics Devices. 2015. Available online: https://www.researchgate.net/publication/325676538_Effect_of_Power_Quality_Issues_in_Power_System_and_Its_Mitigation_by_Power_Electronics_Devices (accessed on 6 June 2021).
14. Zakaria, M.F.; Ramachandaramurthy, V.K. Assessment and Mitigation of Power Quality Problems for Puspati Triga Reactor (RTP). *J. Appl. Phys.* **2017**, 020011. [CrossRef]
15. Gul, O. An Assessment of Power Quality and Electricity Consumer's Rights in Restructured Electricity Market in Turkey. *Electric. Power Qual. Utilis. J.* **2008**, *14*, 29–34.
16. Mindykowski, J. Fundamentals of Electrical Power Quality Assessment. 2003. Available online: https://www.imeko.org/publications/wc-2003/PWC-2003-TC4-027.pdf (accessed on 20 June 2021).
17. Batkiewicz-Pantula, M. The Problem of Selected Parameters of the Power Quality in the Perspective of Tightening Normative Requirements. In Proceedings of the 2019 Modern Electric Power Systems (MEPS), Wroclaw, Poland, 9 September 2019; pp. 1–4.
18. Legarreta, A.E.; Figueroa, J.H.; Bortolin, J.A. An IEC 61000-4-30 Class a-Power Quality Monitor: Development and Performance Analysis. In Proceedings of the 11th International Conference on Electrical Power Quality and Utilisation, Barcelona, Spain, 9–11 October 2011; pp. 1–6.

19. What Is the IEC61000-4-30 Standard for Power Quality Analysers? *Power Qual. Anal.* Available online: https://www.fluke.com/en-gb/learn/blog/power-quality/what-does-the-iec-61000-4-30-class-a-standard-mean-to-me (accessed on 15 June 2021).
20. Bollen, M.H.J.; Milanović, J.V.; Čukalevski, N. CIGRE/CIRED JWG C4.112-Power Quality Monitoring. *Renew. Energy Power Qual. J.* **2014**. [CrossRef]
21. Ferracci, P. Cahier Technique No. 199 Power Quality. *Power Qual.* 2000. Available online: https://eduscol.education.fr/sti/sites/eduscol.education.fr.sti/files/ressources/techniques/3361/3361-ect199.pdf (accessed on 14 July 2021).
22. Karabiber, A. Controllable AC/DC Integration for Power Quality Improvement in Microgrids. *Adv. Electr. Comput. Eng.* **2019**, *19*, 97–104. [CrossRef]
23. Vokas, G.A.; Langouranis, P.A.; Kontaxis, P.A.; Topalis, F.V. Analysis of Power Quality Field Measurements and Considerations on the Power Quality Standard 14. Available online: https://www.researchgate.net/publication/312031589_Analysis_of_power_quality_field_measurements_and_considerations_on_the_power_quality_standard (accessed on 21 June 2021).
24. Sezi, I.T.; Zimmer, I.K.; Lang, J. Power Quality Monitoring and Analysis System. In Proceedings of the 18th International Conference and Exhibition on Electricity Distribution (CIRED 2005), Turin, Italy, 15–18 June 2005; Volume 2005, p. v2-62-v2-62.
25. Nourollah, S.; Moallem, M. A data mining method for obtaining global power quality index. In Proceedings of the 2011 2nd International Conference on Electric Power and Energy Conversion Systems (EPECS), Sharjah, United Arab Emirates, 15–17 November 2011. [CrossRef]
26. Asheibi, A.; Stirling, D.; Robinson, D. Identification of Load Power Quality Characteristics Using Data Mining. In Proceedings of the 2006 Canadian Conference on Electrical and Computer Engineering, Ottawa, ON, Canada, 7–10 May 2006; pp. 157–162.
27. Larose, D.T. *Data Mining Methods and Models*; Wiley: New York, NY, USA, 2006.
28. Larose, D.T. *Discovering Knowledge in Data: An Introduction to Data Mining*; Wiley-Interscience: Hoboken, NJ, USA, 2005; ISBN 978-0-471-66657-8.
29. Ghavidel, S.; Li, L.; Aghaei, J.; Yu, T.; Zhu, J. A Review on the Virtual Power Plant: Components and Operation Systems. In Proceedings of the 2016 IEEE International Conference on Power System Technology (POWERCON), Wollongong, Australia, 28 September–1 October 2016; pp. 1–6.
30. Taylor, K. Oracle Data Mining Concepts 11g Release 2 (11.2). *Doc. E1680807 Oracle 2013.* Available online: https://docs.oracle.com/cd/E11882_01/datamine.112/e16808/title.htm (accessed on 14 July 2021).
31. Ullman, S.; Poggio, T.; Harari, D.; Zysman, D.; Seibert, D. Unsupervised Learn. Slides 2014, Fall 2014 Lecture 13. Available online: http://www.mit.edu/~{}9.54/fall14/Classes/class13.html (accessed on 24 August 2021).
32. Jasiński, M.; Sikorski, T.; Kaczorowska, D.; Rezmer, J.; Suresh, V.; Leonowicz, Z.; Kostyła, P.; Szymańda, J.; Janik, P.; Bieńkowski, J.; et al. A Case Study on Data Mining Application in a Virtual Power Plant: Cluster Analysis of Power Quality Measurements. *Energies* **2021**, *14*, 974. [CrossRef]
33. Dandea, V.; Grigoras, G.; Neagu, B.-C.; Scarlatache, F. K-Means Clustering-Based Data Mining Methodology to Discover the Prosumers' Energy Features. In Proceedings of the 2021 12th International Symposium on Advanced Topics in Electrical Engineering (ATEE), Bucharest, Romania, 25 March 2021; pp. 1–5.
34. Jasiński, M.; Sikorski, T.; Kaczorowska, D.; Rezmer, J.; Suresh, V.; Leonowicz, Z.; Kostyła, P.; Szymańda, J.; Janik, P.; Bieńkowski, J.; et al. A Case Study on a Hierarchical Clustering Application in a Virtual Power Plant: Detection of Specific Working Conditions from Power Quality Data. *Energies* **2021**, *14*, 907. [CrossRef]
35. Jasiński, M.; Sikorski, T.; Leonowicz, Z.; Borkowski, K.; Jasińska, E. The Application of Hierarchical Clustering to Power Quality Measurements in an Electrical Power Network with Distributed Generation. *Energies* **2020**, *13*, 2407. [CrossRef]
36. Neagu, B.-C.; Grigoras, G. A Fair Load Sharing Approach Based on Microgrid Clusters and Transactive Energy Concept. In Proceedings of the 2020 12th International Conference on Electronics, Computers and Artificial Intelligence (ECAI), Bucharest, Romania, 25–27 June 2020; pp. 1–4.
37. Jasilski, M.; Borkowski, K.; Sikorski, T.; Kostvla, P. Cluster Analysis for Long-Term Power Quality Data in Mining Electrical Power Network. In Proceedings of the 2018 Progress in Applied Electrical Engineering (PAEE), Koscielisko, Poland, 18–22 June 2018; pp. 1–5.
38. Jureedi, N.V.V. Karunakar.; Rosalina, K.M.; Prema Kumar, N. Clustering Analysis and Its Application in Electrical Distribution System. *Int. J. Recent Adv. Eng. Technol.* **2020**, *8*, 38–43. [CrossRef]
39. Determining the Optimal Number of Clusters: 3 Must Know Methods. Available online: https://www.datanovia.com/en/lessons/determining-the-optimal-number-of-clusters-3-must-know-methods/ (accessed on 22 June 2021).
40. Patil, C.; Baidari, I. Estimating the Optimal Number of Clusters k in a Dataset Using Data Depth. *Data Sci. Eng.* **2019**, *4*, 132–140. [CrossRef]
41. Aksan, F.F.; Azizah, A.; Prihastomo, E.D. Prediction of Earthquake Magnitude Based on the Clusters in Sulawesi Island, Indonesia. *Int. J. Sci. Res.* **2021**, *7*, 7.
42. Umargono, E.; Suseno, J.E.; Vincensius Gunawan, S.K. K-Means Clustering Optimization Using the Elbow Method and Early Centroid Determination Based on Mean and Median Formula. In *Proceedings of the 2nd International Seminar on Science and Technology (ISSTEC 2019), Yogyakarta, Indonesia, 25 November 2019*; Atlantis Press: Yogyakarta, Indonesia, 2020.
43. EN 50160: Voltage Characteristics of Electricity Supplied by Public Distribution Network. Available online: https://orgalim.eu/position-papers/en-50160-voltage-characteristics-electricity-supplied-public-distribution-system (accessed on 22 June 2021).

44. Scikit-Learn: Machine Learning in Python—Scikit-Learn 0.24.2 Documentation. Available online: https://scikit-learn.org/stable/ (accessed on 21 June 2021).
45. Welcome to Python.Org. Available online: https://www.python.org/ (accessed on 4 May 2021).
46. What Are the Standards for Power Quality Measurements?—Power Quality Analysers. Available online: https://powerqualityanalysers.com/knowledgebase/what-are-the-standards-for-power-quality-measurements/ (accessed on 24 August 2021).
47. Kassambara, A. *Multivariate Analysis 1: Practical Guide To Cluster Analysis in R*, 1st ed.; CreateSpace Independent Publishing Platform: Scotts Valley, CA, USA, 2015; Volume 1, ISBN 1542462703.
48. Rençberoğlu, E. Fundamental Techniques of Feature Engineering for Machine Learning. Available online: https://towardsdatascience.com/feature-engineering-for-machine-learning-3a5e293a5114 (accessed on 19 April 2021).
49. Aggarwal, C.C. *Data Mining*; Springer International Publishing: Cham, Germany, 2015; ISBN 978-3-319-14141-1.
50. Sklearn Preprocessing StandardScaler—Scikit-Learn 0.24.2 Documentation. Available online: https://scikit-learn.org/stable/modules/generated/sklearn.preprocessing.StandardScaler.html (accessed on 16 June 2021).
51. Pandas-Python Data Analysis Library. Available online: https://pandas.pydata.org/ (accessed on 16 June 2021).
52. Sklearn Cluster KMeans—Scikit-Learn 0.24.2 Documentation. Available online: https://scikit-learn.org/stable/modules/generated/sklearn.cluster.KMeans.html (accessed on 30 April 2021).
53. Sklearn Cluster AgglomerativeClustering—Scikit-Learn 0.24.2 Documentation. Available online: https://scikit-learn.org/stable/modules/generated/sklearn.cluster.AgglomerativeClustering.html (accessed on 16 June 2021).
54. Manimaran Clustering Evaluation Strategies. Available online: https://towardsdatascience.com/clustering-evaluation-strategies-98a4006fcfc (accessed on 16 June 2021).
55. Sklearn Metrics Silhouette_score—Scikit-Learn 0.24.2 Documentation. Available online: https://scikit-learn.org/stable/modules/generated/sklearn.metrics.silhouette_score.html (accessed on 14 June 2021).
56. Nanjundan, S.; Sankaran, S.; Arjun, C.R.; Anand, G.P. Identifying the Number of Clusters for K-Means: A Hypersphere Density Based Approach. Available online: https://arxiv.org/abs/1912.00643 (accessed on 22 June 2021).
57. Wei, H. How to Measure Clustering Performances When There Are No Ground Truth? Available online: https://medium.com/@haataa/how-to-measure-clustering-performances-when-there-are-no-ground-truth-db027e9a871c (accessed on 14 June 2021).
58. Milligan, G.W.; Cooper, M.C. An Examination of Procedures for Determining the Number of Clusters in a Data Set. Available online: https://link.springer.com/article/10.1007/BF02294245 (accessed on 22 June 2021).
59. Calinski-Harabasz Index and Boostrap Evaluation with Clustering Methods. Available online: https://ethen8181.github.io/machine-learning/clustering_old/clustering/clustering.html (accessed on 23 June 2021).
60. Kutbay, U. Partitional Clustering. In *Recent Applications in Data Clustering*; Pirim, H., Ed.; InTech: Nappanee, IN, USA, 2018; ISBN 978-1-78923-526-5.
61. Yellowbrick: Machine Learning Visualization—Yellowbrick v1.3.Post1 Documentation. Available online: https://www.scikit-yb.org/en/latest/ (accessed on 16 June 2021).
62. SciPy.Org. Available online: https://www.scipy.org/ (accessed on 16 June 2021).
63. Janik, P.; Sikorski, T. *Control in Electrical Power Engineering*; Wiley: New York, NY, USA, 2009; Volume 168, pp. 1–65.
64. Markiewicz, H. 5.4.2 Standard EN 50160 Voltage Characteristics in Public Distribution Systems. Available online: http://copperalliance.org.uk/uploads/2018/03/542-standard-en-50160-voltage-characteristics-in.pdf (accessed on 22 June 2021).
65. Prabhu, P. *Method for Determining Optimum Number of Clusters*; Social Science Research Network: Rochester, NY, USA, 2012.
66. Wang, X.; Xu, Y. An Improved Index for Clustering Validation Based on Silhouette Index and Calinski-Harabasz Index. *IOP Conf. Ser. Mater. Sci. Eng.* **2019**, *569*, 052024. [CrossRef]
67. Subbalakshmi, C.; Krishna, G.R.; Rao, S.K.M.; Rao, P.V. A Method to Find Optimum Number of Clusters Based on Fuzzy Silhouette on Dynamic Data Set. *Procedia Comput. Sci.* **2015**, *46*, 346–353. [CrossRef]
68. Zhou, S.; Xu, Z.; Liu, F. Method for Determining the Optimal Number of Clusters Based on Agglomerative Hierarchical Clustering. *IEEE Trans. Neural Netw. Learn. Syst.* **2017**, *28*, 3007–3017. [CrossRef] [PubMed]
69. Rahman, M.M.; Masud, M.d.A.; Mazumder, B. Estimation of the Number of Clusters Based on Simplical Depth. In Proceedings of the 2020 2nd International Conference on Sustainable Technologies for Industry 4.0 (STI), Dhaka, Bangladesh, 19 December 2020; pp. 1–5.
70. Visual Studio Code-Code Editing. Redefined. Available online: https://code.visualstudio.com/ (accessed on 15 July 2021).

Article

Adaptive Neuro-Fuzzy Inference System-Based Maximum Power Tracking Controller for Variable Speed WECS

Abrar Ahmed Chhipa [1,*], Vinod Kumar [1], Raghuveer Raj Joshi [1], Prasun Chakrabarti [2], Michal Jasinski [3], Alessandro Burgio [4], Zbigniew Leonowicz [3], Elzbieta Jasinska [5], Rajkumar Soni [2] and Tulika Chakrabarti [6]

[1] College of Technology and Engineering, Maharana Pratap University of Agriculture and Technology, Udaipur 313001, Rajasthan, India; vinodcte@yahoo.co.in (V.K.); rrJoshi_iitd@yahoo.com (R.R.J.)
[2] Techno India NJR Institute of Technology, Udaipur 313003, Rajasthan, India; drprasun.cse@gmail.com (P.C.); rajkumarsoni16@gmail.com (R.S.)
[3] Faculty of Electrical Engineering, Wroclaw University of Science and Technology, 50-370 Wroclaw, Poland; michal.jasinski@pwr.edu.pl (M.J.); zbigniew.leonowicz@pwr.edu.pl (Z.L.)
[4] Independent Researcher, 87036 Rende, Italy; alessandro.burgio.phd@gmail.com
[5] Department of Operations Research and Business Intelligence, Wrocław University of Science and Technology, 50-370 Wrocław, Poland; elzbieta.jasinska@pwr.edu.pl
[6] Department of Basic Sciences, Sir Padampat Singhania University, Udaipur 313601, Rajasthan, India; tulika.chakrabarti20@gmail.com
[*] Correspondence: abrar0613@gmail.com

Citation: Chhipa, A.A.; Kumar, V.; Joshi, R.R.; Chakrabarti, P.; Jasinski, M.; Burgio, A.; Leonowicz, Z.; Jasinska, E.; Soni, R.; Chakrabarti, T. Adaptive Neuro-Fuzzy Inference System-Based Maximum Power Tracking Controller for Variable Speed WECS. *Energies* **2021**, *14*, 6275. https://doi.org/10.3390/en14196275

Academic Editors: Adrian Ilinca and Surender Reddy Salkuti

Received: 21 August 2021
Accepted: 29 September 2021
Published: 1 October 2021

Publisher's Note: MDPI stays neutral with regard to jurisdictional claims in published maps and institutional affiliations.

Abstract: This paper proposes an adaptive neuro-fuzzy inference system (ANFIS) maximum power point tracking (MPPT) controller for grid-connected doubly fed induction generator (DFIG)-based wind energy conversion systems (WECS). It aims at extracting maximum power from the wind by tracking the maximum power peak regardless of wind speed. The proposed MPPT controller implements an ANFIS approach with a backpropagation algorithm. The rotor speed acts as an input to the controller and torque reference as the controller's output, which further inputs the rotor side converter's speed control loop to control the rotor's actual speed by adjusting the duty ratio for the rotor side converter. The grid partition method generates input membership functions by uniformly partitioning the input variable ranges and creating a single-output Sugeno fuzzy system. The neural network trained the fuzzy input membership according to the inputs and alter the initial membership functions. The simulation results have been validated on a 2 MW wind turbine using the MATLAB/Simulink environment. The controller's performance is tested under various wind speed circumstances and compared with the performance of a conventional proportional–integral MPPT controller. The simulation study shows that WECS can operate at its optimum power for the proposed controller's wide range of input wind speed.

Keywords: ANFIS; fuzzy logic; induction generator; MPPT; neural network; renewable energy; variable speed WECS; wind energy conversion system; wind energy

1. Introduction

Electricity is an undeniable source for the development of any nation. Life cannot be imagined without electricity in any sector, whether residential, commercial, or industrial. The generation of electricity depends on fossil fuels such as oil, coal, and natural gases. About 70% of the world's electricity generation is done by coal and other fossil fuels. With the increase in population, the requirement for electricity is also accelerating at an alarming rate, demanding the increased consumption of fossil fuels. As a result, fossil fuel supplies exhaust. All these issues can be eliminated promisingly by renewable energy sources. Wind energy, solar energy, biomass energy, geothermal energy, and tidal energy are some of the well-established and developed renewable energy sources [1,2].

As a clean and green energy source, wind energy is the most effective option for mitigating pollution and meeting energy requirements [3]. Wind energy generation depends on weather conditions, so the power generation from wind energy fluctuates and does not fulfil load demand instantaneously. For reliable operation and performance, an appropriate control strategy should be adopted. Therefore, the operation of the wind energy conversion system (WECS) in obtaining optimal power has become critical due to the intermittent behaviour of wind flow. To resolve this problem and to use WECS more economically and efficiently, MPPT technology needs to be implemented to extract optimal power at variable wind speed conditions.

In the literature, mainly two types of wind energy generation (WEG), variable speed WEG [4], and fixed-speed WEG [5], are available. The variable speed WEG is more advantageous than fixed-speed WEG; it offers wide wind speed range operation, better power-capturing capability, and improved overall efficiency. The doubly fed induction generator (DFIG) is most dominant for variable speed operation applications. It uses reduced capacity power converters, about one-fourth of the system-rated capacity, and is less expensive and easier to maintain [6,7]. The DFIG also provides good damping for the weak grids. The operation and control of wind turbines have been improved today, and the credit goes to the developments in the power electronics industry.

The drawbacks of both the permanent magnet synchronous generator (PMSG) [8] and squirrel cage induction generator (SCIG) [9] WECS are that they need a power converter rated at the total system power rating, which increases their cost. Filters for inverter outputs and EMI are rated for the rated output power, complicating and increasing filter design costs. Additionally, converter efficiency has a significant impact on overall system efficiency throughout the entire operating range.

Most of the WECS are equipped with DFIG using back-to-back power electronic converters in the wind industry [10,11]. Because the converter does not have to transmit the total power generated by the DFIG, its power rating is smaller than the overall machine rating. Such WECS has decreased the cost of inverters because generally, the inverter rating is one fourth of total system power. Additionally, it decreased the cost of inverter and EMI filters since filters are rated for one fourth of total system power, and inverter harmonics represent a smaller proportion of total system harmonics. The higher cost of the wound rotor induction machine than the SCIG is compensated by the smaller size of the power converters and increased energy production [12]. The maximum power point tracking control is implemented utilizing a machine-side control mechanism in such a system.

Maximum power point control (MPPT) algorithms are among the best techniques to extract the maximum possible power at various wind speeds in wind turbine systems. The MPPT algorithms protect the system from overload and various lightening surges [13,14]. Additionally, MPPT assists in stabilizing the output voltage in the presence of higher and lower wind speeds than the rated wind speed.

In the literature [15–17], various MPPT algorithms have extensively been discussed. The hill-climbing search (HCS), optimum relation-based MPPT (ORB), and the incremental conductance (INC) are all classified as direct power control (DPC)-based MPPT algorithms. While optimal torque control (OTC), power signal feedback (PSF), and tip-speed ratio (TSR) algorithms are included in the category of indirect power control (IPC)-based MPPT algorithms. In addition, fuzzy-logic and neural network-based control has been developed [18].

HCS, also known as perturbation and observation, is a resilient, unreliable technique based on previous WT characteristics knowledge. This algorithm provides the local maximal point for the given function [19]. The likelihood of identifying the wrong direction to achieve the most significant power point under a sudden change in wind direction is a disadvantage of this method. Using a modified version of the HCS method, [20,21] was able to address the issue of incorrect direction movement under changing wind speeds.

The ideal relationship between quantities such as WT power output, converter DC voltage, power, current, and speed is required by the ORB-based MPPT algorithm [22,23]. Fast-tracking and no need for sensors are the main advantages of this technique. However, a thorough understanding of the characteristic curves of turbine power and DC current at different wind speeds is required. By observing the optimum current curve, the MPP can be tracked [24].

The INC algorithms are completely independent of sensor needs and the specification of wind turbine and generator parameters. Therefore, systems employed with this algorithm reduce the system's cost and improve reliability [25]. According to the authors in [26,27], the operating point of the MPPT may be determined using the power-speed slope. The disadvantage of this method is that it becomes unstable when the inertia of the turbine varies under a variable speed wind scenario [28]. A new method called as the fractional order INC (FO-INC) is presented in [27] to address the instability issue at different wind speeds. For fast changes, variable step size is used in tracking the MPP under variable wind conditions.

The PSF-based MPPT approach makes use of a power control loop which incorporates information about WT's maximum power curve [12]. While the TSR-based MPPT controller is easy to build and highly efficient, it has a high operating cost. The drawback of this method is it needed optimal power coefficient and optimal tip-speed ratio [29].

The OTC technique involves changing the generator torque based on the most significant power reference torque at any given wind speed [18]. The main advantages of this method are fast response, efficiency, and simplicity. Due to the absence of direct wind speed measurement, changes in wind speed are not reflected in the reference signal [16].

MPPT control techniques based on fuzzy logic offer the benefits of rapid convergence, parameter independence, and acceptance of noisy and incorrect data [30]. The articles [31,32] provide a data-driven design approach for generating a Takagi–Sugeno–Kang (TSK) fuzzy model for MPPT control. Although the fuzzy model offers many advantages over other techniques, the main drawback is that it cannot be used for every issue. Additionally, it necessitates an examination of the parameter used to assign linguistic variables.

The artificial neural network (ANN) is another method to determine the maximum power peak by taking various input variables and processing them to obtain the maximum power [33]. Each neural network (NN) contains an input layer, a hidden layer, and an output layer. There is really no constraint on the number of nodes assigned, and they may vary as per the demand. The ANN-based controller is a more efficient and reliable alternative than conventional controllers for extracting the maximum amount of power from wind's available kinetic energy. The disadvantages of NN include their black box structure, increased computing load, overfitting issue, and empirical nature of model development. This technique necessitates the use of a look-up table containing predefined data [34].

The membership function type, number of rules, and correct selection of parameters of FLC are essential to obtain desired performance in the system. Selecting suitable fuzzy rules, membership functions, and their definitions in the universe of discourse invariable involves painstaking trial-error [35]. The adaptive neuro-fuzzy inference system (ANFIS) is a scheme derived from a synthesis between the neural network and fuzzy inference system [36]. Similar to the method of training a neural network, the membership function parameters have been fine-tuned using adaptive neuro-learning methods. A neural network enhances the adaptability of the model. The primary purpose of using the ANFIS approach is to realize the fuzzy system by using neural network methods automatically. The ANFIS combines the capability of fuzzy reasoning in handling the uncertainties and the capability of ANN in learning from processes [36].

In this paper, an ANFIS MPPT controller is used for maximum power tracking. The generator rotor speed is input to the MPPT controller training input value, and the optimum torque reference is selected as the target value.

The following is the structure of the paper. Section 1 discusses renewable energy importance and the literature of MPPT algorithms employed in wind energy conversion systems. The modeling of doubly fed induction generator-based WECS is given in Section 2. Section 3 deals with rotor side control with maximum power point tracking control. Section 4 defines the ANFIS MPPT controller for achieving maximum power point along with the training process. Section 5 illustrates the performance of the ANFIS approach. A comparison with another conventional approach is also carried out in this section. Finally, Section 6 summarizes the conclusion.

2. Modeling of Doubly Fed Induction Generator-Based WECS

Configuration of the grid-connected DFIG-based WECS considered in this study is shown in Figure 1, which consists of a turbine system connected to the DFIG generator through a gear system. The stator winding of the DFIG is directly connected to the grid. In contrast, the rotor of the DFIG is connected to the grid via a back-to-back power converter. In addition, a MPPT controller is connected to the RSC control of the WECS.

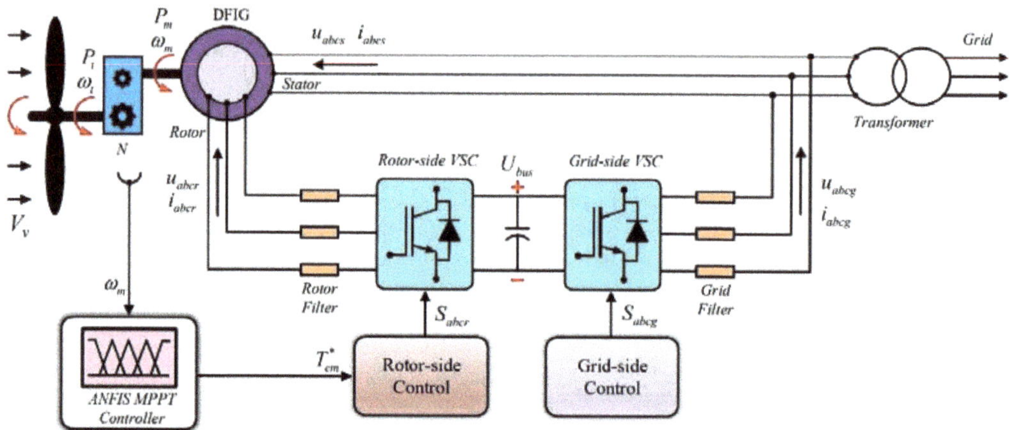

Figure 1. Grid-connected DFIG-based WECS configuration.

2.1. Wind Turbine Modeling

The equation for the amount of power derived by the wind turbine from the wind is expressed in (1) [37]:

$$P_t = \frac{1}{2}\rho\pi R^2 V_v^3 C_p(\lambda, \beta) \tag{1}$$

where ρ is the air density measured in kilograms per cubic meter (kg/m³), R is the rotor radius of the turbine measured in meter (m), V_v presents the wind speed measured in meter per second (m/s), $C_p(\lambda, \beta)$ presents the power coefficient expressed as a function of the tip-speed ratio (λ) and the pitch angle (β).

λ is expressed in (2):

$$\lambda = \frac{R\omega_t}{V_v} \tag{2}$$

where ω_t is the angular rotational speed of the wind turbine rotor (rad/sec).

$C_p(\lambda, \beta)$ is expressed by (3) [37]:

$$C_p(\lambda, \beta) = c_1\left(\frac{c_2}{\lambda_i} - c_3\beta - c_4\beta^{c_5} - c_6\right).e^{\frac{-c_7}{\lambda_i}}. \tag{3}$$

λ_i is given in (4):

$$\lambda_i = \frac{1}{\lambda + 0.02\beta} - \frac{0.003}{\beta^3 + 1} \tag{4}$$

where $c_1 = 0.73$; $c_2 = 151$; $c_3 = 0.58$; $c_4 = 0.002$; $c_5 = 2.4$; $c_6 = 13.2$; $c_7 = 18.4$ The turbine generated torque is expressed in (5) [37]:

$$T_t = \frac{P_t}{\omega_t} \tag{5}$$

2.2. DFIG Modeling

The following equations describe the DFIG model. The voltage vectors of the stator and rotor are expressed in (6) and (7), respectively [36]:

$$\vec{u}_s = R_s \vec{i}_s + \frac{d\vec{\psi}_s}{dt} + j\omega_s \vec{\psi}_s \Rightarrow \begin{cases} u_{ds} = R_s i_{ds} + \frac{d\psi_{ds}}{dt} - \omega_s \psi_{qs} \\ u_{qs} = R_s i_{qs} + \frac{d\psi_{qs}}{dt} + \omega_s \psi_{ds} \end{cases} \tag{6}$$

$$\vec{u}_r = R_s \vec{i}_r + \frac{d\vec{\psi}_r}{dt} + j\omega_r \vec{\psi}_r \Rightarrow \begin{cases} u_{dr} = R_r i_{dr} + \frac{d\psi_{dr}}{dt} - \omega_r \psi_{qr} \\ u_{qr} = R_r i_{qr} + \frac{d\psi_{qr}}{dt} + \omega_r \psi_{dr} \end{cases} \tag{7}$$

where u_{ds}, u_{qs}, u_{dr}, and u_{qr}: Voltages at the stator and rotor in the d–q frame, respectively. i_{ds}, i_{qs}, i_{dr}, and i_{qr}: Currents in the stator and rotor in the d–q frame, respectively. R_r, R_s, ω_s, and ω_r: Stator and rotor phase resistances and angular velocity, respectively.

The flux vectors for the stator and rotor are denoted in (8) and (9), respectively [37]:

$$\vec{\psi}_s = L_s \vec{i}_s + L_m \vec{i}_r \Rightarrow \begin{cases} \psi_{ds} = L_s i_{ds} + L_m i_{dr} \\ \psi_{qs} = L_s i_{qs} + L_m i_{qr} \end{cases} \tag{8}$$

$$\vec{\psi}_r = L_m \vec{i}_s + L_r \vec{i}_r \Rightarrow \begin{cases} \psi_{dr} = L_m i_{ds} + L_r i_{dr} \\ \psi_{qr} = L_m i_{qs} + L_r i_{qr} \end{cases} \tag{9}$$

where $\vec{\psi}_s$, $\vec{\psi}_r$ are the flux vectors for stator and rotor, respectively. ψ_{ds}, ψ_{qs} are the fluxes along the d–q axis stator. ψ_{dr}, ψ_{qr} are the fluxes along with the d–q axis rotor. L_s, L_r: Leakage inductances in the stator and rotor phases, L_m: Mutual inductance between stator and rotor, p: is the generator pole pair count.

The expression of electromagnetic torque is expressed in (10) [37]:

$$T_{em} = \frac{3}{2} p \frac{L_m}{L_s} \left(\psi_{qs} i_{dr} - \psi_{ds} i_{qr} \right) \tag{10}$$

The active and reactive power equations of the stator and rotor are expressed in (11) and (12) [37]:

$$\begin{cases} P_s = \frac{3}{2} \left(u_{ds} i_{ds} + u_{qs} i_{qs} \right) \\ Q_s = \frac{3}{2} \left(u_{qs} i_{ds} - u_{ds} i_{qs} \right) \end{cases} \tag{11}$$

$$\begin{cases} P_r = \frac{3}{2} \left(u_{dr} i_{dr} + u_{qr} i_{qr} \right) \\ Q_r = \frac{3}{2} \left(u_{qr} i_{dr} - u_{dr} i_{qr} \right) \end{cases} \tag{12}$$

where P_s, Q_s presents stator active and reactive power. P_r, Q_r presents rotor active and reactive power. T_{em} is the electromagnetic torque.

3. Rotor Side Control with Maximum Power Point Tracking

The RSC is responsible for the voltage applied to the rotor winding of the DFIG. To derive the voltage equation in dq reference frame, from the DFIG model in the previous section, replacing Equations (8) and (9) in Equation (7) and considering $\psi_{qs} = 0$ we get the following equation as a function of the rotor currents and stator flux [37]:

$$\begin{cases} u_{dr} = R_r i_{dr} + \sigma L_r \frac{di_{dr}}{dt} - \omega_r \sigma L_r i_{qr} + \frac{L_m}{L_s} \frac{d|\vec{\psi}_s|}{dt} \\ u_{qr} = R_r i_{qr} + \sigma L_r \frac{di_{qr}}{dt} + \omega_r \sigma L_r i_{dr} + \omega_r \frac{L_m}{L_s} \frac{d|\vec{\psi}_s|}{dt} \end{cases} \tag{13}$$

where $\sigma = 1 - \frac{L_m^2}{L_s L_r}$. Assuming negligible voltage drop in the stator winding resistance and stator flux are constant because of the constant grid quantities, consequently, zero. It can be seen from Equation (13) that dq component of rotor current can be controlled using regulators. The reactive power proportional–integral (PI) regulator represented as REG-1. The equal PI regulator for both d and q current loop are chosen as REG-2 and REG-3, respectively. Actual values are considered for tunning of the gain parameters of the regulators. The gain parameters for all three regulators are presented in Table 1. The control must be performed on the dq components, so rotor voltage and current are transformed into dq components using abc–dq transform. Θ_s is obtained by first estimate the stator voltage vector and subtracting angle $\pi/2$. The phase-locked loop (PLL) is used for grid synchronization, which also supports in rejection of minor disturbances. The "u" defines the stator-rotor turn ratio that is $1/3$. Figure 2 illustrates the complete vector control of the DFIM with MPPT controller.

Table 1. The gain parameters of PI regulators in rotor side control.

Gains	REG-1	REG-2	REG-3
Proportional	10,160	0.5771	0.5771
Integral	406,400	491.5995	491.5995

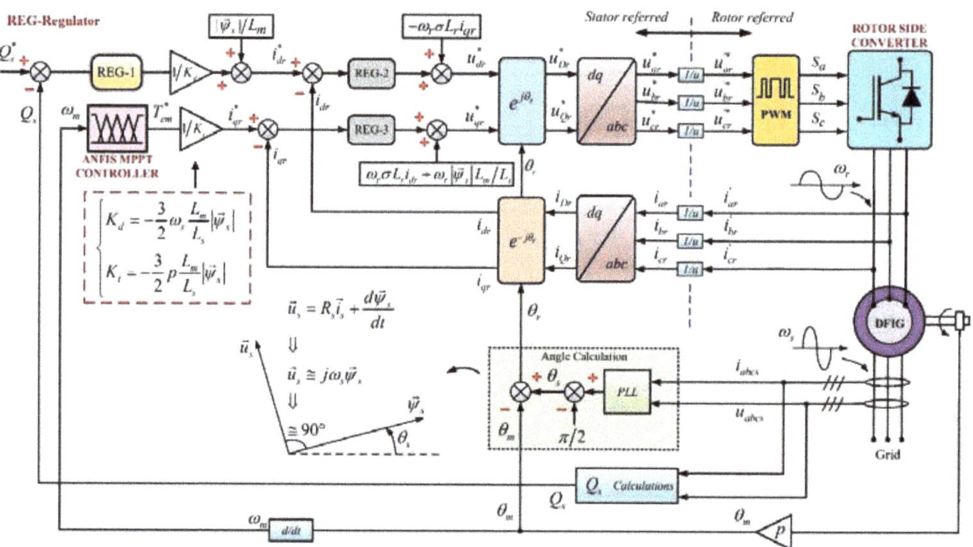

Figure 2. Block diagram of RSC vector control of the DFIG.

The torque expressions in dq frame can be given by:

$$T_{em} = \frac{3}{2} p \frac{L_m}{L_s} \left(\psi_{qs} i_{dr} - \psi_{ds} i_{qr} \right) \Rightarrow T_{em} = -\frac{3}{2} p \frac{L_m}{L_s} \psi_{ds} i_{qr} \Rightarrow T_{em} = K_t i_{qr} \tag{14}$$

The stator reactive power expressions in dq frame can be given by:

$$Q_s = \frac{3}{2} \left(u_{qs} i_{ds} - u_{ds} i_{qs} \right) = -\frac{3}{2} \omega_s \frac{L_m}{L_s} |\vec{\psi}_s| \left(i_{dr} - \frac{|\vec{\psi}_s|}{L_m} \right) \Rightarrow Q_s = K_q \left(i_{dr} - \frac{|\vec{\psi}_s|}{L_m} \right) \tag{15}$$

Equation (14) reveals that the i_{qr} is proportional to the T_{em} to control torque with i_{qr}. Expression in Equation (15) reveals that d component of rotor current i_{dr} controls the Q_s. Therefore, because of the orientation chosen, it can be seen that both rotor current components independently allow us to control the torque and reactive stator power.

4. ANFIS Maximum Power Point Tracking Control

The adaptive neuro-fuzzy inference method is a highly effective technique that incorporates both fuzzy control and artificial neural network concepts [31,38,39]. Due to the combined influence of fuzzy and neural networks, it is an excellent learner and interpreter [40]. The ANFIS controller determines which membership function to use. The general structure of ANFIS consists of five layers, as shown in Figure 3.

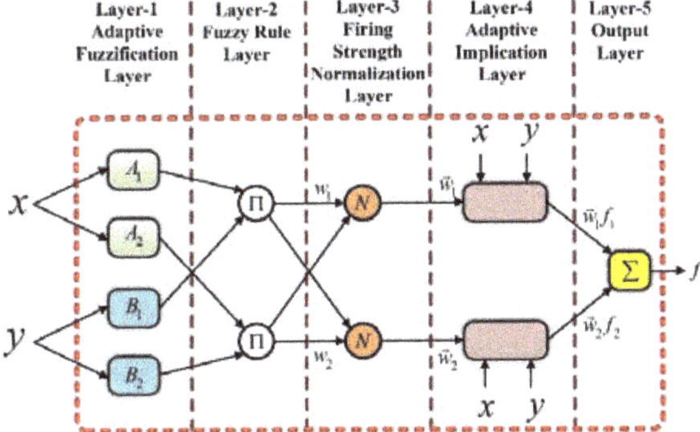

Figure 3. The architecture of the ANFIS controller.

- Layer 1, the adaptive fuzzification layer is composed of user-specified input variables and membership functions (MF).
- Layer 2, the fuzzy rule layer checks the degree of MF, and the corresponding fuzzy set is selected and input to the next layer.
- Layer 3, the firing strength normalization layer evaluates weight for each normalized node.
- Layer 4, the adaptive implication layer outputs values in accordance with inference rules, and each neuron is normalized.
- Layer 5, the output layer adds all of the inputs from layer 4 and transforms the fuzzy values to a crisp value.

The developed ANFIS has single input as rotor speed. The instantaneous torque reference is determined as the output from the ANFIS network. In the developed MPPT controller, the ANFIS first-order Sugeno model as well as with fuzzy IF-THEN rules of Takagi and Sugeno type are used. A backpropagation algorithm trains the ANFIS-based MPPT controller.

Figure 4 illustrates the block diagram of the proposed ANFIS MPPT control. The generated optimal torque (T_{em}^*) is used to determine rotor quadrature current reference (i_{qr}^*) applied to the speed control loop of RSC control that controls the actual rotor speed by adjusting the duty ratio of the RSC. The control objective of the converter is to maximize the output power delivered to the grid.

Figure 5 depicts the architecture of the developed ANFIS controller in MATLAB/Simulink using Neuro-Fuzzy Designer. The ANFIS details are given in Table 2. The trial-and-error method is used for choosing the number and shape of MFs as there is no exact method for choosing the MFs in the literature. Seven Gaussian MFs were selected for this study because they had the lowest root-mean-square error (RMSE) of 0.098280. The primary reason why Gaussian MFs were chosen is that they have the fewest parameters (Only two parameters mean and standard deviation). To define the membership functions and fuzzy rules, the grid partition method is used, generating input membership functions by uniformly partitioning the input variable ranges and creating a single-output Sugeno fuzzy system. Each input membership function combination is represented by a single rule in the fuzzy rule base.

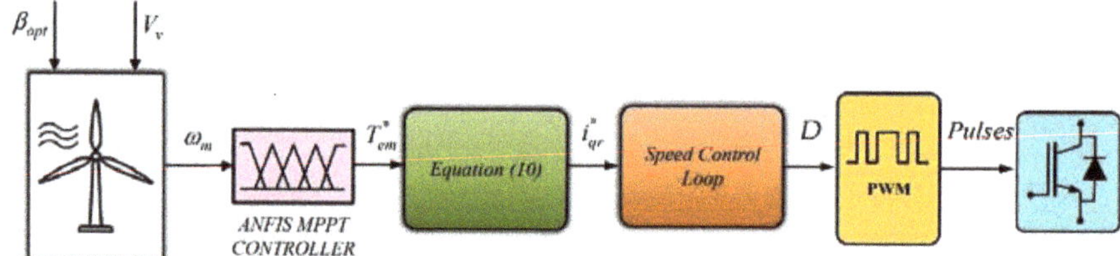

Figure 4. Block diagram of ANFIS MPPT control.

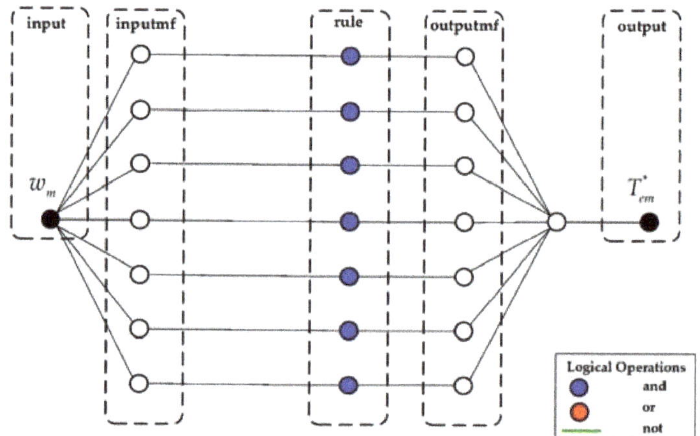

Figure 5. The architecture of the developed ANFIS-based MPPT controller.

Table 2. The ANFIS parameters information.

Parameter	Value	Parameter	Value
Number of nodes	32	Total number of parameters	28
Number of linear parameters	14	Number of training data pairs	10,000,001
Number of nonlinear parameters	14	Number of fuzzy rules	7

Speed is taken as an input to the ANFIS MPPT controller, and it outputs the torque reference. The controller is trained for 1000 epochs. The controller has one input with seven membership functions (MFs). The initial generated input speed membership functions for training are shown in Figure 6, which utilize seven rules. The details of the initial seven input membership functions derived from Equation (13) are presented in Table 3.

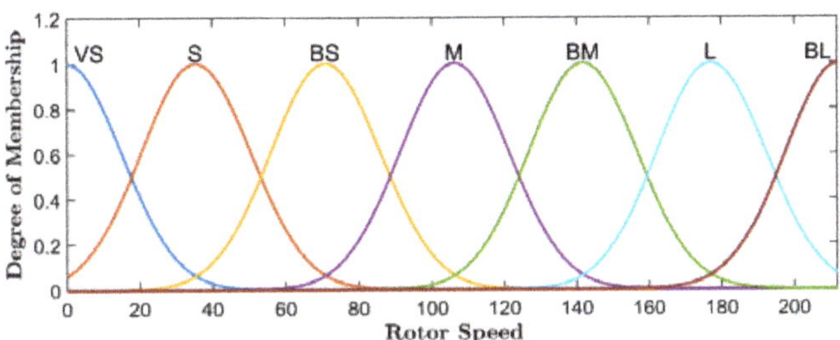

Figure 6. Initial input speed membership functions for training.

Table 3. Initial input membership function details.

Membership Function Name	Type	Parameter	
		Standard Deviation	Mean
Very Small (VS)	Gaussian	15.047	-3.3549×10^{-13}
Small (S)	Gaussian	15.047	35.433
Big Small (BS)	Gaussian	15.047	70.865
Medium (M)	Gaussian	15.047	106.3
Big Medium (BM)	Gaussian	15.047	141.73
Large (L)	Gaussian	15.047	177.16
Big Large (BL)	Gaussian	15.047	212.6

The neural network tuned input speed membership functions is shown in Figure 7. The tunned membership functions details are presented in Table 4. As the output function in the Sugeno fuzzy inference system is selected as a linear function of the input, details of all seven output membership functions are given in Table 5. Figure 8 shows the step size increase/decrease during the training. The root-mean-square error is shown in Figure 9. The expression for the Gaussian membership function is given in Equation (13):

$$f(x, \sigma, c) = e^{\frac{-(x-c)^2}{2\sigma^2}}$$

(16)

where σ is the standard deviation, c is mean, and x is input value.

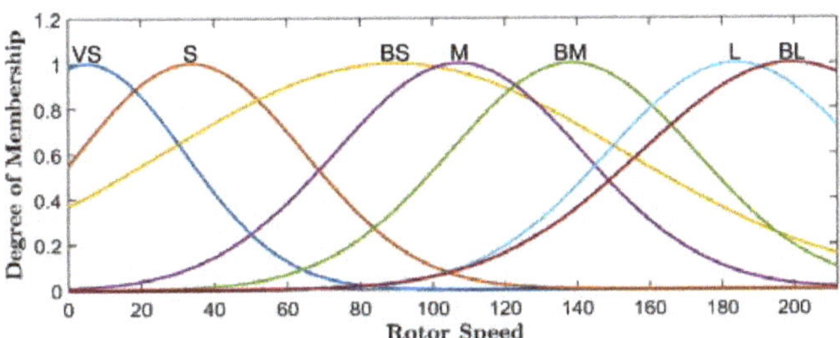

Figure 7. Neural network tuned input speed membership functions.

Table 4. Neural network tunned input membership function details.

Membership Function Name	Type	Parameter	
		Standard Deviation	**Mean**
Very Small (VS)	Gaussian	26.909	5.1594
Small (S)	Gaussian	30.952	33.765
Big Small (BS)	Gaussian	63.437	89.457
Medium (M)	Gaussian	34.224	107.28
Big Medium (BM)	Gaussian	33.967	138.56
Large (L)	Gaussian	35.116	183.63
Big Large (BL)	Gaussian	41.65	199.36

Table 5. Output membership function details.

Membership Function Name	Type	Parameter	
		Lower Limit	**Upped Limit**
Very Small (VS)	Linear	1.9856	1145.1
Small (S)	Linear	7.3276	1244.4
Big Small (BS)	Linear	-4.8424	-5088.7
Medium (M)	Linear	-124.22	9248
Big Medium (BM)	Linear	-143.49	15,022
Large (L)	Linear	-98.038	7647.7
Big Large (BL)	Linear	-149.91	17,393

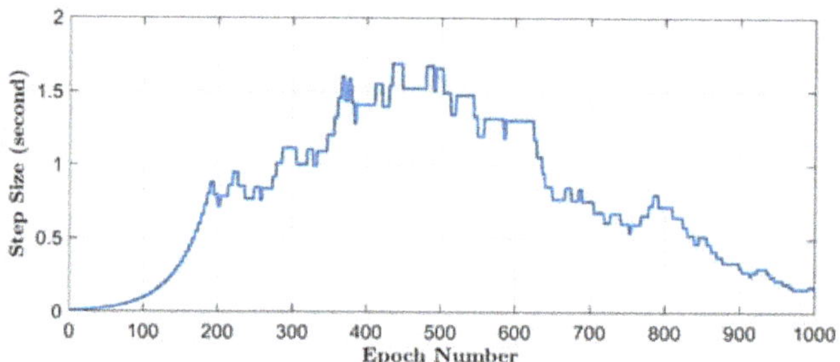

Figure 8. Step size during the training.

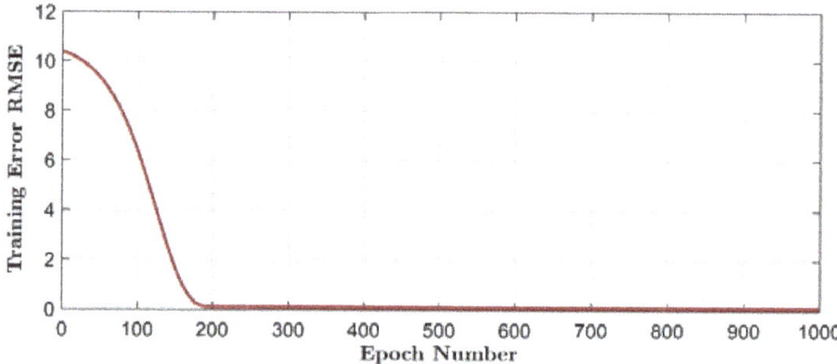

Figure 9. Step size during the training.

- The first layer consists of an input node as the variable. This layer is responsible for transform in input value to the next layer. Here, seven gaussian MFs with minimum = 0 and maximum = 1 are utilized, and corresponding node equations are given (17):

$$O_i^1 = \mu A_i(e_i) \tag{17}$$

where $i = 1, 2, \ldots 7$, O_i is the output of ith node in layer one, A_i is the linguistic label, e_i is the input to the node.

- The second layer verifies the weights of individual MFs. It accepts the first layer's input values and serves as the MF for the corresponding input variables fuzzy sets. The second layer has non-adaptive nodes that multiply incoming signals and output the result as in (18):

$$w_j = \mu A_i(e_1) \times \mu B_i(e_2) \tag{18}$$

where $i = 1, 2, \ldots 7$ and $j = 1, 2, \ldots 7$. The output from each node represents the firing strength of a rule.

- Each node in the third layer computes the activation level of each fuzzy rule, with the number of layers equal to the number of fuzzy rules. Each node of these layers generates the normalized weights. Each node calculates the ratio of the rule's firing strength to the total of all rules' firing strengths, that is, the normalized firing strength given in (19):

$$w_j^* = \frac{w_j}{w_1 + w_2 \ldots w_7} \tag{19}$$

where $j = 1, 2, \ldots 7$.

- The fourth layer contains the output values obtained through rule inference. Node function of the fourth layer is given in (20):

$$O_j^4 = w_j^* f_j = w_j^* \left(p_j e_1 + q_j e_2 + r_j \right) \tag{20}$$

The rule base is given as:

If e_1 is A_1 and e_2 is B_1 then $f_1 = p_1 e_1 + q_1 e_2 + r_1$;

If e_1 is A_2 and e_2 is B_2 then $f_2 = p_1 e_1 + q_2 e_2 + r_2$;

If e_1 is A_7 and e_2 is B_1 then $f_7 = p_7 e_1 + q_7 e_2 + r_7$,

where (p_j, q_j, r_j) is the parameter set and in this layer is referred to as consequent parameters, O_j^4 = output of the ith node in layer-4, A_i, B_i = fuzzy membership function, $i = 1, 2, \ldots 7$ and $j = 1, 2, \ldots 7$.

- The fifth layer is the output layer; it aggregates all of the fourth layer's inputs and converts the fuzzy classification results into a crisp representation. This layer has a non-adaptive nature, having a single node with the output given in (21):

$$y = \sum_{j=1}^{7} w_j^* f_j = \sum_{j=1}^{7} \left(\left(w_j^* e_1 \right) p_j + \left(w_j^* e_2 \right) q_j + \left(w_j^* \right) r_j \right) \tag{21}$$

In practice, the proposed controller can be implemented using the Simulink HDL Coder toolbox that can generate code for the DSP's and FPGA's family chips of different vendors. The complete list of supported chips in the HDL coder and more details on this can be seen from the MathWorks official website [41].

5. Simulation Result and Discussion

The effectiveness of the ANFIS MPPT controller for DFIG-based WECS under variable wind speed operation has been verified in MATLAB/Simulink environment. The simulation comparison results of the proportional–integral (PI) controller and the proposed ANFIS MPPT controller are present in the following figures.

In this work, β is set to zero and designed for the rated wind speed of 11 m/s. The simulated power characteristics at different wind speeds are presented in Figure 10.

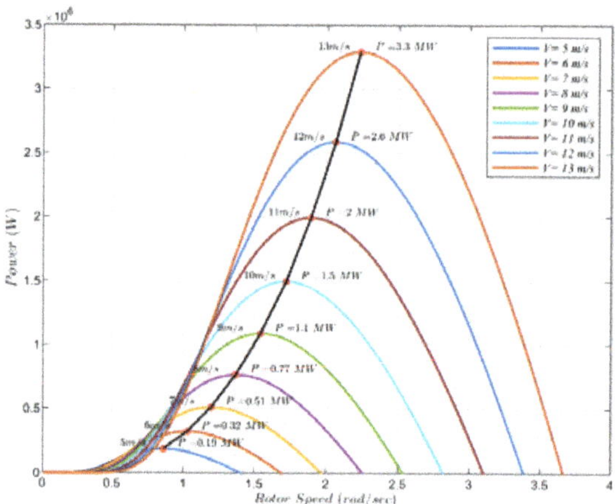

Figure 10. Power and rotor speed characteristics of WT at different wind speed.

$C_p - \lambda_i$ characteristics at different value of β is presented in Figure 9. The design presents that the maximum value $C_{p\,max}$ is 0.4411 and the corresponding λ is 7, as shown in Figure 11. This value $C_{p\,max}$ and λ is the optimum value for capturing peak power from the available wind power. Parameters of the WT are presented in Table 6.

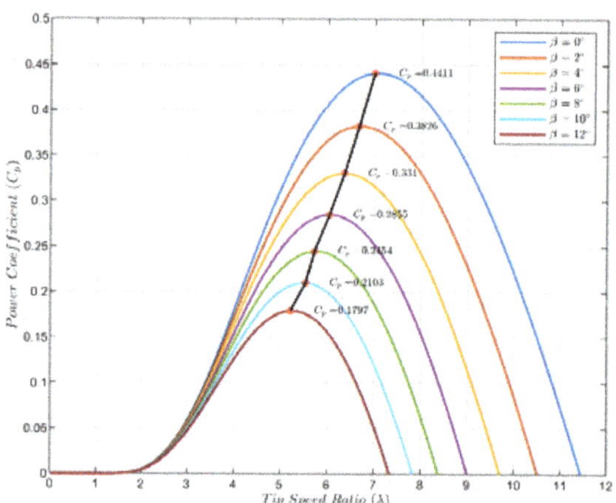

Figure 11. $C_p - \lambda_i$ characteristics with variation in pitch angle (β).

Table 6. WECS parameters.

Parameter	Value	Parameter	Value
Nominal wind speed	11 m/s	Frequency	50 Hz
Air density	1.225 kg/m^3	Rated torque	12,732 N·m
Tip-speed ratio	7	Pole pair	2
Pitch angle	0°	Inertia	127 kg·m^2
Power coefficient	0.4411	Gear ratio	100
Nominal power	2 MW	Radius of turbine	42 m

Three case studies are considered to analyze the performance of the proposed controller. Different input wind speed profiles are considered for all three cases. Table 7 shows an overview of all three cases of wind speed during the simulation study.

Table 7. Input wind speed in all cases.

Time Duration (s)	Input Wind Speed (m/s)		
	Case-I	**Case-II**	**Case-III**
0–5	6	12	8
5–8	7	11	10
8–11	8	10	9
11–14	9	9	12
14–17	10	8	7
17–20	11	7	7
20–23	12	6	7

Simulation time responses of rotor speed (ω_m), electromagnetic torque (T_{em}), stator active power (P_s), DC-link voltage (V_{dc}), stator voltage (V_s), stator current of a-phase (I_{sa}), rotor current of a-phase (I_{ra}), q-axis rotor current component (i_{qr}), and d and q axis rotor voltage component (v_{dr}, v_{qr}) with the change in system input wind speed (V_v) for PI controller and ANFIS controller are presented in following figures below.

5.1. Case-I: Step Increase in Input Wind Speed

In this case, the input wind speed is increased in step manner as shown in Figure 12a and according to data presented in Table 5. The rotor speed tracking of conventional control and the proposed controller is shown in Figure 12b. There was a significant difference in rotor speed response. The electromagnetic torque response is shown in Figure 12c. The stator active power is observed in Figure 12d. Figure 12e and f present the DC-link voltage and stator voltage response, respectively.

The single a-phase stator current response comparison during step wind speed change at $t = 14$ s from 9 m/s to 10 m/s is shown in Figure 12g. The proposed controller stator current response remains sinusoidal without any swell condition in current, whereas; the conventional controller shows the unbalance operation. Figure 12h shows the single a-phase rotor current response comparison. It can be observed at $t = 14$ and 17 s when wind speed increase occurs. The conventional controller shows an unbalance operation along with overshoot. The quadrature axis current component of rotor side converter (i_{qr}) response is shown in Figure 12i. Whenever there is wind speed increase operation, the proposed controller shows a smooth transition, whereas the conventional controller shows oscillation at each change instant. The quadrature and direct axis voltage component of the rotor side converter is shown in Figure 12j.

5.2. Case-II: Step Decrease in Input Wind Speed

In this case, the input wind speed is decreased in a step manner, as shown in Figure 13a and according to data presented in Table 5 Case-II. The rotor speed tracking of conventional control and the proposed controller are shown in Figure 13b. The electromagnetic torque response is shown in Figure 13c. The stator active power is observed in Figure 13d. Figure 13e and f present the DC-link voltage and stator voltage response, respectively.

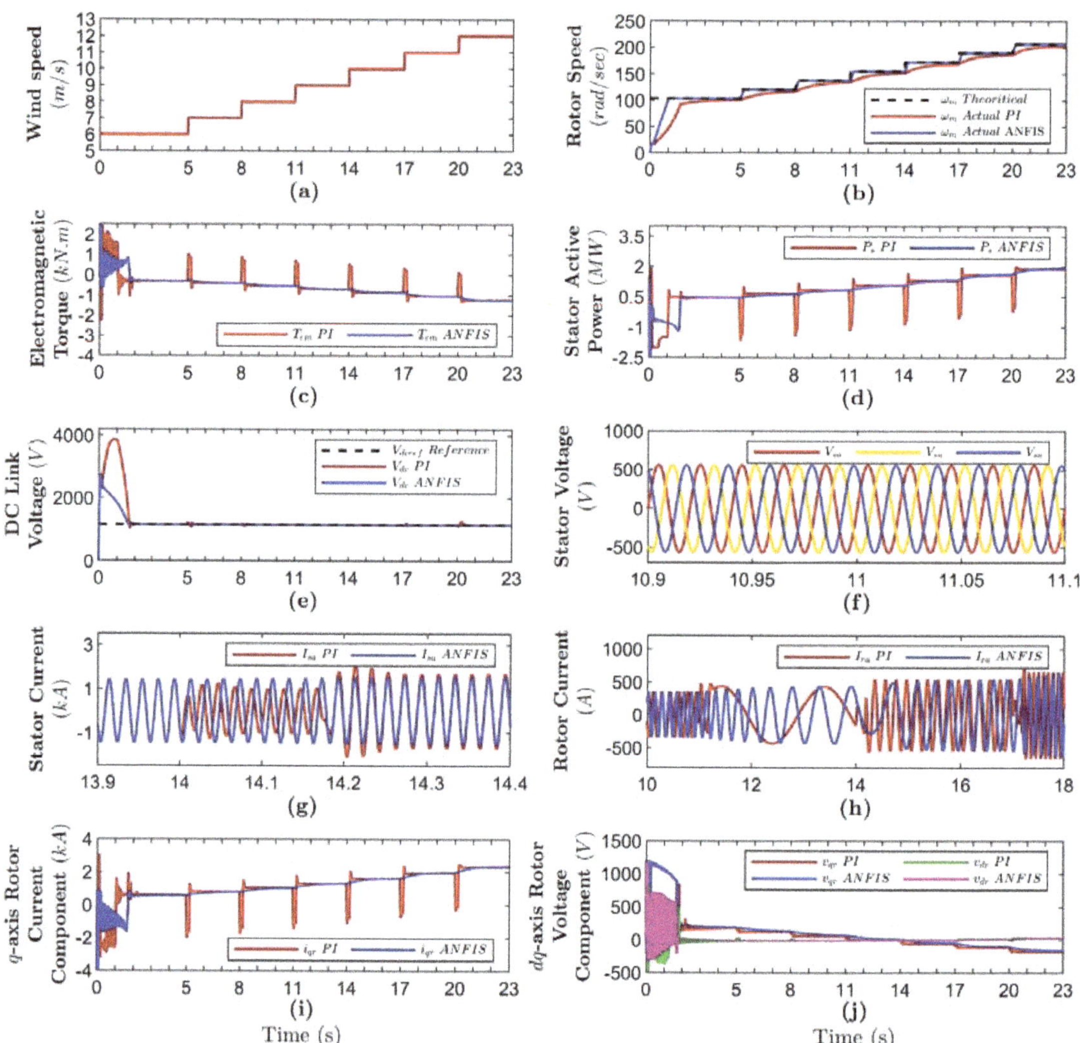

Figure 12. Simulated response under step wind speed increase: (**a**) input wind speed, (**b**) rotor speed, (**c**) electromagnetic torque, (**d**) stator active power, (**e**) DC-link voltage, (**f**) stator voltage, (**g**) stator current of a-phase, (**h**) rotor current of a-phase, (**i**) *q*-axis rotor current component, and (**j**) *d* and *q* axis rotor voltage component.

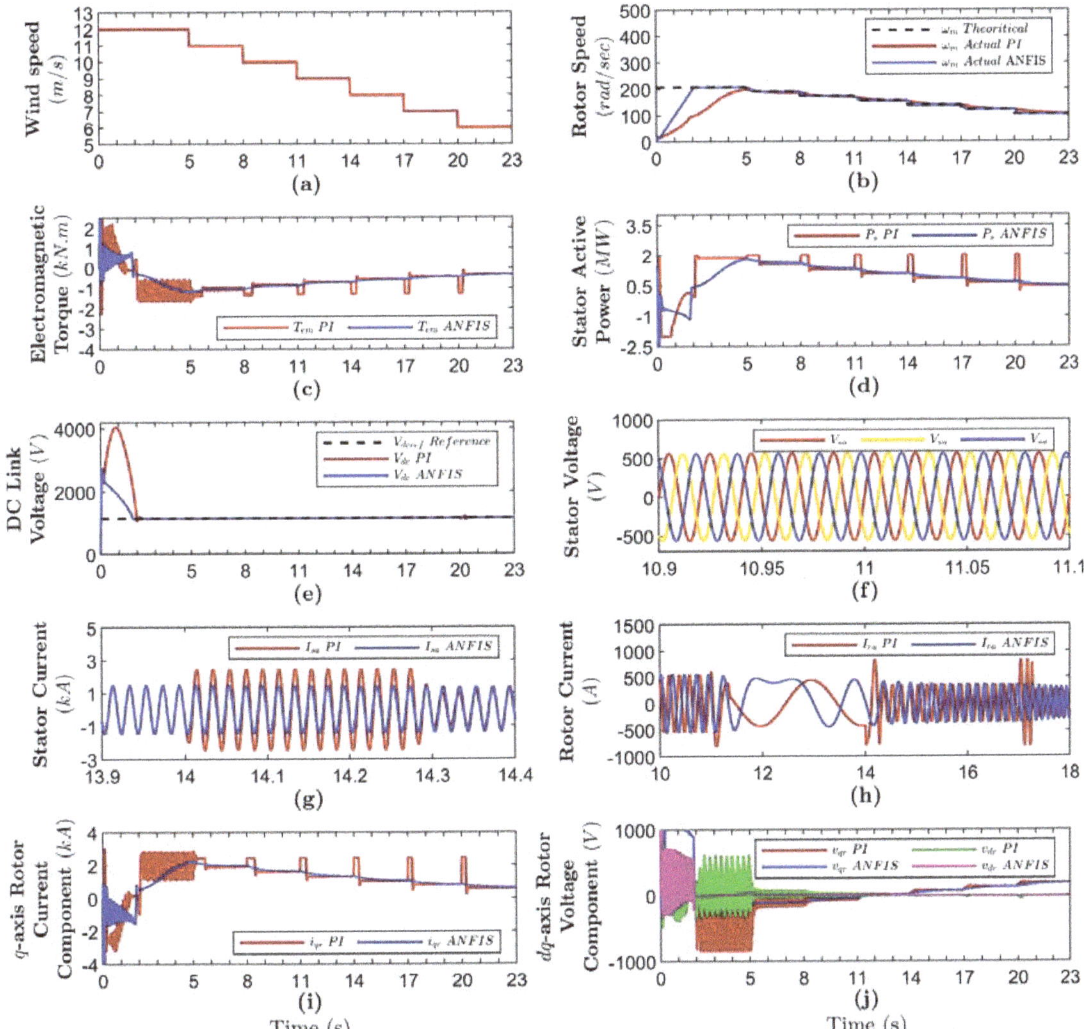

Figure 13. Simulated response under step wind speed decrease: (**a**) Input wind speed. (**b**) Rotor speed. (**c**) Electromagnetic torque. (**d**) Stator active power. (**e**) DC-link voltage. (**f**) Stator voltage. (**g**) Stator current of a-phase. (**h**) Rotor current of a-phase. (**i**) q-axis rotor current component. (**j**) d and q axis rotor voltage component.

The single a-phase stator current response comparison during step wind speed decrease at $t = 14$ s from 9 m/s to 8 m/s is shown in Figure 13g. The proposed controller stator current response remains sinusoidal without any swell condition in current whereas, the conventional controller shows current swell. Figure 13h shows the single a-phase rotor current response comparison. It can be observed at $t = 10$ and 18 s when wind speed decrease occurs, the conventional controller shows unbalanced operation and overshoot. The quadrature axis current component of rotor side converter (i_{qr}) response is shown in Figure 13i. Whenever there is a decrease in wind speed operation, the proposed controller shows a smooth transition, whereas the conventional controller shows oscillation at each change instant. The quadrature and direct axis voltage component of the rotor side converter is shown in Figure 13j.

5.3. Case-III: Intermittent Change in Input Wind Speed

In this case, the input wind speed is intermittent and shown in Figure 14a, according to data presented in Table 5 Case-III. The rotor speed tracking of conventional control and the proposed controller is shown in Figure 14b. The electromagnetic torque response is shown in Figure 14c. The stator active power is observed in Figure 14d. Figure 14e,f present the DC-link voltage and stator voltage response, respectively.

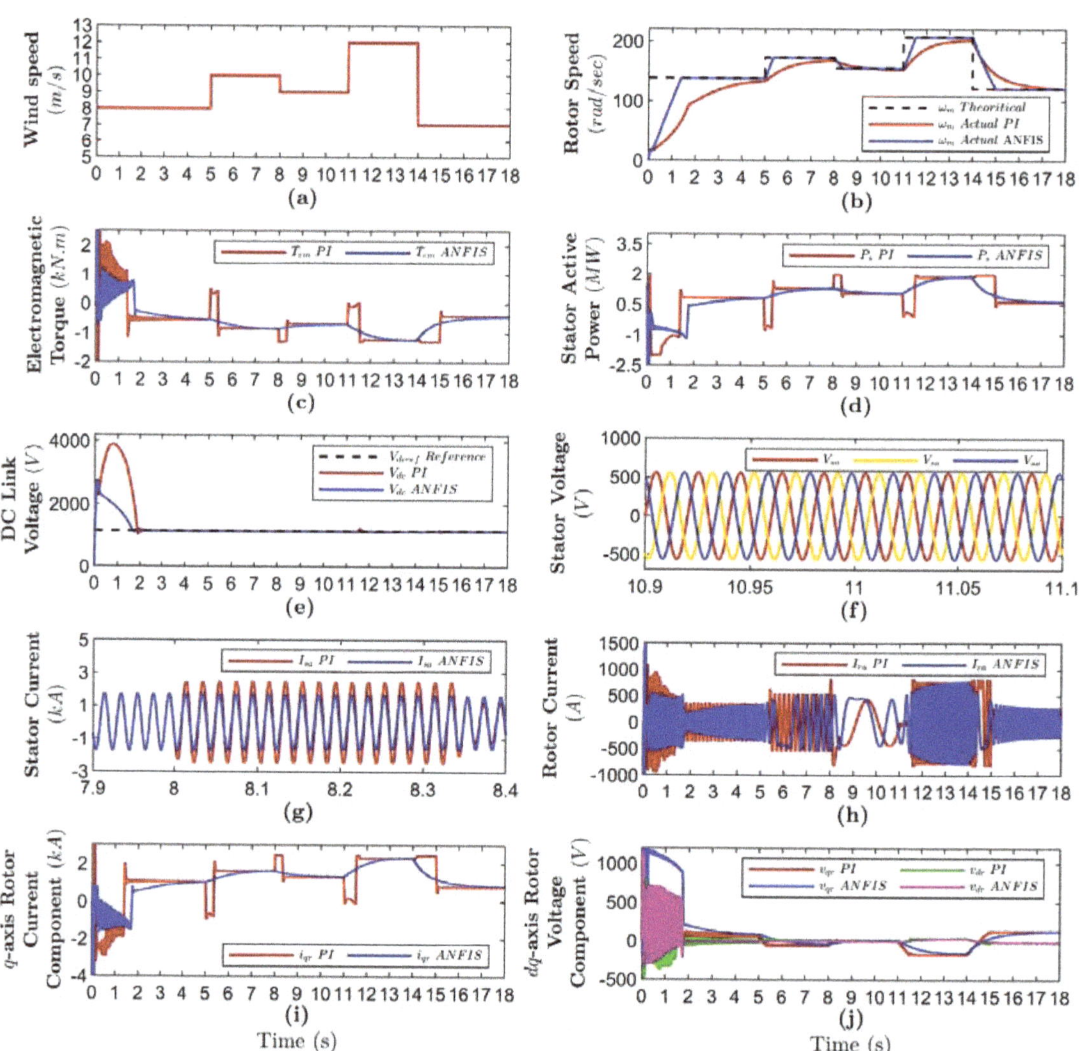

Figure 14. Simulated response under step increase wind speed: (**a**) input wind speed, (**b**) rotor speed, (**c**) electromagnetic torque, (**d**) stator active power, (**e**) DC-link voltage, (**f**) stator voltage, (**g**) stator current of a-phase, (**h**) rotor current of a-phase, (**i**) q-axis rotor current component, and (**j**) d and q axis rotor voltage component.

The single a-phase stator current response comparison during step wind speed decrease at $t = 8$ s from 10 m/s to 9 m/s is shown in Figure 14g. The proposed controller stator current response remains sinusoidal without any swell condition in current, whereas the conventional controller shows the current swell. Figure 14h shows the single a-phase rotor current response comparison. It can be observed at t = 5, 8, 11, and 14 s when wind speed change occurs, and the conventional controller shows unbalance operation along with overshoot. The quadrature axis current component of rotor side converter (i_{qr}) response is shown in Figure 14i. Whenever there is a change in wind speed operation, the proposed controller shows smooth transition, whereas the conventional controller shows oscillation at each change instant. The quadrature and direct axis voltage component of the rotor side converter is shown in Figure 14j. The performance comparison of ANFIS and PI controller is presented in Table 8 considering rotor speed and stator active power. Considering all three cases, there is 3.28% improvement in stator active power.

Table 8. Performance comparison of ANFIS and PI controller.

Case Study	Simulation Time Instant (sec)	Wind Speed	Rotor Speed (rad/sec)		Stator Active Power (MW)		Percentage Improvement in Power (%)
			PI	ANFIS	PI	ANFIS	
	4	6	100.6	103.7	0.4786	0.4878	1.89
	7	7	115.9	120.9	0.6261	0.6372	1.74
	10	8	133.4	138.2	0.8386	0.8491	1.24
Case-I	13	9	151.1	155.5	1.0562	1.0683	1.13
	16	10	168.2	172.8	1.3150	1.3320	1.28
	19	11	185.4	190.2	1.5860	1.6140	1.73
	22	12	202.5	207.3	1.8830	1.9200	1.93
	4	12	195.2	207.3	1.7732	1.8740	5.38
	7	11	187.3	190.2	1.5860	1.6380	3.17
	10	10	170.5	172.8	1.3150	1.3620	3.45
Case-II	13	9	153.7	155.5	1.0680	1.1120	3.96
	16	8	137.1	138.2	0.8479	0.8894	4.67
	19	7	120.2	120.9	0.6533	0.6928	5.70
	22	6	101.1	103.7	0.4939	0.5236	5.67
	4	8	129.1	138.2	0.7889	0.8478	6.95
	7	10	165.9	172.8	1.3127	1.3265	1.04
Case-III	10	9	153.8	155.5	1.0680	1.0970	2.64
	13	12	199.9	207.3	1.8830	1.9070	1.26
	17	7	120.9	123.4	0.6531	0.7071	7.64

In terms of stabilizing the stator power output, the suggested controller outperforms the PI controller. In contrast, the PI solution exhibits power oscillations at speed change instant, while the ANFIS response exhibits smooth tracking.

The voltage reference for the DC-link is 1150 volts. The ANFIS controller DC-link voltage response is constant during operation compared with the ANFIS response; the PI response demonstrates that the DC-link voltage oscillates at the instant of speed change and overshoots at around 3980 volts max, which is 37% greater than the ANFIS response.

6. Conclusions

This paper proposed an ANFIS controller for maximum power extraction from the wind for grid-connected DFIG-based WECS. The controller has implemented an ANFIS controller for peak power point tracking. For training, an ANFIS includes input and target data; in this case, the rotor speed is used as the input data, and the torque reference is used as the target or output data. The proposed controller has implemented a 2 MW variable speed wind turbine in MATLAB/Simulink subject to variable wind speed conditions. The simulation study shows that the proposed ANFIS MPPT controller approach exhibited good dynamic performance and quick response for wind speed change while ensuring peak power point tracking. Comparison analysis with a conventional proportional–integral controller approach showed that the ANFIS approach resulted in smoother power tracking and reduced chattering than the conventional approach with a wide range of wind speed changes. The proposed ANFIS MPPT controller shows a 3.28 percent improvement in stator active power than the proportional–integral controller.

This research may be further explored, taking into consideration the controller with multivariable input. The system's performance may be improved in the future if the changing pitch angle and the actual power generated are taken into consideration, along with rotor speed. Furthermore, performance may be compared with that of other intelligent controllers.

Author Contributions: Conceptualization, A.A.C. and V.K.; methodology, A.A.C.; investigation, A.A.C. and M.J.; formal analysis, A.B., Z.L. and E.J.; validation, A.A.C., V.K. and R.R.J.; resources and data curation, A.A.C., A.B., R.S. and T.C.; writing—original draft preparation, A.A.C.; writing—review and editing, V.K., P.C. and M.J.; visualization, A.A.C. and M.J.; supervision, A.A.C., V.K., A.B., Z.L. and P.C.; project administration, A.A.C.; funding acquisition, M.J., Z.L. and E.J. All authors have read and agreed to the published version of the manuscript.

Funding: K38W05D02 Wroclaw University of Science and Technology.

Institutional Review Board Statement: Not applicable.

Informed Consent Statement: Not applicable.

Data Availability Statement: Data are available on request to abrar0613@gmail.com.

Conflicts of Interest: The authors declare no conflict of interest.

References

1. Qazi, A.; Hussain, F.; Rahim, N.A.B.D.; Hardaker, G.; Alghazzawi, D.; Shaban, K.; Haruna, K. Towards Sustainable Energy: A Systematic Review of Renewable Energy Sources, Technologies, and Public Opinions. *IEEE Access* **2019**, *7*, 63837–63851. [CrossRef]
2. Chhipa, A.A.; Vyas, S.; Kumar, V.; Joshi, R.R. Role of Power Electronics and Optimization Techniques in Renewable Energy Systems. In *Intelligent Algorithms for Analysis and Control of Dynamical Systems*; Kumar, R., Singh, V.P., Akhilesh, M., Eds.; Springer: Singapore, 2021; pp. 167–175. [CrossRef]
3. Wang, H.; Di Pietro, G.; Wu, X.; Lahdelma, R.; Verda, V.; Haavisto, I. Renewable and Sustainable Energy Transitions for Countries with Different Climates and Renewable Energy Sources Potentials. *Energies* **2018**, *11*, 3523. [CrossRef]
4. Chen, Z.; Yin, M.; Zou, Y.; Meng, K.; Dong, Z.Y. Maximum Wind Energy Extraction for Variable Speed Wind Turbines with Slow Dynamic Behavior. *IEEE Trans. Power Syst.* **2017**, *32*, 3321–3322. [CrossRef]
5. Rodríguez-Amenedo, J.L.; Arnaltes, S.; Rodríguez, M.A. Operation and Coordinated Control of Fixed and Variable Speed Wind Farms. *Renew. Energy* **2008**, *33*, 406–414. [CrossRef]
6. Portillo, R.C.; Prats, M.Á.M.; Leon, J.I.; Sánchez, J.A.; Carrasco, J.M.; Galván, E.; Franquelo, L.G. Modeling Strategy for Back-to-Back Three-Level Converters Applied to High-Power Wind Turbines. *IEEE Trans. Ind. Electron.* **2006**, *53*, 1483–1491. [CrossRef]
7. Valenciaga, F.; Puleston, P.F.; Battaiotto, P.E. Variable structure system control design method based on a differential geometric approach: Application to a wind energy conversion subsystem. *IEE Proc. Control. Theory Appl.* **2004**, *151*, 6–12. [CrossRef]
8. Zhou, F.; Liu, J. A Robust Control Strategy Research on PMSG-Based WECS Considering the Uncertainties. *IEEE Access* **2018**, *6*, 51951–51963. [CrossRef]
9. Mesemanolis, A.; Mademlis, C.; Kioskeridis, I. Optimal Efficiency Control Strategy in Wind Energy Conversion System with Induction Generator. *IEEE J. Emerg. Sel. Top. Power Electron.* **2013**, *1*, 238–246. [CrossRef]
10. Akhmatov, V. Variable-Speed Wind Turbines with Doubly-Fed Induction Generators Part III: Model with the Back-to-Back Converters. *Wind Eng.* **2003**, *27*, 79–91. [CrossRef]
11. Naidu, N.K.S.; Singh, B. Grid-Interfaced DFIG-Based Variable Speed Wind Energy Conversion System with Power Smoothening. *IEEE Trans. Sustain. Energy* **2017**, *8*, 51–58. [CrossRef]
12. Cheng, M.; Zhu, Y. The State of the Art of Wind Energy Conversion Systems and Technologies: A Review. *Energy Convers. Manag.* **2014**, *88*, 332–347. [CrossRef]
13. Lin, W.M.; Hong, C.M. Intelligent Approach to Maximum Power Point Tracking Control Strategy for Variable-Speed Wind Turbine Generation System. *Energy* **2010**, *35*, 2440–2447. [CrossRef]
14. Hosseinzadeh, M.; Salmasi, F.R. Analysis and detection of a wind system failure in a micro grid. *J. Renew. Sustain. Energy* **2016**, *8*, 043301–043317. [CrossRef]
15. Tripathi, S.M.; Tiwari, A.N.; Singh, D. Grid-Integrated Permanent Magnet Synchronous Generator Based Wind Energy Conversion Systems: A Technology Review. *Renew. Sustain. Energy Rev.* **2015**, *51*, 1288–1305. [CrossRef]
16. Abdullah, M.A.; Yatim, A.H.M.; Tan, C.W.; Saidur, R. A Review of Maximum Power Point Tracking Algorithms for Wind Energy Systems. *Renew. Sustain. Energy Rev.* **2012**, *16*, 3220–3227. [CrossRef]

17. Kumar, D.; Chatterjee, K. A Review of Conventional and Advanced MPPT Algorithms for Wind Energy Systems. *Renew. Sustain. Energy Rev.* **2016**, *55*, 957–970. [CrossRef]
18. Apata, O.; Oyedokun, D.T.O. An Overview of Control Techniques for Wind Turbine Systems. *Sci. Afr.* **2020**, *10*, e00566. [CrossRef]
19. Lalouni, S.; Rekioua, D.; Idjdarene, K.; Tounzi, A. Maximum Power Point Tracking Based Hybrid Hill-Climb Search Method Applied to Wind Energy Conversion System. *Electr. Power Compon. Syst.* **2015**, *43*, 1028–1038. [CrossRef]
20. Hua, A.C.-C.; Cheng, B.C.-H. Design and Implementation of Power Converters for Wind Energy Conversion System. In Proceedings of the 2010 International Power Electronics Conference—ECCE ASIA, Sapporo, Japan, 21–24 June 2010; pp. 323–328. [CrossRef]
21. Femia, N.; Granozio, D.; Petrone, G.; Spagnuolo, G.; Vitelli, M. Predictive Amp; Adaptive MPPT Perturb and Observe Method. *IEEE Trans. Aerosp. Electron. Syst.* **2007**, *43*, 934–950. [CrossRef]
22. Urtasun, A.; Sanchis, P.; San Martín, I.; López, J.; Marroyo, L. Modeling of Small Wind Turbines Based on PMSG with Diode Bridge for Sensorless Maximum Power Tracking. *Renew. Energy* **2013**, *55*, 138–149. [CrossRef]
23. Xia, Y.; Ahmed, K.H.; Williams, B.W. Wind Turbine Power Coefficient Analysis of a New Maximum Power Point Tracking Technique. *IEEE Trans. Ind. Electron.* **2013**, *60*, 1122–1132. [CrossRef]
24. Abdullah, M.A.; Yatim, A.H.M.; Tan, C.W. An Online Optimum-Relation-Based Maximum Power Point Tracking Algorithm for Wind Energy Conversion System. In Proceedings of the 2014 Australasian Universities Power Engineering Conference, AUPEC 2014—Proceedings, Perth, WA, Australia, 28 September–1 October 2014. [CrossRef]
25. Bendib, B.; Belmili, H.; Krim, F. A Survey of the Most Used MPPT Methods: Conventional and Advanced Algorithms Applied for Photovoltaic Systems. *Renew. Sustain. Energy Rev.* **2015**, *45*, 637–648. [CrossRef]
26. Hosseini, S.H.; Farakhor, A.; Haghighian, S.K. Novel Algorithm of Maximum Power Point Tracking (MPPT) for Variable Speed PMSG Wind Generation Systems through Model Predictive Control. In Proceedings of the ELECO 2013—8th International Conference on Electrical and Electronics Engineering, Bursa, Turkey, 28–30 November 2013; pp. 243–247. [CrossRef]
27. Yu, K.N.; Liao, C.K. Applying Novel Fractional Order Incremental Conductance Algorithm to Design and Study the Maximum Power Tracking of Small Wind Power Systems. *J. Appl. Res. Technol.* **2015**, *13*, 238–244. [CrossRef]
28. Hohm, D.P.; Ropp, M.E. Comparative Study of Maximum Power Point Tracking Algorithms. *Prog. Photovolt. Res. Appl.* **2003**, *11*, 47–62. [CrossRef]
29. Pagnini, L.C.; Burlando, M.; Repetto, M.P. Experimental Power Curve of Small-Size Wind Turbines in Turbulent Urban Environment. *Appl. Energy* **2015**, *154*, 112–121. [CrossRef]
30. Hiloowala, R.M.; Sharaf, A.M. A Rule-Based Fuzzy Logic Controller for a PWM Inverter in a Stand Alone Wind Energy Conversion Scheme. *IEEE Trans. Ind. Appl.* **1996**, *32*, 57–65. [CrossRef]
31. Galdi, V.; Piccolo, A.; Siano, P. Designing an Adaptive Fuzzy Controller for Maximum Wind Energy Extraction. *IEEE Trans. Energy Convers.* **2008**, *23*, 559–569. [CrossRef]
32. Galdi, V.; Piccolo, A.; Siano, P. Exploiting Maximum Energy from Variable Speed Wind Power Generation Systems by Using an Adaptive Takagi-Sugeno-Kang Fuzzy Model. *Energy Convers. Manag.* **2009**, *50*, 413–421. [CrossRef]
33. Ata, R. Artificial Neural Networks Applications in Wind Energy Systems: A Review. *Renew. Sustain. Energy Rev.* **2015**, *49*, 534–562. [CrossRef]
34. Thongam, J.S.; Bouchard, P.; Ezzaidi, H.; Ouhrouche, M. Artificial Neural Network-Based Maximum Power Point Tracking Control for Variable Speed Wind Energy Conversion Systems. In Proceedings of the IEEE International Conference on Control Applications, St. Petersburg, Russia, 8–10 July 2009; pp. 1667–1671. [CrossRef]
35. Patyra, M.J.; Grantner, J.L.; Koster, K. Digital Fuzzy Logic Controller: Design and Implementation. *IEEE Trans. Fuzzy Syst.* **1996**, *4*, 439–459. [CrossRef]
36. Jang, J.-S.R. ANFIS: Adaptive-Network-Based Fuzzy Inference System. *IEEE Trans. Syst. Man. Cybern.* **1993**, *23*, 665–685. [CrossRef]
37. Abad, G.; López, J.; Rodríguez, M.A.; Marroyo, L.; Iwanski, G. *Doubly Fed Induction Machine: Modeling and Control for Wind Energy Generation*; Institute of Electrical and Electronics Engineers: Piscataway, NJ, USA, 2011. [CrossRef]
38. Biswas, D.; Sahoo, S.S.; Tripathi, P.M.; Chatterjee, K. Maximum Power Point Tracking for Wind Energy System by Adaptive Neural-Network Based Fuzzy Inference System. In Proceedings of the 2018 4th International Conference on Recent Advances in Information Technology (RAIT), Dhanbad, India, 15–17 March 2018; pp. 1–6. [CrossRef]
39. Asghar, A.B.; Liu, X. Adaptive Neuro-Fuzzy Algorithm to Estimate Effective Wind Speed and Optimal Rotor Speed for Variable-Speed Wind Turbine. *Neurocomputing* **2018**, *272*, 495–504. [CrossRef]
40. Kumar, A.; Giribabu, D. Performance Improvement of DFIG Fed Wind Energy Conversion System Using ANFIS Controller. In Proceedings of the 2016 2nd International Conference on Advances in Electrical, Electronics, Information, Communication and Bio-Informatics (AEEICB), Chennai, India, 27–28 February 2016; pp. 202–206. [CrossRef]
41. Mathworks. Available online: https://www.mathworks.com/products/hdl-coder.html (accessed on 21 August 2021).

energies

MDPI

Article

The Influence of Power Network Disturbances on Short Delayed Estimation of Fundamental Frequency Based on IpDFT Method with GMSD Windows

Józef Borkowski [1], Mirosław Szmajda [2,*] and Janusz Mroczka [1]

[1] Faculty of Electronics, Photonics and Microsystems, Wroclaw University of Science and Technology, 50-372 Wroclaw, Poland; jozef.borkowski@pwr.edu.pl (J.B.); janusz.mroczka@pwr.edu.pl (J.M.)

[2] Faculty of Electrical Engineering, Automatic Control and Informatics, Division of Control Science and Engineering, Opole University of Technology, 45-758 Opole, Poland

* Correspondence: m.szmajda@po.edu.pl

Abstract: This paper presents an application of the IpDFT spectrum interpolation method to estimate the fundamental frequency of a power waveform. Zero-crossing method (ZC) with signal prefiltering was used as a reference method. Test models of disturbances were applied, based on real disturbances recorded in power networks, including voltage harmonics and interharmonics, transient overvoltages, frequency spikes, dips and noise. It was determined that the IpDFT method is characterized by much better dynamic parameters with better estimation precision. In an example, in the presence of interharmonics, the frequency estimation error was three times larger for the reference method than that for the IpDFT method. Furthermore, during the occurrence of fast transient overvoltages, the IpDFT method reached its original accuracy about three times faster than the ZC method. Finally, using IpDFT, it was possible to identify the type of disturbances: impulsive, step changes of frequency or voltage dips.

Keywords: frequency estimation; spectrum interpolation; power network disturbances; power quality

Citation: Borkowski, J.; Szmajda, M.; Mroczka, J. The Influence of Power Network Disturbances on Short Delayed Estimation of Fundamental Frequency Based on IpDFT Method with GMSD Windows. *Energies* **2021**, *14*, 6465. https://doi.org/10.3390/en14206465

Academic Editor: Tomonobu Senjyu

Received: 12 September 2021
Accepted: 2 October 2021
Published: 9 October 2021

Publisher's Note: MDPI stays neutral with regard to jurisdictional claims in published maps and institutional affiliations.

1. Introduction

Monitoring the frequency of voltage waveforms in power supply networks is one of the most important parameters from the point of view of, e.g., power quality measurements [1,2] and power network protection automation devices, especially those occurring in inverters used in the photovoltaic industry [3–5]. In the first industry, incorrect estimation of the fundamental frequency of the waveform results in a so-called spectrum leakage, and consequently, in an increased uncertainty of measurement of spectral parameters such as harmonics and interharmonics. On the other hand, incorrect estimation of the fundamental frequency in the second industry results in large power losses in the case of frequency mismatch between two different power systems and their phase synchronization [6]. This is important both in the case of cooperation among large power systems and also when connecting a large number of micro photovoltaic power plants to the power grid [4]. Inverters, which are a key component of such systems, must be equipped with disturbance-tolerant algorithms that estimate the instantaneous frequency of the electrical grid. According to [3], the requirements for power generation modules require the use of the so-called limited frequency sensitive mode at overfrequency (LFSM-O) and limited frequency sensitive mode at underfrequency (LFSM-U), which consist in a dynamic change of the generated energy as a function of changes in the grid frequency. Moreover, the aim of the protection automatics in photovoltaic systems is to protect the systems against unwanted switching between grid-connected and islanding operating modes [4,7]. For this purpose, the rate-of-change-of-frequency (ROCOF) method is used, among others. This method requires calculating the frequency over a number of cycles and comparing with a specified trigger threshold [3,8].

There are many algorithms for estimating the fundamental frequency of power waveforms. These methods can be divided into time, filter, frequency and other methods (e.g., time–frequency methods).

One of the basic temporal methods is the zero-crossing method. This method, due to its sensitivity to noise found in real power waveforms, is supplemented by prefiltering or regression least mean squares (LMS) modeling methods [9]. These models [10,11] and their extended least squares (ELS) extensions [12] are also used independently of the zero-crossing method.

Filtering methods include the use of Kalman filters [12–15]. In the ROCOF method, Kalman filters have been used, for example, to reduce the nondetection zone (NDZ) [8]. Furthermore, a combination of Kalman filtering and the least squares (LS) method is used to improve the dynamic properties of accurate frequency estimation [16].

Spectral methods provide the estimation of the fundamental frequency of the power waveform based on the spectrum processing obtained by discrete Fourier transform (DFT). In order to improve the accuracy of the spectrum estimation, its interpolation IpDFT (which will be discussed later in this paper) or other modifications, such as the so-called smart DFT (SDFT) with the additional use of the complex-valued least squares (CLS-SDFT) method [17], are used. The DFT transformation is also used repeatedly [18]. Spectrum interpolation can also be based on non-Fourier methods, e.g., Prony's method to model the signal using exponentially damped sinusoids [19] or the Aboutanios and Mulgrew's (HAM) method [20] and the cooperative weighted least squares (WLS) method representing HAM-WLS [20]. Spectrum estimation is also performed using time–frequency analyses e.g., discrete wavelet transform (DWT) [21] or Wigner–Ville transform [22].

Finally, the methods can be applied to the analysis of single- and three-phase waveforms [23–25].

IpDFT spectrum interpolation methods deserve special attention; the first one was developed by Rife and Vincent in 1970 [26], but their rapid development has occurred over the past two decades. These methods combine the advantageous properties of using nonrectangular time windows, a FFT algorithm and further processing of the resulting DFT spectrum to reduce errors due to the discrete nature of the resulting spectrum. The use of such time windows as Hann (often referred to as Hanning windows), Hamming, Kaiser, Chebyshev and many others, allows a significant reduction of the phenomenon of spectrum leakage, originating from harmonics and interharmonics in the conditions of nonsynchronous (noncoherent) sampling with the period of the measured signal. On the other hand, the use of the FFT algorithm in a modern DSP system with a signal processor that is optimized for fast execution of such an algorithm makes the computation time much smaller than the signal measurement time, while maintaining a low cost of the system. The time window method combined with the FFT algorithm belongs to classical Fourier analysis and has a reputation for being fast but not very accurate due to the so-called picket fence effect. This effect is the cyclic nature of the frequency and amplitude estimation errors when these parameters are determined from the local maximum of the raw DFT spectrum. Then, the maximum error of frequency estimation is half of the DFT computational resolution (i.e., $\pm 0.5/NT$—half the inverse of the measurement time) and does not depend on the applied data window. For typical time windows, the maximum amplitude estimation errors caused by picket fence effect range from -9% to -22% [27]. The essence of IpDFT methods is to further process the raw DFT spectrum to eliminate errors caused by the discrete nature of the DFT spectrum (picket fence effect) without significantly increasing the computation time. In this way, IpDFT methods combine the beneficial properties of classical Fourier analysis methods (high speed and low cost) and non-Fourier methods (usually much more accurate, but much slower). The latest strands of IpDFT methods are those that take into account the conjugate component in the spectrum, developed especially for applications with short measurement times—of the order of a few signal periods [28–31]. Their latest version [31] is further discussed in a more detailed way in Section 3 and its basic flowchart for a real-time grid-monitoring system is presented in Figure 1.

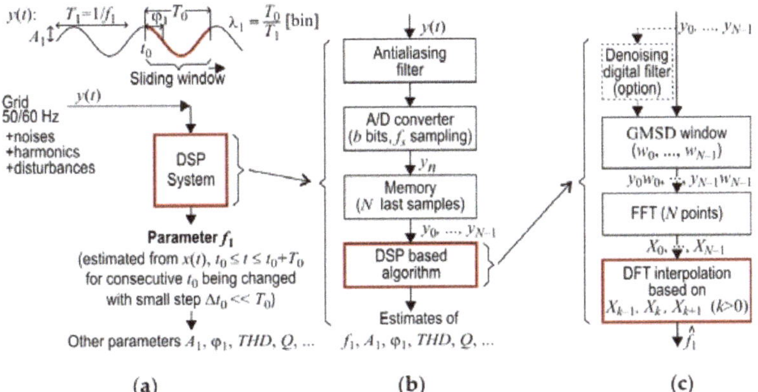

Figure 1. The real-time grid-monitoring system with a sliding time window: (**a**) the general system scheme with the definition of the sliding window; (**b**) the DSP block diagram; (**c**) the algorithm to estimate the fundamental frequency f_1 of the grid signal $y(t)$.

The method presented in [31] and in Figure 1 has the following advantages:

- It allows for the estimation in short measurement times with high accuracy because the conjugate component in the spectrum (i.e., a component with a negative frequency resulting from the mathematical properties of the Fourier transform) is taken into account during derivation of the estimating formula;
- The estimation formula used in the calculations only applies a GMSD window, the FFT algorithm and a simple interpolation formula, which involves three points of the FFT-resulting spectrum;
- It allows for a cheap implementation, because the calculation time depends mainly on the calculation time of the FFT algorithm, hence modern cheap signal processors and microcontrollers optimized for FFT are applicable;
- It can be used for signals with a large THD coefficient, because GMSD windows significantly eliminate its influence on the accuracy of the estimation.

The basic comparison of the presented method with parametric methods is presented in Table 1. More quantitative details are provided, e.g., in [29,30]. It shows that the presented method has very good properties for short, delayed estimation of the fundamental frequency, as for $\lambda_1 \approx 1 \dots 3$ it still maintains its high accuracy and at the same time it does not require many calculations.

The paper focuses on the influence of the most important disturbances on the estimation results. It is the first systematical study of this aspect of the method [31]. The models of grid disturbances are presented in Section 2 and the interpolated discrete Fourier transform is summarized in detail in Section 3. The important role of the proper choice of the method's parameters is described in detail in Section 4, as well as a short description of the characteristics of the zero-crossing (ZC) method used in the paper as a reference method. The simulations performed in the MATLAB software environment are presented in Section 5, and they include detailed information about the features of the method when a grid signal contains various disturbances, as is usually the case. The conclusions are presented in Section 6.

Table 1. Qualitative comparison of the IpDFT method with GMSD windows (which takes into account the conjugate component during derivation of the estimation formula) vs. parametric methods depending on the normalized frequency λ_1 (the number of signal periods in the measurement window).

Measurement Window Duration	IpDFT-Based Frequency Estimation Methods		Parametric Methods
	IpDFT Method Which Takes into Account Conjugate Component	**IpDFT Methods Which Do Not Take into Account Conjugate Component**	**Parametric Methods: Prony LS, TLS (Total Least Squares), ESPRIT**
Very short window ($\lambda_1 < 1$)	Applicable only for low level of noise	Not applicable due to great number of systematic errors	Applicable, especially for high resolution methods (e.g., ESPRIT)
Short window ($\lambda_1 \approx 1 \dots 3$)	Applicable Good accuracy (systematic errors below the level caused by noise) Cheap implementation	Not applicable due to great number of systematic errors	Applicable, but expensive in practice due to the great number of calculations required (even by a few orders for N > 1000)
Long window ($\lambda_1 \gg 3$)	Applicable Good accuracy Cheap implementation	Applicable Good accuracy Cheap implementation	Applicable Good accuracy Expensive implementation

2. Disturbances and Its Models in Power Networks

2.1. Disturbances in Power Networks and Power Quality

The voltage electrical waveform in power networks is by definition a strictly deterministic signal. It is modeled by a single sinusoidal waveform with nominal rms value, symmetric with respect to the 0 V level and nominal frequency. In three-phase networks, these waveforms are additionally shifted in individual phases by 120°.

In fact, this waveform additionally contains disturbances of deterministic and non-stationary character. The first group includes cyclic or permanent disturbances that are specific to the supply network, for example, the continuous presence of the 5th harmonic, long-lasting voltage underdeviation or asymmetry. The second group includes disturbances appearing every indefinite time, or whose character evolves in time in an unpredictable manner, e.g., voltage overvoltages caused by lightning, momentary changes of the fundamental frequency under the influence of rapid changes of the load or oscillatory disturbances occurring during activation of capacitor banks to compensate reactive power.

Some of the disturbances occurring in the supply voltage waveform have a direct impact on the power quality and on the operation and safety of electrical equipment. These include all kinds of voltage fluctuations, harmonics, interharmonics, frequency fluctuations and supply voltage asymmetry. The remainder at least affects the user comfort of electrical equipment. These disturbances can include, inter alia: the phenomenon of flickering caused by a series of changes in the rms value of the voltage of the light source supply, dips and interruptions in power supply. Definitions and acceptable levels of most of the mentioned disturbances are defined in relevant standards of power quality, such as EN501610 [32], and electromagnetic compatibility from the IEC 61000 series [33].

Based on the authors' own research, various types of interference were acquired, which can be divided into three groups:

- Temporary disturbances—of incidental nature;
- Sustained disturbances—occurring in the power network for a longer period of time;
- Mixed disturbances—containing the two above mentioned types of disturbances.

Figure 2 as well as Figures 3 and 4 present examples of disturbances from the first group: transient overvoltage, voltage fluctuations and dips and instantaneous frequency changes, respectively.

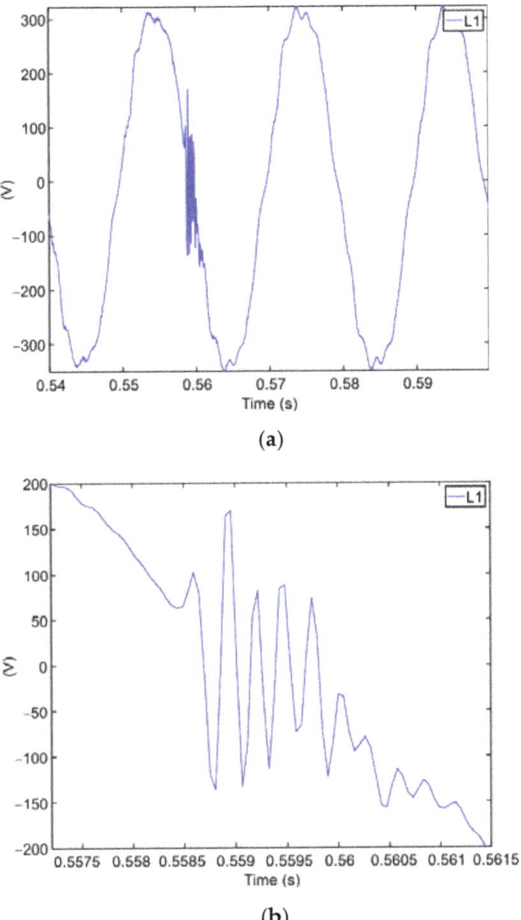

Figure 2. High-frequency damped sinusoidal fast transient overvoltage with oscillation frequency of 4 kHz: (**a**) signal with the disturbance; (**b**) expanded disturbance.

The duration of the first one is about 3 ms and it has an oscillatory character with a frequency of about 4 kHz.

The second disturbance is an example of a voltage drop of about 2.4% and two consecutive dips of 12% and 24%.

The third disturbance represents changes in the instantaneous frequency of the voltage. These changes are often of a sustained disturbance nature, which include long-term deviations of the power network frequency within ±1% for 99.5% of the year [32]. However, dynamic fluctuations of the fundamental frequency are also encountered, caused by rapid changes in the power network load, which include the frequency fluctuation of 0.3% shown in Figure 3.

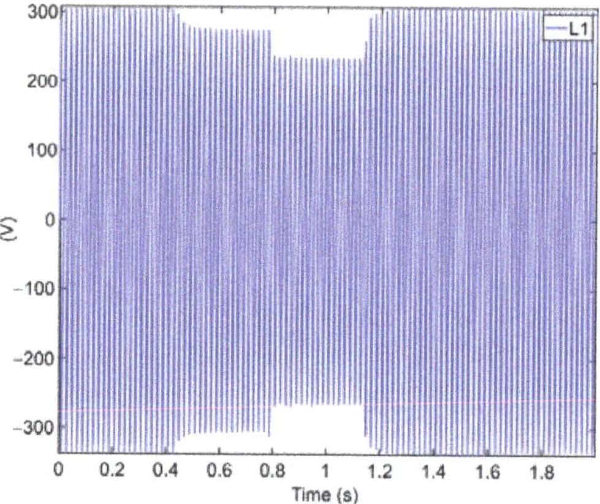

Figure 3. Waveform with voltage variations: 0–0.4 s voltage drop 2.4%; 0.4–0.8 s voltage dip 12%; 0.8–1.2 s voltage dip 24%; after 1.2 s voltage drop 2%.

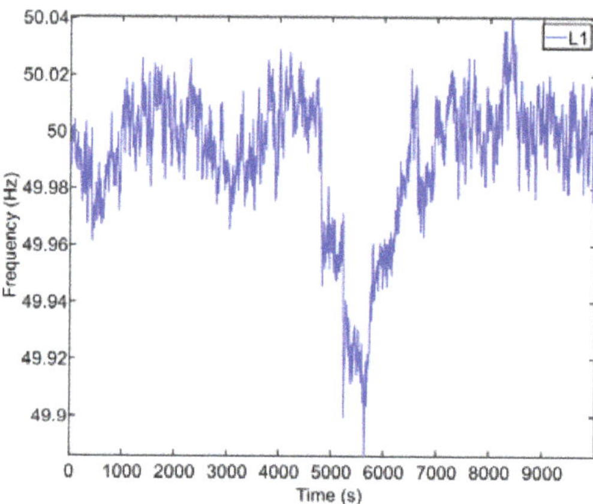

Figure 4. Temporary changes in fundamental frequency.

The second group of disturbances—sustained disturbances—often involves, among other things, the occurrence of harmonics and interharmonics, which are generated by nonlinear loads, in particular by voltage converters, available, for example, in any IT equipment. These disturbances are parameterized, for example, by so-called spectral power quality parameters, which include [34]:

- Total distortion factor of harmonic subgroups—Equation (1),

$$THDS_U = \sqrt{\sum_{h=h_{min}}^{h_{max}} \left(\frac{U_{sg,h}}{U_{sg,1}} \right)^2} \tag{1}$$

where:

$$U_{sg,h}^2 = \sum_{k=-1}^{1} U_{C,(N \wedge h)+k}^2 \tag{2}$$

$U_{sg,h}^2$—harmonic subgroup of the h order.
$U_{sg,1}^2$—first-harmonic subgroup.
$U_{C,l}$—rms value of the spectral voltage component of the l order.

N—number of periods of the fundamental component.
h—harmonic order.
h_{min}—minimum harmonic order: 2.
h_{max}—maximum harmonic order: 50.

- Centered subgroups of interharmonics—Equation (3),

$$U^2_{isg,h} = \sum_{k=2}^{N-2} U^2_{C,(N\times h)+k'} \tag{3}$$

where the interpretation of individual parameters is as for Equation (2).

Examples of harmonic disturbances registered in an office building are presented in Figure 5. The values of particular harmonic subgroups are shown in Table 2. Deformation of the waveform in the area of minima and maxima is characteristic for this type of disturbances. Moreover, this type of disturbance usually has a quasi-stationary character and lasts for a minimum of several dozen periods or persists permanently. As a rule, the values of even harmonics are minimal, while the dominant ones are the 5th, 7th and 11th harmonics. The total distortion factor of the harmonic subgroups should remain below 8% for 95% of the week [32]. The mentioned example represents the exceedance of this value: 10.3%. When interharmonics are present, the distortion factor of the centered interharmonic subgroups has not yet been defined in the relevant standard.

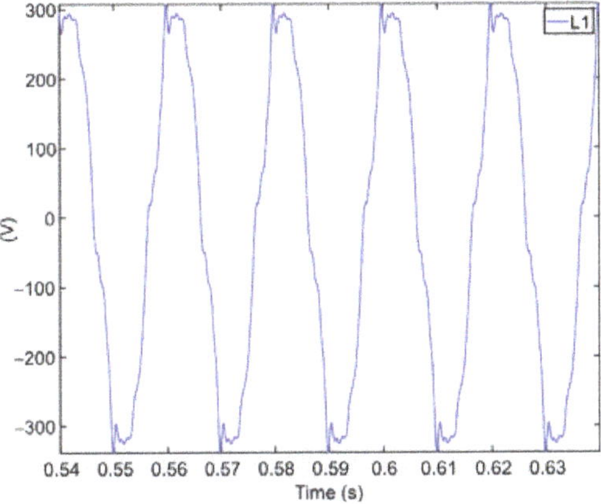

Figure 5. Harmonic disturbance.

Table 2. Parameters of measured harmonics from $THDS_U = 10.3\%$.

Parameter	i—Harmonic Number											
	1	3	5	7	9	11	13	15	17	19	21	23
$f_i = i \cdot f_1\,[\text{Hz}]$	50	150	250	350	450	550	650	750	850	950	1050	1150
$U_{sg,i}$ rms $[\text{V}]$	225	0.6	15.2	14.5	1.2	6.4	4.1	0.8	4.6	3.6	0.9	0.6
$U_{sg,i}$ rms $[\%]$	100	0.29	6.74	6.47	0.56	2.91	1.82	0.37	2.04	1.65	0.41	0.26

Noise is also a disturbance that affects the distortion of the waveform spectrum. Figure 6 shows an example of recorded noise-like interference with a signal-to-noise (SNR) of 27 dB. Disturbances of this type are also usually of sustained character.

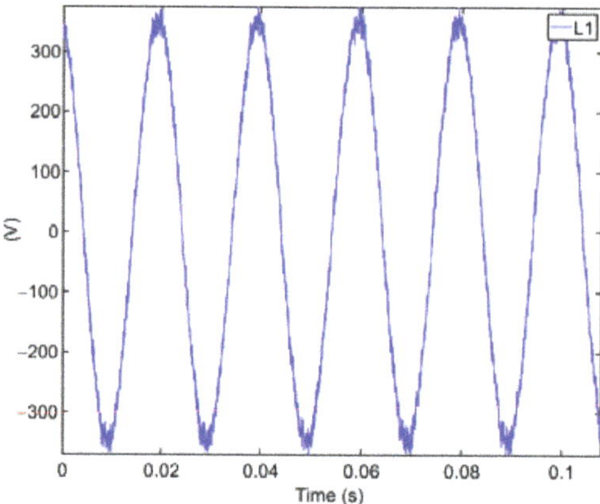

Figure 6. Noise waveform with SNR of 27 dB.

An example of the third group of disturbances, concerning a combination of both transient and sustained disturbances, is presented in Figure 7. It shows a harmonic disturbance with superimposed transient overvoltage of oscillatory nature.

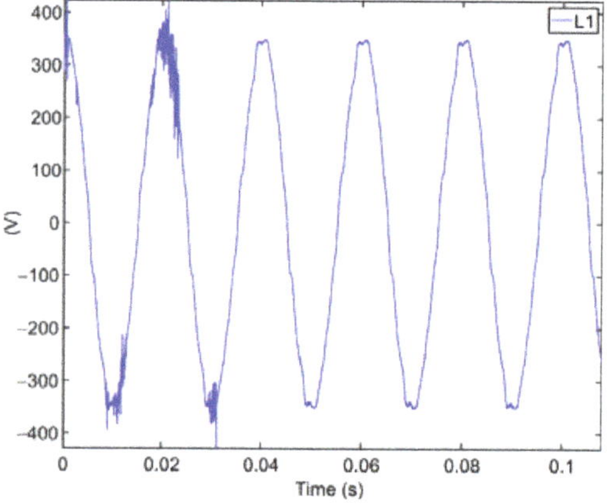

Figure 7. Waveform with complex disturbances: harmonics and oscillations.

2.2. Disturbance Modeling

Based on the real disturbances occurring in power grids and on the test procedures included in the standards for power quality and electromagnetic compatibility, as well as in the relevant documents related to the requirements imposed on inverters in PV microgrids, the following disturbance models are proposed. These models will allow us to investigate the influence of certain types of disturbances on the accuracy of the instantaneous frequency measurement.

2.2.1. Harmonics and Interharmonics

The parameters of the signal models representing the harmonic and interharmonic disturbances are presented in Tables 3 and 4, respectively. Due to the specificity of the power waveforms, harmonics are practically nonexistent and they have not been taken into account. The amplitudes of the individual harmonics and interharmonics have been selected based on actual tests. The $THDS_U$ parameter exceeds the acceptable threshold defined in the standards—8% [32].

Table 3. Harmonic parameters of the signal from $THDS_U$ = 10.3%.

Parameter	*i*—Harmonic Number											
	1	3	5	7	9	11	13	15	17	19	21	23
$f_i = i \cdot f_1 \,[\mathrm{Hz}]$	50	150	250	350	450	550	650	750	850	950	1050	1150
$U_{sg,i}\,[\mathrm{V}]$	225	0.6	15.2	14.5	1.2	6.4	4.1	0.8	4.6	3.6	0.9	0.6
$\varphi_i\,[^\circ]$	0	0	0	0	0	0	0	0	0	0	0	0

Table 4. Interharmonic signal parameters.

Parameter	*i*—Interharmonic Number								
	1	3.2	3.3	3.4	3.5	3.6	3.7	3.8	3.9
$f_i = i \cdot f_1\,[\mathrm{Hz}]$	50	160	165	170	175	180	185	190	195
$U_{ig,i}\,[\mathrm{V}]$	230	1.5	2	4.5	12	17	4	3	2
$\varphi_i\,[^\circ]$	0	0	0	0	0	0	0	0	0

Graphical representations of the waveforms with harmonic and interharmonic disturbance models are shown in Figures 8a and 8b, respectively.

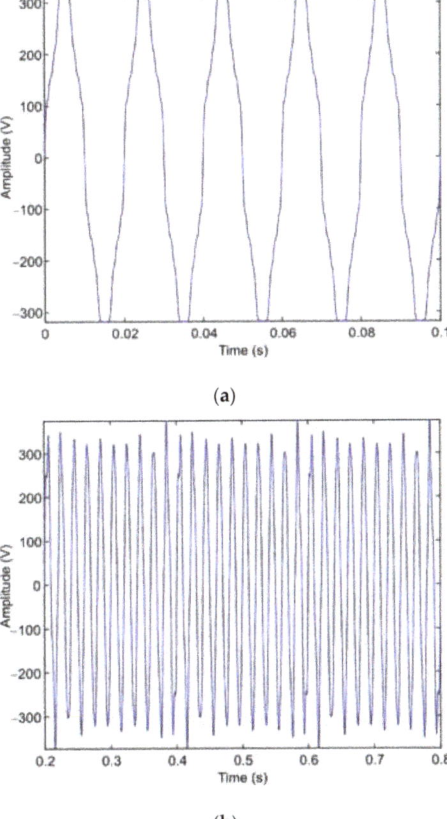

(a)

(b)

Figure 8. Models of (**a**) harmonic and (**b**) interharmonic disturbances.

2.2.2. Transient Overvoltages

The transient overvoltages were modeled using exponentially damped oscillatory disturbances. Two models were proposed: short duration and long duration. These models were expressed as (4) and (5):

$$s_{\mathrm{I}}(t) = 2000\mathrm{e}^{-0.0583t}\sin(2\pi 10^{6}t) \tag{4}$$

$$s_{\mathrm{II}}(t) = 500\mathrm{e}^{-0.00029278t}\sin(2\pi 10^{3}t) \tag{5}$$

In the first case, the oscillation frequency is equal to 1 MHz, the amplitude is 2000 V, and the duration after which the waveform is damped 100 times is 15 µs.

In the second case, these parameters are, respectively: 1 kHz, 500 V and 6 ms.

In the case of a short-duration disturbance, its amplitude was limited to 400 V due to the fact that the input systems of the measuring equipment are practically unable to transfer such a steep disturbance and in fact, the final waveform to be acquired has an amplitude of 400 V.

Figure 9a shows the signal with the disturbance expressed by (4), and Figure 9b shows the superimposed disturbance on the power waveform. Figure 10 shows the above for the disturbance defined by Equation (5).

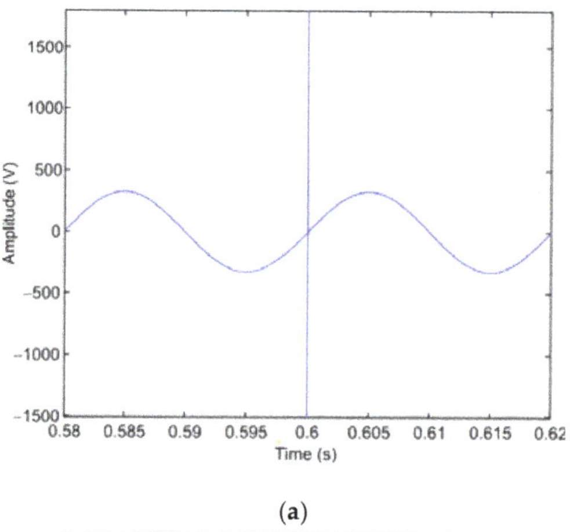

(a)

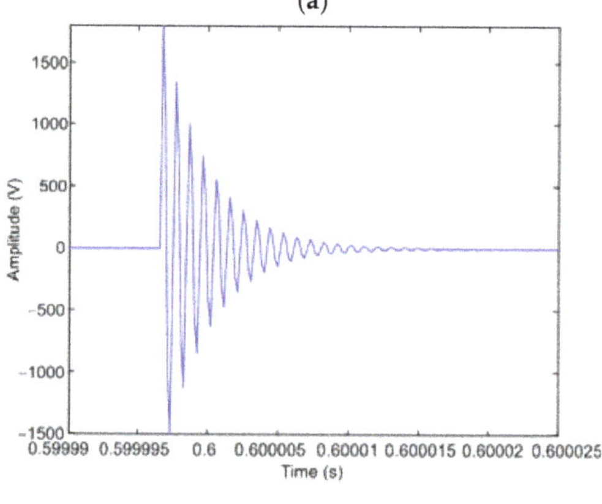

(b)

Figure 9. Model of fast transient disturbance with frequency oscillation of 1 MHz: (**a**) power signal and (**b**) expanded disturbance.

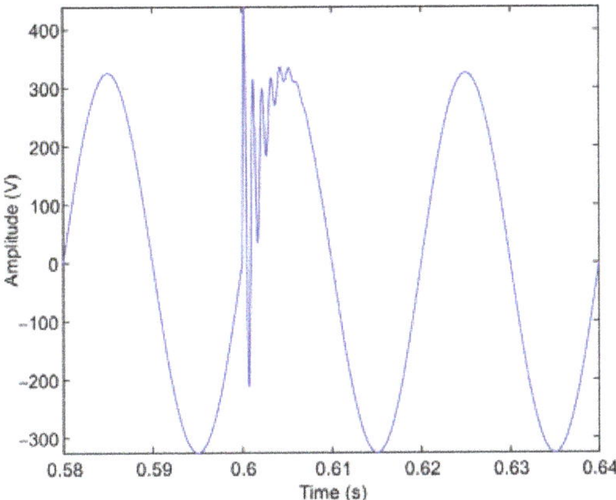

Figure 10. Model of fast transient disturbance with frequency oscillation of 1 kHz.

2.2.3. Changes in Fundamental Frequency

The model of the signal containing fluctuations of the basic component of the power waveform consists of five waveforms with the rms value of 230 V and the following frequencies (6):

$$
y(t) = \begin{cases}
42.50 \text{ Hz} & \text{for } t < 0.353 \text{ [s]} \\
46.25 \text{ Hz} & \text{for } 0.353 < t < 0.677 \text{ [s]} \\
50.00 \text{ Hz} & \text{for } 0.677 < t < 0.977 \text{ [s]} \\
53.375 \text{ Hz} & \text{for } 0.977 < t < 1.258 \text{ [s]} \\
57.50 \text{ Hz} & \text{for } 1.258 < t < 1.600 \text{ [s]}
\end{cases}
\tag{6}
$$

Each waveform contains 16 complete periods at the frequencies specified above.

2.2.4. Multicomponent Disturbances: Harmonics, Dips, Transient Overvoltages

Finally, a model consisting of harmonics, two transient overvoltages and three voltage dips was used to illustrate the response of the tested frequency measurement method in the presence of many different types of disturbances. Detailed parameters are shown in Table 5. Throughout the test run, the fundamental frequency was constant at 50 Hz.

Table 5. Multicomponent disturbance parameters.

Type of Disturbance	Parameter	Occurrence Time (s)
Harmonics	In accordance with Table 2	0–1.6
Transient overvoltages	In accordance with Formula (5)	0.6–1.2
Voltage dips	Urms = 230 V	0–0.7
	Urms = 215 V	0.7–0.9
	Urms = 181 V	0.9–1.1
	Urms = 162 V	1.1–1.3
	Urms = 230 V	1.3–1.6

Figure 11 presents the signal with disturbances, as described in Table 5.

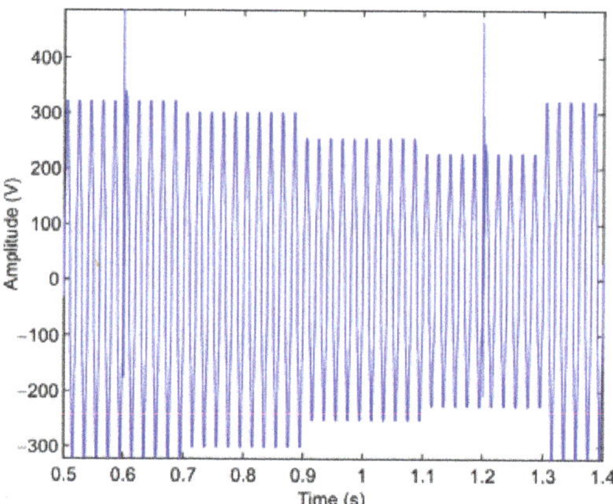

Figure 11. Model of multicomponent disturbance.

3. Interpolated Discrete Fourier Transform with the Use of GMSD Windows

The characteristics of the power network signal presented in Section 2 allow us to define the stationary part of this signal as:

$$y(t) = \sum_{m=1}^{K} A_m \sin(2\pi f_m t + \varphi_m) \tag{7}$$

where A_1, f_1, and φ_1 are the parameters (amplitude, frequency and phase, respectively) of the fundamental component, and A_i, f_i, and φ_i are the harmonic and interharmonic parameters. Such a signal is found in many fields of science and technology and is often referred to as a multifrequency signal because it consists of multiple sinusoidal components. For obvious reasons, such a signal is further disturbed by the noise present in each measurement and a number of additional distortions, which for the case of the mains signal are described in Section 2. Rms values $U_{i\text{ rms}}$ of the sinusoidal components are obtained from the relation $U_{i\text{ rms}} = A_i / \sqrt{2}$. For a complete description of the power network signal, estimation of all A_i, f_i and $]\varphi_i$ parameters is required for all i, but the most important of these is the value of f_1, i.e., the frequency of the power network signal. It is nominally 50 Hz or 60 Hz, and in practice it is a value varying within the range allowed by legal regulations or exceeding this range in cases of control system failures in the process of electricity generation. Precise determination of the parameter f_1 determines the remaining parameters of the power network signal.

Many methods have been developed, parametric and nonparametric, allowing estimation of the parameters of the multifrequency signal modeled by Equation (7). Particularly noteworthy are spectrum interpolation methods (IpDFT), whose development and range of applications is constantly increasing. The application of IpDFT methods for the determination of f_1 is also studied in the context of power network signal parameter estimation for the following reasons:

- In IpDFT methods, the computation time is much smaller than the signal measurement time, which significantly reduces the total estimation time of f_1 (the total estimation time is equal to the sum of the signal measurement time and the duration of calculations necessary to be performed after the measurement is completed);
- IpDFT methods achieve high estimation accuracy of f_1, even with short signal measurement times, i.e., between one and three signal periods; this allows for a quick response in systems controlling power generation or disconnecting devices from the supply network;
- In IpDFT methods, the computational complexity of the estimation algorithm is much lower than in parametric methods, which lowers the cost of the DSP system

for determining the parameters of the power network signal; this is particularly important for small power generation systems (e.g., small photovoltaic installations) or for power quality monitoring by consumers.

The latest IpDFT methods are based on the complex values of the DFT spectrum obtained by the FFT algorithm and include in their solution a conjugate component, i.e., a component with a negative frequency, resulting from the mathematical properties of the Fourier transform. The use of the complex values of the DFT spectrum eliminates the drawbacks of earlier IpDFT methods, which used the DFT spectrum modules. On the other hand, the inclusion of the conjugate component allows for a significant reduction in signal measurement time, relative to earlier IpDFT methods, without increasing the systematic error values. The paper [31] presents the latest version of the IpDFT method taking into account the conjugate component and using the complex values of the DFT spectrum. This method uses windows of $\sin^m$ type, which are, as shown in [31], generalized MSD windows already used in earlier IpDFT methods. The IpDFT method presented in [31] uses GMSD windows, which are described by two, mutually equivalent, relations:

$$w_n = \sum_{h=0}^{H-1} (-1)^h a_h \cos\left[\frac{(2h+z)\pi n}{N} + \frac{z\pi}{2}\right], \; n = 0, \ldots, N-1 \tag{8}$$

$$w_n = \sin^m\left(\frac{\pi n}{N}\right) = \sin^{2r+z}\left(\frac{\pi n}{N}\right), \; n = 0, \ldots, N-1 \tag{9}$$

where w_n are the values of the GMSD window function for indices $n = 0, \ldots, N-1$, N is the number of samples of the GMSD window, $m = 2r + z$ is a natural number, $H = r + 1$, $z = 0$ for even m and $z = 1$ for odd m, and a_h are appropriately chosen coefficients to ensure that the GMSD window has the largest sidelobes damping for a given H. Relationships (8) and (9) are equivalent, as shown in [31]. To derive the solution of the method in [31], (8) is used, and for practical use, the more convenient form (9).

The signal $y(t)$ from (7) is processed in an A/D converter operating at a sampling rate f_s ($T = 1/f_s$ is the sampling period) and N signal samples $y_n = y(nT)$ are obtained at the A/D output. After introducing normalized frequencies $\lambda_m = f_m NT$ in (bins) or (*cycles in range (CiR)*) (and in general, $\lambda = fNT$) the estimation of f_1 is equivalent to the estimation of λ_1. The use of λ in place of f and λ_m in place of f_m simplifies the mathematical model and allows a direct frequency reference of f_1 from the power network signal to the duration of the signal measurement (time window duration). For example, the value of $\lambda_1 = 2.5$ bin means that a time window of NT (N samples obtained every sampling period T) contains 2.5 periods of the power network signal.

Applying the time window (8) and (9) to the signal samples (7) reduces to multiplying y_n by w_n for all n and then the discrete-time Fourier transform (DtFT) is calculated, i.e., X_λ defined by:

$$X_\lambda = X(\lambda) = \sum_{n=0}^{N-1} y_n w_n e^{-j2\pi n\lambda/N} \text{ for any } \lambda \tag{10}$$

The spectrum X_λ for natural values of λ is a commonly known discrete Fourier transform (DFT), which is usually calculated with the FFT algorithm. However, the IpDFT method from [31] requires knowledge of the spectrum $X(\lambda)$ for $\lambda = 0, 0.5, 1, 1.5$, etc. (i.e., with a step of 0.5 bin), which is obtained by the FFT algorithm in so-called double zero padding technique [27].

The most important result in [31] is the dependencies on $\hat{\lambda}_1$ (and consequently on $\hat{f}_1 = \hat{\lambda}_1/NT$):

$$\hat{\lambda}_1^2 = l_1^2 + \text{Re}\{\varepsilon_1\} \tag{11}$$

$$\varepsilon_1 = (m+2)\frac{(m+2+4l_1)X_{l_1+1} + 2mX_{l_1} + (m+2-4l_1)X_{l_1-1}}{4(X_{l_1+1} - 2X_{l_1} + X_{l_1-1})} \tag{12}$$

where:

$$l_1 = 1, \ 1.5, \ 2, \ 2.5, \ \dots \tag{13}$$

This solution means that the spectrum interpolation procedure uses three DtFT spectrum points obtained with a step of 1 bin. The value of l_1 should be close to the value of λ_1, as this condition minimizes the variance of the estimator and hence the error due to noise in the signal.

The value $\hat{\lambda}_1$ differs from λ_1 due to the presence of systematic errors (which even occur for pure sine wave), the influence of harmonics and interharmonics, the presence of noise in a real measured signal and other signal disturbances. The determination of the impact of typical disturbances occurring in the power network on the obtained frequency estimation accuracy f_1 is the goal of the research presented in Section 5 in accordance with the methodology presented in Section 4.

4. Research Methodology

4.1. Assumptions for IpDFT

4.1.1. Measurement Time (Time Window Duration) NT

Based on the properties of the IpDFT method presented in [31], it was assumed in this study that the NT time of the power network signal measurement with frequency $f_1 = 50$ Hz is $NT = 50$ ms. This means that the normalized frequency $\lambda_1 = 2.5$ bin, i.e., that the measurement window (time window) covers 2.5 periods of the power network signal. The value of the measurement time $NT = 50$ ms was chosen in the study as a compromise value between aiming for the shortest possible measurement time and aiming for the highest possible accuracy. The IpDFT method of [31] allows accurate signal measurement for several times smaller measurement times, but then the effect of signal harmonics on the estimation accuracy increases, causing a significant reduction in accuracy. For such smaller measurement times (of the order of 1–2 signal periods) it is necessary, to maintain high accuracy, to use filtering that reduces the influence of harmonics, and this would degrade the dynamic properties of the method. As demonstrated by the studies presented in Section 5 adoption of $NT = 50$ ms ($\lambda_1 = 2.5$ bin) maintains high estimation accuracy without the need for additional filtering at the input of the method. In practice the time window will not cover exactly 2.5 signal periods due to the power network frequency variation f_1 and because of the characteristics of the A/D converters used, for which a change in sampling frequency $f_s = 1/T$ is not possible continuously, but only among a limited set of values. However, the most important advantage of IpDFT methods is that there is no need to precisely synchronize the duration of the measurement window with the current frequency f_1 of the supply network. This is an essential feature of IpDFT methods that distinguishes them from methods with so-called coherent sampling. The adoption of $NT = 50$ ms is also the condition furthest from coherent sampling, since in coherent sampling, the duration of the measurement window must be equal to an integer multiple of the signal period. The lack of significant sensitivity of the IpDFT method tested in Section 5 to changing the duration of the time window is shown in Section 5.1, where the accuracy of the method was tested for $NT = 50$ ms and for two values differing by ca. $\pm 3\%$: $NT = 48.44$ ms and $NT = 51.65$ ms. In the other studies (Sections 5.2–5.7), it was also found that a similar change in NT measurement time does not significantly affect the estimation results, and therefore only the case of $NT = 50$ ms was included in the presented graphs to maintain the readability of the presented figures.

4.1.2. Number of Signal Samples N and Sampling rate $f_s = 1/T$

For most studies in Section 5, the number of samples was assumed to be equal to $N = 1024$ and $N = 2048$. This means, with measurement time $NT = 50$ ms, that the sampling rate is, respectively $f_s = 20{,}480$ Hz and $f_s = 40{,}960$ Hz. The accuracy of the method for wide-spectrum signal disturbances (Section 5.3) is also shown for $f_s = 2{,}621{,}440$ Hz ≈ 2.62 MHz (for $N = 2^{17}$), to show that the error for smaller values of f_s is caused by the fact that the conditions of Shannon's sampling theorem regarding the minimum value of the sampling frequency are not satisfied. The increase in the error due to the failure to satisfy the conditions of Shannon's sampling theorem is eliminated in practice by the use of an antialiasing filter at the input of the A/D converter and partly by the limited spectrum of the coupling systems between the network and the measuring system. With the limited frequency spectrum of the power network signal, it is possible to use smaller values of N and f_s. For the studies presented in Sections 5.1, 5.2 and 5.5, the following are presented in addition to the two fundamental cases $N = 1024$ and 2048 ($f_s = 20{,}480$ Hz and 40,960 Hz providing $NT = 50$ ms) and the results for $N = 128$, 256 and 512 ($f_s = 2560$ Hz, 5120 Hz and 10,240 Hz providing $NT = 50$ ms). The presentation of results for five different values of N from 128 to 2048 provided a better representation of how the parameters N and f_s at the given NT affect the accuracy of the estimation. In many cases, low values of N and f_s allow to obtain satisfactory estimation accuracy under the condition of limited power network signal bandwidth (e.g., due to the use of antialiasing low-pass filter) and fulfillment of the condition of Shannon's theorem for the minimum sampling frequency. It should also be remembered that increasing the number of samples of the signal N (i.e., increasing the sampling frequency of f_s at a constant NT measurement time value) reduces the effect of noise on the estimation accuracy, as shown in Section 5.7.

4.1.3. The m Parameter of GMSD Window

The basic parameter of the IpDFT method from [31] is, in addition to N and f_s, the m parameter of the GMSD window defined by (8) and (9). In the studies presented in Section 5 a fixed value of $m = 3$ was adopted as a compromise between the desire to suppress the sidelobes as much as possible (long-range spectral leakage) and to widen the main lobe width as little as possible. An increase in the value of m also results in an increase in the variance of the estimator, i.e., the error due to noise in the signal. The authors' studies have shown that the value of $m = 3$ is the optimum case for many applications of power network parameter estimation and can be treated as a universal condition when there is no more specific other rationale for adopting a different value of m. A study of the effect of other values of m in the presence of the disturbances considered in Section 5, in the opinion of the authors, should be studied for individual target applications, which is beyond the scope of this paper, which aims to present the basic properties of the method for a wide class of applications.

4.1.4. The Effect of Signal Phase and Sliding Window

The phase φ_1 of the power network signal described by (7) has important implications on the accuracy of the estimation. In the case of continuous observation of the signal (Figure 12), as is the case in network parameter monitoring systems, the phase of the signal changes over time as the measurement window of duration NT includes the last N samples of the signal relative to the current instant of time (Figure 12b). Each such set of consecutive N samples can be referred to as a data frame (Figure 12c). As shown in [31], the computation time required in a typical DSP system to determine the frequency f_1 based on a single data frame is much smaller than the NT time of measuring that frame, which is a well-known advantage of IpDFT methods. Therefore, in this study, it is assumed that the computation time is negligibly small and that the current estimate f_1 appears in the output of the method with the last sample of a given frame, i.e., with the end of the measurement time of that frame. The next estimate f_1 is calculated after a sampling period T and it takes into account the last N samples of the signal. This updating of the estimate f_1 with each successive signal sample was adopted in the study in Section 5, in order to accurately demonstrate the effect of the phase φ_1 value, which changes with a small step for successive data frames. In practice, the calculation of the frequency f_1 can be performed less frequently because successive data frames differ by only one sample, i.e., they overlap to a large extent. Frequency calculation of f_1 with every kth sample reduces the computational power requirements of the DSP system by k times. For example, in the

DSP system used in [31], the frequency f_1 is updated every second sample of the signal, i.e., every $2T \approx 90.7$ μs, where $T = 1/f_s \approx 45.4$ μs ($f_s = 22.05$ kHz).

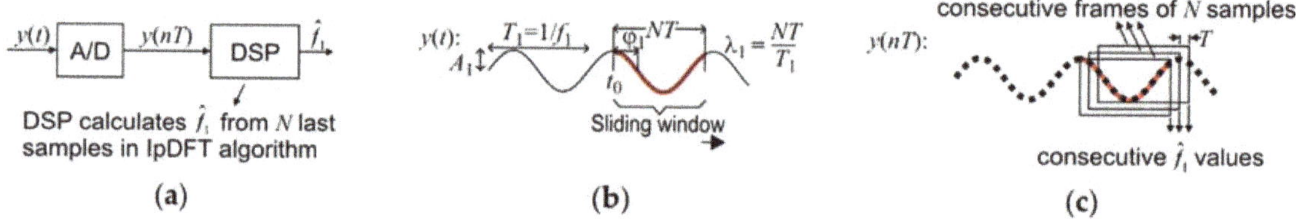

Figure 12. Determination of the frequency estimate f_1 with IpDFT method at every signal sample based on consecutive data frames: (**a**) estimation of f_1 based on (11) and (12) in the DSP system; (**b**) sliding window and phase definition φ_1 of the fundamental component relative to the origin point t_0 of the window; (**c**) updating of the estimate f_1 every sampling period T for successive data frames.

4.2. Assumptions for Zero Crossing Method

The zero-crossing (ZC) method was used as a reference method for the investigated IpDFT method from [31], since it is one of the most popular methods for estimating the frequency of a power network signal. This frequency is calculated in the ZC method based on the last two transitions of the signal through zero, thus by determining the duration of one half-period of the signal (Figure 13). It is possible to reduce the estimation error by taking into account many half-periods in one measurement instead of one, but the essential properties of the ZC method do not change significantly, and additionally the dynamic properties of the method deteriorate. Therefore, this paper is limited to the determination of a single signal half-period. In order to precisely determine the moment when the signal passes through zero, linear interpolation was used (Figure 13b), as is usually the case in the ZC method.

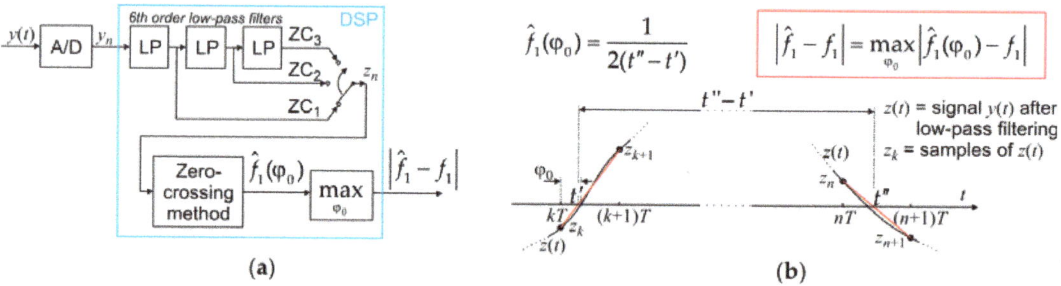

Figure 13. Determining the frequency estimate f_1 by zero-crossing (ZC) method and the maximum estimation error $\left| \hat{f}_1 - f_1 \right|$ with low-pass prefiltering: (**a**) three prefiltering options (ZC$_1$, ZC$_2$ and ZC$_3$); (**b**) estimation of f_1 in ZC method with linear interpolation.

In the ZC method, low-pass filtering of the input signal is necessary, because without such filtering any high-frequency (impulse) disturbance introduces too large errors into the frequency estimation result. A digital filter of order 4–6 is most commonly used; however, results presented in Section 5.2 showed that higher-order filters are necessary to significantly reduce the errors of the zero-crossing method for disturbances due to interharmonics. In the studies presented in Section 5, three versions of a digital Butterwoth low-pass filter were adopted, each consisting of 1, 2, or 3 sections, respectively, of a 6th-order Butterworth filter with a cutoff frequency of $f_c = 60$ Hz (Figure 13a). The signal obtained as a result of such filtering is labeled in Section 5 with the labels ZC$_1$, ZC$_2$ and ZC$_3$ (Figures 14b and 15b), and this filtering introduces a signal delay that depends on the filter order (Figure 14b). The results of the frequency estimation of f_1 obtained using these three types of filtering are also labeled in Section 5 with the labels ZC_1, ZC_2 and ZC_3 (Figures 14f, 15d–f, 16e, 17e, 18d, 19e,g,i and 20b,c).

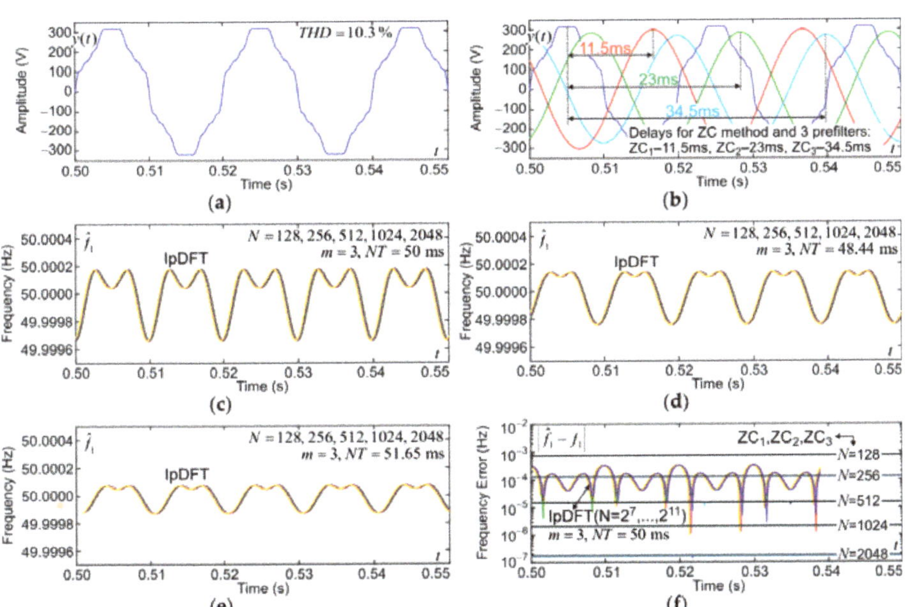

Figure 14. Effect of harmonics on estimation results: (**a**) 50 Hz power network signal model of $THDS_U = 10.3\%$; (**b**) delays brought on by 6th-order (ZC_1), 12th-order (ZC_2) and 18th-order (ZC_3) filters for ZC; (**c–e**) frequency estimation by IpDFT for $m = 3$ and of N samples at measurement times of 50 ms, 48.44 ms and 51.65 ms; (**f**) comparison of the accuracy of the IpDFT method for $m = 3$ and $NT = 50$ ms with the ZC method for $N = 128, 256, 512, 1024$ and 2048.

The ZC method is sensitive to the phase shift φ_0 of the first sample considered in the calculations with respect to the point where the signal passes through zero (Figure 13b). Therefore, in the studies in Section 5 for the zero-crossing method, the maximum value of the estimation error was determined from the set of values calculated for multiple values of φ_0.

5. Results

5.1. The Influence of Harmonics

The study of harmonics influence was carried out for a signal containing 12 consecutive odd harmonics with parameters defined in Table 2 and with $THDS_U = 10.3\%$ (Figure 9a). For the duration of the measurement window of 50 ms, the maximum estimation error depends on the phase φ_1 of the power network signal relative to the beginning of the time window, which results in a variation of the estimated value of f_1 on the time axis. Changing the measurement time by a few (or even by several percent) does not significantly affect the estimation accuracy (Figure 14c–e illustrates this effect for $NT = 48.44$ ms, 50 ms and 51.65 ms). In addition, the change in f_1 (as is the case in practice) at a constant NT value will not result in a significant change in accuracy, since the decisive factor in both of the above cases (i.e., changing NT at a constant f_1 and the change of f_1 at constant NT) is the value $\lambda_1 = f_1 NT = NT/T_1$, and thus the ratio of the NT window duration to the signal period of T_1. The number of samples N does not affect the error values due to harmonics in the signal (Figure 14c–e). This property occurs when the condition of the sampling theorem is satisfied and when the dominant component of the estimation is the error due to harmonics in the signal. For the ZC method, a different situation occurs because the estimation error depends on the number of samples N. For $N = 128$. the estimation error is larger than for the IpDFT method, for $N = 256$ it is, depending on the phase φ_1, smaller or larger, and for $N \geq 512$, the estimation error for the ZC method is almost always smaller than for the IpDFT method (Figure 14f). This ability to decrease the estimation error with increasing N for the ZC method is a beneficial property of the method, but it comes at the expense of the additional delay (Figure 14b) introduced by the low-pass filtering shown in Figure 13a. In the case of errors due to harmonics, the order of the filter used (6 for ZC_1, 12 for ZC_2 and 18 for ZC_3) does not affect the results of the ZC estimation (Figure 14f). For the IpDFT method and for the adopted parameters of this method, the estimation error does not exceed, depending on the quotient NT/T_1 the value, ca. 0.0003 Hz (Figure 14c), ca. 0.0002 Hz (Figure 14d) and ca. 0.0001 Hz (Figure 14e).

5.2. The Influence of Interharmonics

The study of the effect of harmonics was performed for a signal containing a group of interharmonics with parameters defined in Table 4 (Figure 15a). The influence of the interharmonic group on estimation accuracy of f_1 is larger than for the harmonics group. For the IpDFT method, the estimation error is less than ca. 0.0015 Hz and it does not depend, as in Section 5.1, on the number of samples N (Figure 15c). For the ZC method, in order to eliminate the influence of interharmonics, a 12th-order low-pass filter is necessary (ZC_2, Figure 15e) for $N < 512$, because the filter of order 6 causes a significant increase in estimation error (ZC_1, Figure 15d). For $N \geq 512$, a full minimization of errors caused by interharmonics in the ZC method requires a filter order of 18 (ZC_3, Figure 15f), which is done at the expense of a higher signal delay (the signal with interharmonics after filtering is shown in Figure 15b, and the values of delays contributed by the filters ZC_1, ZC_2 and ZC_3 are provided in Figure 14b).

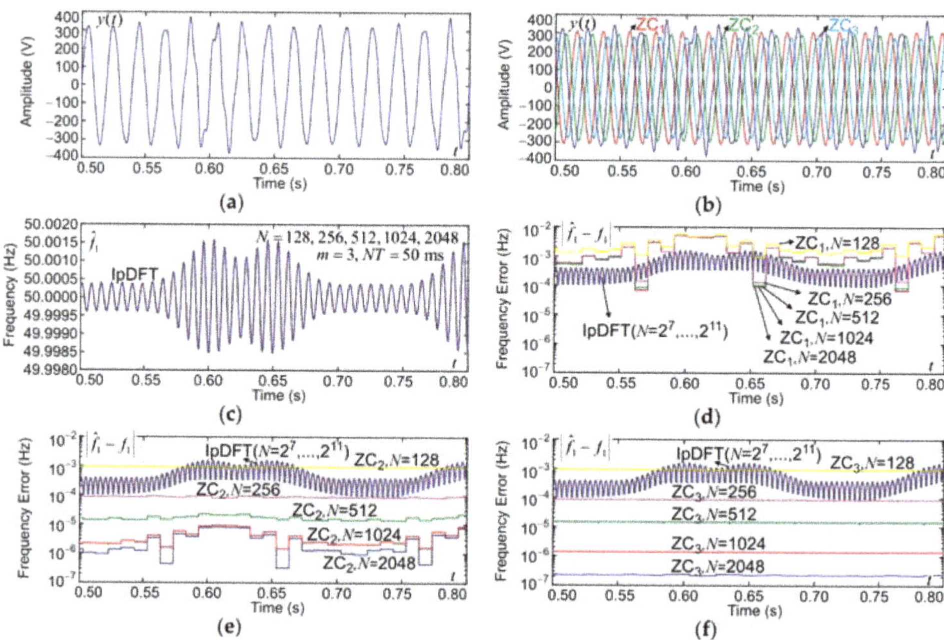

Figure 15. Effect of interharmonics on estimation results: (**a**) 50 Hz power network signal disturbed by interharmonics; (**b**) signals at the output of 6th order (ZC_1), 12th order (ZC_2) and 18th order (ZC_3) filters for ZC; (**c**) IpDFT frequency estimation for $m = 3$ at a measurement time of 50 ms and for $N = 128, 256, 512, 1024$ and 2048; (**d–f**) comparison of the accuracy of the IpDFT method for $m = 3$ and $NT = 50$ ms with the ZC method for $N = 128, 256, 512, 1024$ and 2048 and for the 6th-order (ZC1), 12th-order (ZC2) and 18th-order (ZC3) filters.

5.3. The Influence of Transient Overvoltages—High-Frequency Pulse

To study the effect of transient overvoltages of short duration, i.e., wide frequency band, the model (4) with oscillations of 1 MHz was used. Such a signal (Figure 16a) was added to the power network signal (Figure 16b) and the signal amplitude was limited to $\pm$ 400 V, due to the saturation effect of the input circuits and the A/D converter itself that occurs in practice. For N = 1024 (f_s = 20.48 kHz for NT = 50 ms), the maximum estimation error of f_1 is 0.1 Hz (Figure 16c). An increase of N by a factor of two to 2048 (i.e., increase by a factor of two of f_s) causes the reduction of the maximum error by ca. two times, i.e., to a value of ca. 0.05 Hz (Figure 16c). However, only adopting $N = 2^{17}$, $f_s > 2 \times 1$ MHz eliminates errors caused by failure to satisfy Shannon's sampling theorem (Figure 16d). The comparison of the accuracy of the IpDFT method with the ZC method shows (Figure 16e) that the maximum errors are at a similar level, but the return to very low values of the estimation errors after the occurrence of the high-frequency pulse is much faster for the IpDFT method than for the ZC method. Taking into account the principle of the ZC method, it is obvious that it is not possible to use this method for this kind of disturbances without low-pass filtering at the input of the method. For the IpDFT method, however, the presented results indicate that it is advisable to apply an antialiasing filter at the input of the A/D converter in order to reduce the impact of such interference on the result of the estimation, but this is not a necessary condition—when it is not met, the maximum error is ca. 0.1 Hz (Figure 16c).

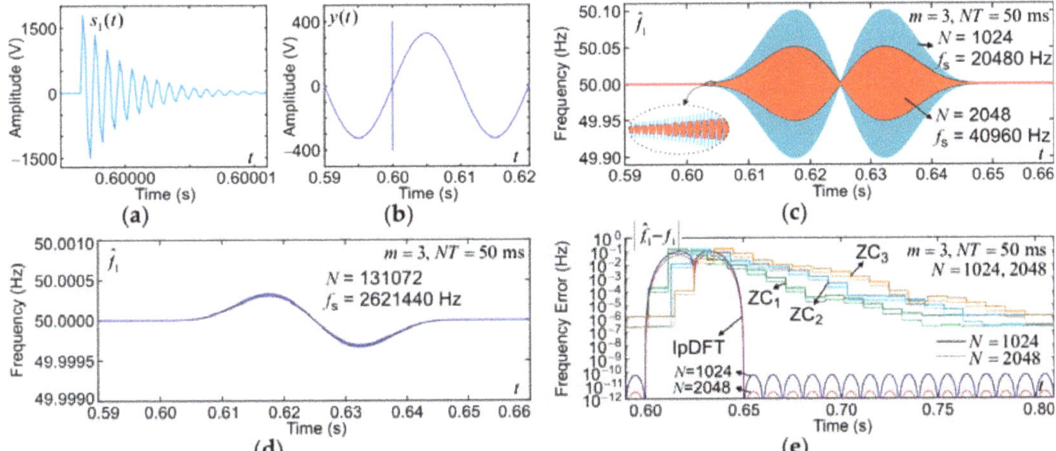

Figure 16. Effect of high-frequency pulse disturbance on estimation results: (**a**) disturbing pulse; (**b**) 50 Hz power network signal disturbed by the pulse and limited in amplitude to $\pm$ 400V; (**c**,**d**) estimation result of IpDFT method for m = 3 at 50 ms measurement time and for N = 1024, 2048 and 131072; (**e**) comparison of accuracy of IpDFT method for m = 3 and NT = 50 ms with ZC method for N = 1024 and 2048.

5.4. The Influence of Transient Overvoltages—Low-Frequency Pulse

To study the effect of transient overvoltages with long duration, i.e., narrow frequency band, model (5) with 1 kHz oscillations was used. Such a signal (Figure 17a) was added to the power network signal (Figure 17b). For N = 1024 and 2048, the conditions of Shannon's theorem for sampling frequency are satisfied, and the maximum error of estimation of f_1 in the IpDFT method does not exceed 0.5 Hz and practically does not depend on N (Figure 17c). The return to very low estimation errors after a low-frequency pulse is much faster for the IpDFT method than for the ZC method (Figure 17d,e). The pulse modeled by (5) results in larger estimation errors than that modeled by (4), even though the condition of Shannon's theorem on sampling frequency is satisfied, because the energy contributed by pulse (5) is much larger than that by pulse (4). Similar to the study in Section 5.3, it is clear from the performance of the ZC method that it is not possible to apply this method to low-frequency pulse disturbances without low-pass filtering at the input of the method. On the other hand, for the IpDFT method, the results presented show that the error rate does not exceed ca. 0.5 Hz with a rapid return to low values after the cessation of the disturbance.

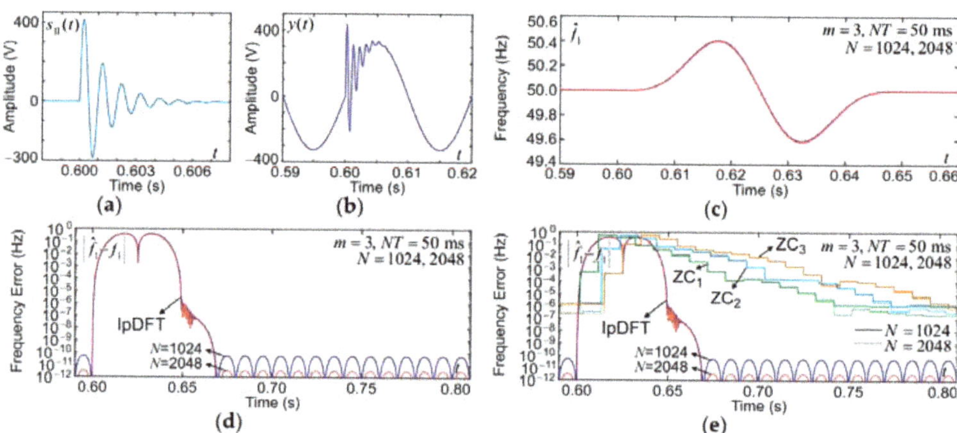

Figure 17. Effect of low-frequency pulse interference on estimation results: (**a**) disturbing pulse; (**b**) 50 Hz power network signal disturbed by pulse; (**c**,**d**) estimation result of IpDFT method for m = 3 at 50 ms measurement time and for $N = 1024$ and 2048; (**e**) Comparison of accuracy of IpDFT method for $m = 3$ and $NT = 50$ ms with ZC method for $N = 1024$ and 2048.

5.5. The Influence of Frequency Changes

The impact of a step change in f_1 on the estimation results was examined for a signal for which the frequency changed in steps from 42.5 Hz every ca. 0.25–0.35 s (Figure 18a). Since the IpDFT method determines f_1 based on the last N samples of the signal, then after time NT the estimation error caused by the frequency step decreases to zero (Figure 18b,c). In this case, the IpDFT method also recovers much faster after a frequency step to very low values than the ZC method (Figure 18d).

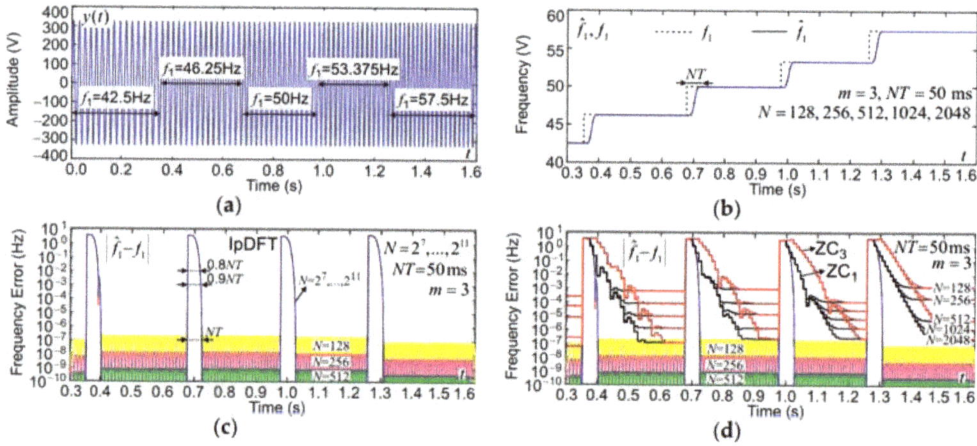

Figure 18. Effect of frequency stepping on the estimation results: (**a**) frequency change from 42.5 Hz to 57.5 Hz every 15 periods; (**b**,**c**) estimation result of IpDFT method for m = 3 at 50 ms measurement time and for $N = 128, 256, 512, 1024$ and 2048; (**d**) comparison of accuracy of IpDFT method for $m = 3$ and $NT = 50$ ms with ZC method for $N = 128, 512, 1024$ and 2048.

5.6. The Composition of Harmonics, Pulse and Voltage Fall and Rise

To compare the influence of different types of disturbances, the estimation accuracy was determined for the signal in which harmonics, transient overvoltages and voltage dips occur, with the parameters defined in Table 5 (Figure 19a). In the IpDFT method, there are large changes in the estimation result (Figure 19d) due to the low-frequency pulse (at time instants 0.6 s and 1.2 s) and smaller changes due to the amplitude step change (at time instants 0.7, 0.9, 1.1 and 1.3 s). The determination of the type of disturbance only based on changes in the estimation result of f_1 does not seem simple, but in combination with the estimation of the signal amplitude (Figure 19f) there appears, as an effect of the application of the IpDFT method, an additional possibility of real-time detection and classification of the types of disturbances of the power network signal. Even the simplest estimation of the power network signal amplitude seems to be sufficient to perform such detection and classification. For example, Figure 19f shows the result obtained after determining the value of the complex spectrum for the determined f_1 (Figure 19d). For this purpose, one of the methods of determining one point of the spectrum for a given frequency, e.g., Goertzel's algorithm, can be used. The plot of the estimation error on the logarithmic scale (Figure 19h) shows good dynamic properties of the IpDFT method, i.e., a fast return to very small estimation errors after the disturbance disappears.

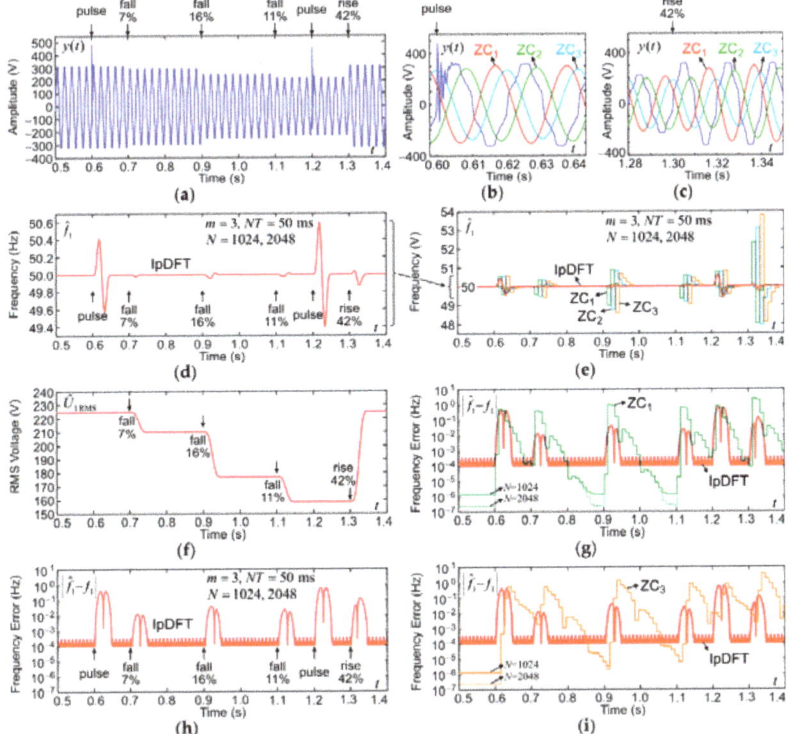

Figure 19. Effect of the sum of distortions by harmonics, low-frequency pulse, voltage dip, and voltage rise on the estimation results: (**a**) time domain signal distorted with harmonics and two pulses (at 0.6 s and 1.2 s) and four voltage steps (at 0.7 s, 0.9 s, 1.1 s and 1.3 s); (**b,c**) filter output signals for ZC method; (**d,h**) estimation result of IpDFT method for $m = 3$ at 50 ms measurement time and for $N = 1024$ and 2048; (**f**) rms measurement of fundamental component for voltage step detection; (**e,g,i**) comparison of accuracy of IpDFT method for $m = 3$ and $NT = 50$ ms with ZC method (for ZC_1 and ZC_3) for $N = 1024$ and 2048.

The ZC method gives more than ten times larger estimation errors of f_1 than the IpDFT method for dip-type disturbances (Figure 19e). The ZC method also has significantly worse dynamic properties in this case (Figure 19g,i).

5.7. The Influence of Noise

The effect of Gaussian white noise on frequency estimation errors in the IpDFT method of [31] is described in more detail in [31]. For comparison with the ZC method, sample noise was added to a signal composed of several disturbances and shown in Section 5.6 and Figure 19a. The estimation results of such a signal with example noise are shown in Figure 20. According to the properties of the method described in [31] for a pure sinusoid with noise, the IpDFT method is unbiased, i.e., the error component due to noise is dominant. Moreover, the properties of the method in [31] show that the variance of the frequency estimator decreases with increasing number of N samples. The results in Figure 20 confirm these properties. At the moments of interference mentioned in Section 5.6, the estimation errors caused by these disturbances are larger than those caused by noise in the signal (Figure 20a). The properties of the IpDFT method for the noisy signal are also clearly more favorable than in the ZC method (Figure 20b,c) for both values of N (1024 and 2048) and regardless of the filter order in the ZC method.

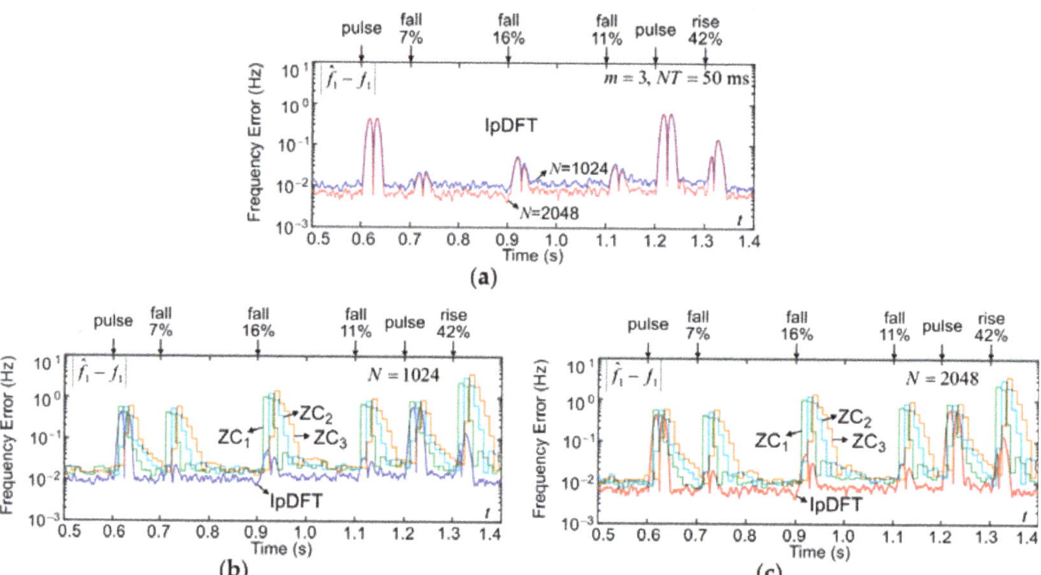

Figure 20. The effect of white noise with normal distribution and $\sigma = 1$ V on the estimation results in the presence of signal noise in Figure 14a: (**a**) estimation result of IpDFT method for $m = 3$ at 50 ms measurement time and for $N = 1024$ and 2048; (**b**,**c**) comparison of accuracy of IpDFT method for $m = 3$ and $NT = 50$ ms with ZC method for $N = 1024$ and 2048.

6. Conclusions

The paper presents the use of the IpDFT spectrum interpolation method for estimating the fundamental frequency of power waveforms. The zero-crossing method with prefiltering by three types of low-pass filters was used as a reference method. The waveforms created on the basis of recorded disturbances occurring in low-voltage power networks were used as test models of signals. The models included: voltage harmonics and inter-harmonics, oscillatory transient overvoltages exponentially damped with high and low frequency pulses, frequency spikes and disturbances consisting of harmonics, voltage dips, transient overvoltages and noise. The results of frequency estimation in the presence of modeled disturbances were analyzed as a function of the number of samples—sampling frequency, measurement window length and in the case of the ZC method for three types of filters.

Conducted research indicates that there is no significant influence of nonsynchronous sampling in the case of the IpDFT method, as it is in case of the ZC method. For a larger number of samples (above 512), the frequency estimation error for the ZC method is smaller than for IpDFT. A simplified summary of the obtained results are demonstrated in Table 6.

Table 6. Maximum frequency estimation level errors for the grid disturbances described in Section 5. For two cases of N (1024 and 2048), NT = 50ms and for the presented IpDFT method with the use of GMSD window with parameter m = 3 and the reference zero-crossing method (ZC1 for one 6th-order filter section, ZC2 and ZC3—for two and three, respectively, filter sections).

Type of Disturbance	Type of Method	For Number of Samples N		Data Source
		1024	**2048**	
The influence of harmonics (*THDSu* = 10.3%)	IpDFT	3.4×10^{-4} Hz		Figure 14f
	ZC1 … 3	2.0×10^{-6} Hz	1.8×10^{-7} Hz	Figure 14f
The influence of interharmonics	IpDFT	0.0016 Hz		Figure 15d–f
	ZC1	0.0052 Hz		Figure 15d
	ZC2	9.2×10^{-6} Hz	8.1×10^{-6} Hz	Figure 15e
	ZC3	1.4×10^{-6} Hz	2.4×10^{-7} Hz	Figure 15f
The influence of transient overvoltages—high-frequency pulse	IpDFT	0.10 Hz in pulse (above 0.001 Hz for 43 ms [1])	0.051 Hz in pulse (above 0.001 Hz for 41 ms [1])	Figure 16e
	ZC1	0.14 Hz in pulse (above 0.001 Hz for 60 ms [1])	0.068 Hz in pulse (above 0.001 Hz for 50 ms [1])	Figure 16e
	ZC2	0.13 Hz in pulse (above 0.001 Hz for 80 ms [1])	0.067 Hz in pulse (above 0.001 Hz for 70 ms [1])	Figure 16e
	ZC3	0.14 Hz in pulse (above 0.001 Hz for 90 ms [1])	0.071 Hz in pulse (above 0.001 Hz for 80 ms [1])	Figure 16e
The influence of transients overvoltages—low-frequency pulse	IpDFT	0.41 Hz in pulse (above 0.001 Hz for 45 ms [1])		Figure 17d,e
	ZC1	0.55 Hz in pulse (above 0.001 Hz for 60 ms [1])		Figure 17e
	ZC2	0.54 Hz in pulse (above 0.001 Hz for 90 ms [1])		Figure 17e
	ZC3	0.58 Hz in pulse (above 0.001 Hz for 100 ms [1])		Figure 17e
The influence of frequency changes	IpDFT	3.8 Hz in pulse (above 0.001 Hz for 45 ms [1])		Figure 18c,d
	ZC1	4.1 Hz in pulse (above 0.001 Hz for 100 ms [1])		Figure 18d
	ZC3	5.0 Hz in pulse (above 0.001 Hz for 150 ms [1])		Figure 18d
The influence of voltage fall and rise	IpDFT	0.13 Hz in pulse (above 0.001 Hz for 41 ms [1])		Figure 19g–i
	ZC1	2.3 Hz in pulse (above 0.001 Hz for 81 ms [1])		Figure 19g
	ZC3	3.8 Hz in pulse (above 0.001 Hz for 120 ms [1])		Figure 19i
The influence of noise	IpDFT	0.014 Hz	0.008 Hz	Figure 20a
	ZC1 … 3	0.020 Hz	0.012 Hz	Figure 20b,c

[1] The time in milliseconds of the temporary increase of the error above the level 0.001 Hz in the transient state caused by the disturbance. This time allows us to present the dynamic property of the compared methods.

The effect of harmonics (Section 5.1) in the ZC method is minimal, due to the use of input prefilters, while for the IpDFT method it does not exceed 0.34 mHz for the harmonic disturbance model used. The precision of fundamental frequency estimation is similar for both methods.

In the presence of interharmonic interference in the signal model (Section 5.2), the frequency estimation error increases from 0.34 mHz to a maximum of 1.6 mHz in the case of IpDFT. For the ZC method, better accuracies are obtained with higher filter orders (minimum 12th-order) and larger sample numbers (minimum 256). For example, for the filter order 6 (ZC1), the estimation precision is about 5.2 mHz, representing over three times higher frequency estimation error than for IpDFT.

In the presence of high-frequency damped oscillatory pulses (Section 5.3), the frequency estimation errors are at a comparable level when the disturbance occurs. In contrast, once the disturbance disappears, the IpDFT method reaches its initial accuracy much faster. The IpDFT method reaches its initial accuracy after only 43 ms ($N = 1024$), while the ZC method with the lowest-order filter reaches it after about 60 ms ($N = 1024$ ZC1) or 90 ms ($N = 1024$ ZC3). It is also noticeable that there is a characteristic short oscillation in the frequency estimated waveform for IpDFT immediately after the disturbance, which can be used to identify the disturbance.

Low-frequency pulses (Section 5.4) have a greater impact on the fundamental frequency estimation than do high-frequency pulses. In the former case, the estimation error is 0.1 Hz ($N = 1024$), compared to 0.41 Hz ($N = 1024$) in the latter case. The value of the error depends primarily on the energy of the disturbance. Comparing with the ZC method, one observes, similarly to the high-frequency pulse, a significantly increased recovery time to the original accuracy after the occurrence of the pulse. The situation is very similar when stepping the frequency (Section 5.5). The time to return to the initial accuracy is at least three times faster for the IpDFT method than for the ZC method.

Finally, the IpDFT method has better dynamic performance under all types of disturbances (Section 5.6) with good estimation of the fundamental frequency. Moreover, based on the analysis of frequency estimation error variations and instantaneous amplitude variations of the fundamental frequency component, it is possible to determine the type of disturbances: pulse type, rms variations and frequency variations. Tests in the presence of noise (Section 5.7) also prove better dynamic properties of the IpDFT method in comparison to ZC, and the frequency estimation accuracy is higher for the first method.

In conclusion, the IpDFT method may find application in protection systems and measuring devices monitoring rapid changes of the fundamental frequency, even in the presence of significant disturbances of the power waveform voltage.

Author Contributions: Conceptualization, M.S., J.B. and J.M.; methodology, M.S., J.B. and J.M.; software, M.S. and J.B.; validation, M.S., J.B. and J.M.; formal analysis, M.S. and J.B.; investigation, M.S. and J.B.; resources, M.S. and J.B.; data curation, M.S. and J.B.; writing—original draft preparation, M.S. and J.B.; writing—review and editing, M.S. and J.B.; visualization, M.S. and J.B.; supervision, J.M.; project administration, M.S.; funding acquisition, M.S. All authors have read and agreed to the published version of the manuscript.

Funding: This research received no external funding.

Institutional Review Board Statement: Not applicable.

Informed Consent Statement: Not applicable.

Data Availability Statement: Data and MATLAB scripts available upon request from the corresponding author.

Conflicts of Interest: The authors declare no conflict of interest.

References

1. Martinek, R.; Bilik, P.; Baros, J.; Brablik, J.; Kahankova, R.; Jaros, R.; Danys, L.; Rzidky, J.; Wen, H. Design of a Measuring System for Electricity Quality Monitoring within the SMART Street Lighting Test Polygon: Pilot Study on Adaptive Current Control Strategy for Three-Phase Shunt Active Power Filters. *Sensors* **2020**, *20*, 1718. [CrossRef]
2. Ramos, P.M.; Cruz Serra, A. Comparison of frequency estimation algorithms for power quality assessment. *Measurement* **2009**, *42*, 1312–1317. [CrossRef]
3. ENTSO-E Network Code for Requirements for Grid Connection Applicable to all Generators, Commission Regulation (EU) 2016/631, April 2016. Available online: https://eur-lex.europa.eu/legal-content/EN/TXT/?uri=CELEX%3A32016R0631 (accessed on 1 September 2021).
4. Mastromauro, R.A. Grid Synchronization and Islanding Detection Methods for Single-Stage Photovoltaic Systems. *Energies* **2020**, *13*, 3382. [CrossRef]

5. Jeong, S.; Lee, J.; Yoon, M.; Jang, G. Energy Storage System Event-Driven Frequency Control Using Neural Networks to Comply with Frequency Grid Code. *Energies* **2020**, *13*, 1657. [CrossRef]

6. Gozdowiak, A. Faulty synchronization of salient pole synchronous hydro generator. *Energies* **2020**, *13*, 5491. [CrossRef]

7. Nömm, J.; Rönnberg, S.K.; Bollen, M.H.J. An analysis of frequency variations and its implications on connected equipment for a nanogrid during Islanded operation. *Energies* **2018**, *11*, 2456. [CrossRef]

8. Grebla, M.; Yellajosula, J.R.A.K.; Hoidalen, H.K. Adaptive Frequency Estimation Method for ROCOF Islanding Detection Relay. *IEEE Trans. Power Deliv.* **2020**, *35*, 1867–1875. [CrossRef]

9. Pawłowski, E. Power Grid Frequency Estimation Based on Zero Crossing Technique Using Least Squares Method to Approximate Sampled Voltage Signal Around Zero Level. *Lect. Notes Electr. Eng.* **2018**, *548*, 248–268.

10. Abdollahi, A.; Matinfar, F. Frequency estimation: A least-squares new approach. *IEEE Trans. Power Deliv.* **2011**, *26*, 790–798. [CrossRef]

11. Pradhan, A.K.; Routray, A.; Basak, A. Power system frequency estimation using least mean square technique. *IEEE Trans. Power Deliv.* **2005**, *20*, 1812–1816. [CrossRef]

12. Subudhi, B.; Ray, P.K.; Mohanty, S.R.; Panda, A.M. A comparative study on different power system frequency estimation techniques. *Int. J. Autom. Control* **2009**, *3*, 202–215. [CrossRef]

13. Dash, P.K.; Pradhan, A.K.; Panda, G. Frequency estimation of distorted power system signals using extended complex Kalman filter. *IEEE Trans. Power Deliv.* **1999**, *14*, 761–766. [CrossRef]

14. Routray, A.; Pradhan, A.K.; Rao, K.P. A novel Kalman filter for frequency estimation of distorted signals in power systems. *IEEE Trans. Instrum. Meas.* **2002**, *51*, 469–479. [CrossRef]

15. Huang, C.H.; Lee, C.H.; Shih, K.J.; Wang, Y.J. Frequency estimation of distorted power system signals using a robust algorithm. *IEEE Trans. Power Deliv.* **2008**, *23*, 41–51. [CrossRef]

16. Agha Zadeh, R.; Ghosh, A.; Ledwich, G. Combination of Kalman filter and least-error square techniques in power system. *IEEE Trans. Power Deliv.* **2010**, *25*, 2868–2880. [CrossRef]

17. Xia, Y.; He, Y.; Wang, K.; Pei, W.; Blazic, Z.; Mandic, D.P. A Complex Least Squares Enhanced Smart DFT Technique for Power System Frequency Estimation. *IEEE Trans. Power Deliv.* **2017**, *32*, 1270–1278. [CrossRef]

18. Nam, S.R.; Kang, S.H.; Kang, S.H. Real-time estimation of power system frequency using a three-level discrete fourier transform method. *Energies* **2015**, *8*, 79–93. [CrossRef]

19. Zygarlicki, J. Fast second order original Prony's method for embedded measuring systems. *Metrol. Meas. Syst.* **2017**, *24*, 721–728. [CrossRef]

20. Sun, J.; Aboutanios, E.; Smith, D.B. Iterative weighted least squares frequency estimation for harmonic sinusoidal signal in power systems. *Eur. Signal Process. Conf.* **2018**, *2018*, 176–180.

21. Peng, Z.; Li, H.-B. Power system frequency estimation algorithm for electric energy metering of nonlinear loads. *Metrol. Meas. Syst.* **2012**, *19*, 307–320. [CrossRef]

22. Szmajda, M.; Górecki, K.; Mroczka, J. Gabor transform, spwvd, gabor-wigner transform and wavelet transform-tools for power quality monitoring. *Metrol. Meas. Syst.* **2010**, *17*, 6. [CrossRef]

23. Zečević, Ž.; Krstajić, B.; Popović, T. Improved frequency estimation in unbalanced three-phase power system using coupled orthogonal constant modulus algorithm. *IEEE Trans. Power Deliv.* **2017**, *32*, 1809–1816. [CrossRef]

24. Xia, Y.; Qiao, L.; Yang, Q.; Pei, W.; Mandic, D.P. Widely linear adaptive frequency estimation for unbalanced three-phase power systems with multiple noisy measurements. *Int. Conf. Digit. Signal Process. DSP* **2017**, *2017*, 142.

25. Sun, J.; Aboutanios, E.; Smith, D.B.; Fletcher, J.E. Robust Frequency, Phase, and Amplitude Estimation in Power Systems Considering Harmonics. *IEEE Trans. Power Deliv.* **2020**, *35*, 1158–1168. [CrossRef]

26. Rife, D.C.; Vincent, G.A. Use of the Discrete Fourier Transform in the Measurement of Frequencies and Levels of Tones. *Bell Syst. Tech. J.* **1970**, *49*, 197–228. [CrossRef]

27. Borkowski, J.; Mroczka, J. LIDFT method with classic data windows and zero padding in multifrequency signal analysis. *Measurement* **2010**, *43*, 1595–1602. [CrossRef]

28. Chen, K.F.; Cao, X.; Li, Y.F. Sine wave fitting to short records initialized with the frequency retrieved from Hanning windowed FFT spectrum. *Meas. J. Int. Meas. Confed.* **2009**, *42*, 127–135. [CrossRef]

29. Borkowski, J.; Kania, D.; Mroczka, J. Interpolated-DFT-based fast and accurate frequency estimation for the control of power. *IEEE Trans. Ind. Electron.* **2014**, *61*, 7026–7034. [CrossRef]

30. Borkowski, J.; Kania, D.; Mroczka, J. Comparison of sine-wave frequency estimation methods in respect of speed and accuracy for a few observed cycles distorted by noise and harmonics. *Metrol. Meas. Syst.* **2018**, *25*, 283–302.

31. Borkowski, J.; Mroczka, J.; Matusiak, A.; Kania, D. Frequency Estimation in Interpolated Discrete Fourier Transform with Generalized Maximum Sidelobe Decay Windows for the Control of Power. *IEEE Trans. Ind. Inform.* **2021**, *17*, 1614–1624. [CrossRef]

32. European Std Committee EN 50160-2010. *Voltage Characteristics of Electricity Supplied by Public Distribution Systems*; CENELEC: Brussels, Belgium, 2010.

33. International Electrotechnical Commission (IEC); Electromagnetic Compatibility (EMC). *Part 4-30: Testing and Measurement Techniques—Power Quality Measurement Methods*. IEC 61000-4-30 Ed.3. 2015. Available online: https://webstore.iec.ch/publication/68642 (accessed on 1 September 2021).
34. International Electrotechnical Commission (IEC); Electromagnetic Compatibility (EMC). *Part 4–7: Testitng and Measurement Techniques-General Guide on Harmonics and Interharmonics Measurements and Instrumentation, for Power Supply Systems and Equipment Connected Thereto*. IEC 61000-4-7:2002+AMD1. 2009. Available online: https://standards.iteh.ai/catalog/standards/iec/c4e1c16d-4eec-481b-a45f-67dd8fee0e1a/iec-61000-4-34-2005-amd1-2009 (accessed on 1 September 2021).

Article

Empirical Analysis of the Impact of COVID-19 Social Distancing on Residential Electricity Consumption Based on Demographic Characteristics and Load Shape

Minseok Jang, Hyun Cheol Jeong, Taegon Kim, Dong Hee Suh and Sung-Kwan Joo *

The School of Electrical Engineering, Korea University, Seoul 02841, Korea; blue7xa7@korea.ac.kr (M.J.); jeonghc93@korea.ac.kr (H.C.J.); xorhs98@korea.ac.kr (T.K.); dhsuh@korea.ac.kr (D.H.S.)
* Correspondence: skjoo@korea.ac.kr; Tel.: +82-2-3290-4820

Abstract: Since January 2020, the COVID-19 pandemic has been impacting various aspects of people's daily lives and the economy. The first case of COVID-19 in South Korea was identified on 20 January 2020. The Korean government implemented the first social distancing measures in the first week of March 2020. As a result, energy consumption in the industrial, commercial and educational sectors decreased. On the other hand, residential energy consumption increased as telecommuting work and remote online classes were encouraged. However, the impact of social distancing on residential energy consumption in Korea has not been systematically analyzed. This study attempts to analyze the impact of social distancing implemented as a result of COVID-19 on residential energy consumption with time-varying reproduction numbers of COVID-19. A two-way fixed effect model and demographic characteristics are used to account for the heterogeneity. The changes in household energy consumption by load shape group are also analyzed with the household energy consumption model. There some are key results of COVID-19 impact on household energy consumption. Based on the hourly smart meter data, an average increase of 0.3% in the hourly average energy consumption is caused by a unit increase in the time-varying reproduction number of COVID-19. For each income, mid-income groups show less impact on energy consumption compared to both low-income and high-income groups. In each family member, as the number of family members increases, the change in electricity consumption affected by social distancing tends to decrease. For area groups, large area consumers increase household energy consumption more than other area groups. Lastly, The COVID-19 impact on each load shape is influenced by their energy consumption patterns.

Keywords: COVID-19; time-varying reproduction number; social distancing; load profile; demographic characteristic; household energy consumption; demand-side management; energy management

Citation: Jang, M.; Jeong, H.C.; Kim, T.; Suh, D.H.; Joo, S.-K. Empirical Analysis of the Impact of COVID-19 Social Distancing on Residential Electricity Consumption Based on Demographic Characteristics and Load Shape. *Energies* **2021**, *14*, 7523. https://doi.org/10.3390/en14227523

Academic Editors: Zbigniew Leonowicz and Michał Jasinski

Received: 6 October 2021
Accepted: 9 November 2021
Published: 11 November 2021

Publisher's Note: MDPI stays neutral with regard to jurisdictional claims in published maps and institutional affiliations.

1. Introduction

By the end of December 2020, the total number of COVID-19 confirmed cases worldwide was 80 million, with the death toll reaching 1.7 million. The global pandemic caused by the rapid and widespread proliferation of COVID-19 affected national policies, industries, and even the daily lives of people worldwide [1]. In South Korea, the total number of confirmed cases reached 60,000 by the end of December since the first COVID-19 case in Korea was identified on 20 January 2020, with the number of confirmed cases increasing by approximately 1000 cases every day. To cope with this situation, the Korean government began implementing and strengthening a policy based on social distancing. As COVID-19 affected the entire country, authorities across the world had to analyze their current situations and implement policies to respond to the pandemic. The US government imposed a lockdown from March 2020 to prevent the spread of COVID-19 in the US, and energy consumption declined by 20–40% in April and May 2020 [2]. Energy consumption dropped to weekend levels under lockdown, with dramatic reductions in the services and industry sectors only partially offset by higher residential use [3,4]. After shutting down on

22 March 2020 in New York City, weekday energy consumption in apartments increased by 7%, and weekend use increased by 4%. A load shift was observed during the weekday morning hours. A ramp-up started at 6 am, rising throughout the morning and early afternoon after reaching its pre-full morning peak at 9 am, highlighting the impact of this new stay-at-home behavior. This load shift for weekday mornings and afternoons was also seen in overall urban level consumption patterns observed by New York Independent System Operator (NYISO) [5]. India's energy consumption declined by 9.75% in March–April 2020, decreasing further in May to 14.16%. Unlike the overall Indian energy system, India's household energy consumption increased during the COVID-19 pandemic [6]. An independent electricity system operator reported that in Canada, Ontario's residential energy consumption increased during the spring and summer months of 2020 compared to the previous year [7].

In [8], the autoregressive distributed lag (ARDL) model developed by Pesaran et al. [9] was used to analyze the relationship between COVID-19 and regional energy consumption. India was segmented into five regions, and the effect in GWh was estimated using COVID-19 data for each region. However, reference [8] only analyzed the impact of COVID-19 on regional and not residential energy consumption. In [10], the impact of COVID-19 on household energy consumption was analyzed in Bhutan based on the number of household members. In 2020, Bhutan's household energy consumption increased by 6.8% compared to 2019 and 18% compared to 2018. However, reference [10] has limitations in comparing the total energy consumption used by the household members rather than estimating the coefficient of the COVID-19 impact. Reference [11] showed that in Brazil, the increase in home office arrangements caused by the COVID-19 pandemic resulted in residential energy consumption increasing by 10–13%. Moreover, after dividing Brazil into five regions, which are north, northeast, southeast-midwest, south, and all areas, the total energy consumption before/after COVID-19 by region was only analyzed using the *t*-test to determine the significance of the reduction in energy consumption using statistical methods. However, reference [11] only demonstrated the increase in residential energy consumption due to COVID-19. Apart from lifestyle changes in people's lives during the pandemic, other factors affected energy load consumption, such as weekdays, holidays, weather, etc. [12,13].

In [14], energy consumption of HVAC in the housing sector increased by as much as 16% during the lockdown period. In addition, the hourly load of lighting and home appliances at home decreased slightly in the morning and increased significantly from daytime to evening. There is a limitation as this study only compared the average daily energy consumption during 23–31 March 2020 and the average during March 2017–2019. For Australian households, daytime energy use increased by almost 15% compared with February–May 2019 and the same period in 2020 [15]. In [16], based on the measurement data from about 7000 dwellings in Warsaw household, the result shows that the average daily energy consumption has increased during the lockdown of the COVID-19 pandemic compared to the period of the year before the pandemic. However, refs [15,16] also only compared the average daily usage pattern during the period of the year before the pandemic and the lockdown period. This comparison method has the potential problem of not distinguishing between the impact changed by COVID-19 and the impact changed by external effects. In addition, there is a disadvantage in that it is not possible to perform an analysis that reflects the demographic characteristics and energy usage patterns of each household.

This study analyzes the impact of COVID-19 on household energy consumption with demographic information of individual residential customers to capture the different impacts on each group. The model used in this study better analyzes the impact of COVID-19 by using a two-way fixed effect that captures household heterogeneity. In particular, this study performs clustering of households for each load shape. The impact of COVID-19 based on consumers' pre-social distancing lifestyle patterns is analyzed. The results of this study have implications for the planning and operating of Korea's electrical utility power systems in response to the COVID-19 pandemic.

Most previous studies have only analyzed the impact of COVID-19 using a dummy variable. This study proposes the use of the time-varying reproduction number of COVID-19 to check the robustness and persistence of the impact of COVID-19 on household energy consumption in Korea.

The remainder of this paper is organized as follows. Section 2 describes data used in this study. Section 3 introduces the method for handling missing data, clustering the load pattern algorithm, and estimating the COVID-19 impact on household energy consumption. Section 4 presents the results of the analysis using the proposed method in this study. Section 5 discuss the results of the study. Finally, Section 6 concludes the study.

2. Data and Description

Household energy consumption data are extracted through smart meters. The reproduction number is used to determine the impact of COVID-19 during the social distancing period. The mean infection rate of a population at a certain point in time is defined as the time-varying reproduction number—i.e., R_j, which can be repeatedly calculated over time. R_j is the time-varying reproduction number of COVID-19 for week j in 2020. The instantaneous reproductive index is defined as the number of secondary infections expected to occur at time divided by the number of infected individuals and infectivity at time and is useful for estimating the number of retrospective or real-time transmission. Korea's R_j data show that R_j is high during the eight week of social distancing policy implementation in Korea as shown in Table 1, but R_j exhibits a continuously declining trend during the social distancing period [17,18].

Table 1. Estimated Reproduction Number During the Period March to April 2020 in Korea [18].

Week Index on 2020	R_j	Issue
6 (2.2–2.8)	1.5	
7 (2.9–2.15)	0.81	
8 (2.16–2.22)	9.35	1st regional infection in Daegu, Gyeongbuk (2.20–2.29)
9 (2.23–2.29)	5.66	
10 (3.1–3.7)	1.94	First social distancing (2.29–3.21)
11 (3.8–3.14)	0.67	
12 (3.15–3.21)	0.45	
13 (3.22–3.28)	0.76	Strengthened national social distancing (3.22–4.19)
14 (3.29–4.4)	0.85	
15 (4.5–4.11)	0.67	
16 (4.12–4.18)	0.53	
17 (4.19–4.25)	0.58	
18 (4.26–5.2.)	0.59	

For weather data matched with each household, for a total of 1133 houses, the average temperature and humidity in hour units extracted from the data of the Korea Meteorological Agency's Automated Surface Observing System (ASOS) in January–April 2020, which is closest to the area of the sampled household. The weather data is extracted from nine meteorological observatories. The household energy consumption extracted in 15-min intervals on January–April 2020 from the Korea Electric Power Corporation (KEPCO) smart metering and resampled the average hourly energy consumption. The average electricity price is calculated based on progressive rate design. If 200 kWh or less is used in a month, the electricity rate is 88.3 KRW/kWh. If 400 kWh above is used in a month, the electricity rate is 275.6 KRW/kWh and except for the two preceding conditions, the electricity rate is 182.9 KRW/kWh. A description and statistic of the time varying data is presented in Table 2 and demographic characteristic is shown in Table 3.

Table 2. Summary of time variant variable statistics.

Variable	Unit	Mean	Standard Deviation	Min	Max	Observation [1]
Use	kWh	0.039	0.67	0.003	6.81	3,290,232
Temp	°C	6.09	6.13	−13	26.8	3,290,232
Humid	%	59.7	22.4	4	100	3,290,232
Price	KRW/kWh	99.5	13.6	93.3	234.5	3,290,232

Note: [1] *Use* is extracted from smart meters, *Temp* and *Humid* are extracted from ASOS, and *Price* is calculated using *Use* for each household in January–April 2020.

Table 3. Summary of demographic characteristic.

Characteristic	Unit	Description	Percentage
Income	KRW	≤2.4 million	24%
		2.41–5.50 million	48%
		≥5.51 million	28%
Family member	person	≤Two	38%
		Three	24%
		Four	28%
		≥Five	10%
House area	m^2	33.1–62.8 (i.e., very small)	7%
		66.1–95.6 (i.e., small)	21%
		99.2–128.9 (i.e., medium)	54%
		≥132.2 (i.e., large)	18%

3. Methods for Analysis of the Impact of Social Distancing on Household Energy Consumption

In this section, a two-way fixed effect model is presented. In addition, a simple hybrid-imputation method, which combines linear and statistical methods, and a clustering algorithm based on the load shape is introduced.

3.1. The Method for Estimating the Impact of Social Distancing Using Household Energy Consumption Model

The proposed analysis method used to identify COVID-19 factors that affected air pollution [19]. In the previous paper on electricity usage, household energy consumption is modeled using various factors that affect residential energy usage, such as airborne particulate matter (PM) with a diameter of 10 microns or less, called air pollutants such as PM10 and PM2.5, and temperature, [20–22]. The proposed method estimates the energy consumed by residential customers with time-varying reproduction number during the COVID-19 pandemic. Furthermore, the proposed model considers demographic characteristics, such as house area, income quantile, and the number of household members.

The household energy consumption method is often used to predict energy consumption or estimate factors that may affect energy consumption. This model is used for two reasons. First, this econometric approach provides flexibility in controlling differences in weather, seasonality, and other factors. Second, using household-specific fixed-effect, panel data analysis can control time-invariant but unobservable characteristics per household, which may cause bias in the impact estimation results.

Subsequently, using the household energy consumption model in Equation (1), a model is created to measure the impact of COVID-19 on household energy consumption:

$$ln(Use_{ith}) = \alpha R_{j,t} + \beta_1 temp_{ith} + \beta_2 temp_{ith}^2 + \beta_3 humid_{ith} + \beta_4 price_{ith} + \beta_5 holiday_t + \delta_{it} + \delta_{th} + \varepsilon_{ith} \tag{1}$$

where Use_{ith} is the energy consumption of household i used in hour h on t day from January to April 2020, and $R_{j,t}$ is the time-varying reproduction number of COVID-19 that represents the risk of infection at week t for each week (from the 6th week 2020, when the local infection began in Daegu at that time). α is the parameter of most interest, —that is, a value that estimates the effect of the COVID-19 on the average household energy consumption, which multiplied by 100 gives the rate of change. This study uses α to estimate the impact of COVID-19, which enforce people to stay indoors in the prohibition period, on the household energy consumption. The people may use less electricity because they go out to be active. Similarly, a decline in outdoor activity due to social distancing may result in increase of household electricity usage. $temp_{ith}$ and $humid_{ith}$ are weather variables, which are respectively temperature and humidity, of the closed location of the household i in hour h on t day. In [20–22], the temperature has the relationship between electricity consumption and temperature shows a U-shape. Therefore, if Equation (1) also has the U-shape relationship, the robustness of Equation (1) could be checked. In [23], the residential customer is more sensitive to average price rather than marginal or expected marginal price. Therefore, $price_{ith}$ is the average electricity price of each household. $holiday_t$ is the dummy variable for holidays. β_{1-5} are coefficients for each exogenous variable of residential energy consumption for description and summary statistics of the data described in Table 2, respectively. In addition, δ_{it} is the individual fixed effects of household, household-month, and household-weekday-month to check the robustness of household energy consumption model. δ_{th} is the time fixed effects of hour-weekday, which captures weekday specific change of hourly energy consumption. The standard errors were clustered by each household's daily level in Equation (1). Equation (1) is used to estimate the impact of social distancing by each demographic characteristic group.

Furthermore, another important topic is an impact on each groups' energy usage pattern. Therefore, the extended models are used to capture the impact of COVID-19 on each group's hourly energy consumption. G_h^j are dummies for capturing impact on residential hourly usage in each group. The same notation in Equation (1) is used in Equation (2). The models are given as follows:

$$ln(Use_{ith}) = \sum_{h=1}^{j=24} \sum_j G_h^j \left(\alpha_{1,h} + \alpha_{2,h} R_{j,t} \right) + \beta_1 temp_{ith} + \beta_2 temp_{ith}^2 + \beta_3 humid_{ith} + \tag{2}$$
$$\beta_4 price_{ith} + \beta_5 holiday_t + \delta_{it} + \varepsilon_{ith}$$

The household cluster level of standard error is used in these models. A parametric bootstrapping algorithm is conducted to make mean and confidence intervals of the model coefficients [21]. The average of hourly consumption in hour h and calculated mean and 95% confidence intervals are used to estimate the hourly impact of the time-varying reproduction number 1.94 from 1 March to 7 March 2020, which is the first social distancing week as shown in Table 1. The model for estimating the hourly impact is given as follows:

$$Impact_h = Use_h \left(1 + e^{1.94 \times \alpha_{2,h}} \right) \tag{3}$$

3.2. The Method for Hybrid-Imputation Model for Missing Data

To fill in missing data values, a hybrid method combining statistical and linear method is used. Household energy consumption is driven by human behavior. Therefore, household energy consumption can be divided into two categories: first, because the consumption is consistent over a short period of time, the short missing intervals can be replaced with the available surrounding measurements; second, a person's behavior can be characterized during specific time intervals such as commute/work. Various methods exist to replace the missing variables. The linear interpolation is a commonly used method and is often used in short missing intervals that characterize continuous use. y^L is the energy consumption of the time c to be replaced, and y_{h_a} and y_{h_b} are the energy consumptions at the time when there is no missing value adjacent to h_c, and if the trailing time of h_a, h_b, and h_c are the preceding times. The linear interpolation method is as follows:

$$y_c^L = y_{h_b} + \frac{y_{h_a} - y_{h_b}}{h_a - h_b} (h_c - h_b), \quad h_b < h_c < h_a \tag{4}$$

The linear imputation is fast and simple. However, the linear interpolation method has a disadvantage in that the performance decreases when the missing interval becomes longer [24]. To compensate for this shortcoming, the missing values in long intervals are replaced by utilizing statistical values in an hour, day of the week, or month. Therefore, a statistical imputation method is required. Statistical imputation is a method by which the distribution of the time at which the month and day of the time to be replaced is the same, and the missing value can be replaced by the median value of the corresponding time [25]. For example, in the case of a person living alone, because they go to work around 8 a.m. on weekdays, there is no special energy consumption. Consequently, the user's behavior pattern of consumption data is contained, and the accuracy of the model can be improved by performing substitution using the median value. Here, the median value is used because it is less affected by anomalous data than the average value. The statistical method is as follows:

$$y_h^S = median(z), \; where \; z \in y_j, j \in \theta_h \tag{5}$$

where θ_h represents the set of previous h hours that have the same month and day, and z is the energy consumed during θ_h time. Hybrid imputation adopts a method that introduces the Monte Carlo scenario [26] to find optimal point of the above two methods. After selecting a household with no missing values for a specific month, the missing scenario is generated randomly, and the RMSE of the two imputations algorithms is calculated to determine the optimal replacement method.

3.3. The Method for Clustering of Household Load Shape

As an initial stage for grouping, the representative load pattern of each household energy consumption was derived from median value of hourly energy consumption [27]. Normalization is performed using the maximum and minimum normalization of the representative load pattern for each household [28]:

$$Normalized\_L_{i,h} = \frac{L_{i,h} - min(L_i)}{max(L_i) - min(L_i)} \tag{6}$$

where $Normalized\_L_{i,h}$ is the normalized energy consumption of customer i at time h, $L_{i,h}$ is the average energy consumption of customer i at time h, and L_i is the load profile of customer i. To use the clustering method for the household load profile, there are various clustering methods such as Kohonen self-organizing map (SOM), hierarchical clustering, K-medoids, and Fuzzy C-means. However, there are no statistically significant improvements compared to K-means in clustering of load shape of household energy consumption data. Therefore, in this study, the K-means algorithm is chosen for the sake of simplification [29,30]. The K-means algorithm is used to determine the optimal number of groups. Moreover, there is no significant difference between the elbow method and the silhouette score in finding the optimal number K [31]. Therefore, the elbow method is adopted in this study. The number of groups K is optimized using the elbow method— the elbow method optimizes the number of groups by using the difference in distance between the household in the center of each group and the households in the group. It can determine the optimal point using the sum of squared error (SSE), which is calculated as follows [32,33]:

$$SSE = \sum_{k=1}^{n} \sum_{h=0}^{23} (a_{i,h} - c_{k,h})^2 \; where \; a_i \in c_k \tag{7}$$

where n is the number of clusters, $c_{k,h}$ is the normalized energy consumption of the kth cluster center point at time h, a_i is the household in c_k and $a_{i,h}$ is the normalized energy consumption of customer i at time h. If the value of K increases, the SSE decreases.

3.4. Procedure of Proposed Method

The generic procedure for analyzing the impact of social distancing enforced by COVID-19 on household energy consumption, shown in Figure 1, can be described as follows:

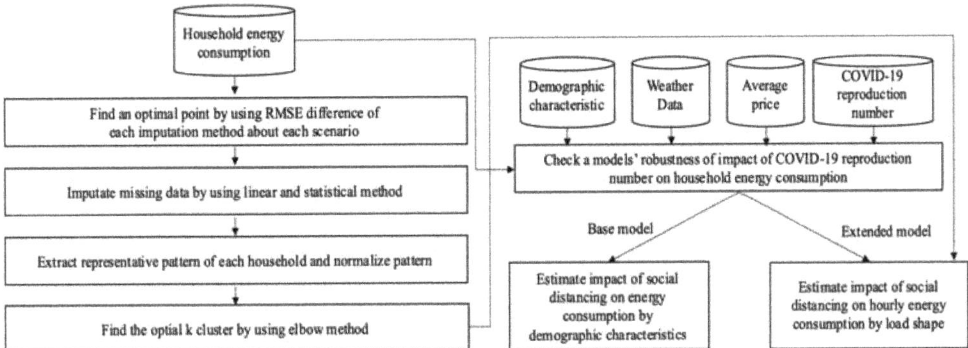

Figure 1. Flowchart of the proposed procedure for analyzing the impact of social distancing on household energy consumption.

Step (1) Find optimal point of each imputation method.

Step (2) Impute the missing data by using the proposed hybrid method.

Step (3) Cluster household usage pattern by using k-means.

Step (4) Estimate the impact of the time-varying reproduction number of COVID-19 on household energy consumption by using a two-way fixed effect.

Step (5) Check the robustness of the impact of the time-varying reproduction number on household energy consumption.

Step (6) Estimate the impact of COVID-19 on household energy consumption by each demographic characteristic groups.

Step (7) Estimate the impact of COVID-19 on the hourly household energy consumption by load shape.

4. Empirical Results

4.1. Results of Hybrid Imputation and Clustering of Household Load Shape

The Monte Carlo scenario is performed by extracting 200 households with no missing data and COVID-19 from the energy consumption data for January–February 2020. Figure 2 shows the results indicating the algorithm is more advantageous when it comes to handling missing values. In intervals that are not grayed out, which means the number of missing values is less than nine per day, the linear imputation method performs better than the statistical imputation method. However, the longer the missing interval, the more accurate the statistical imputation method.

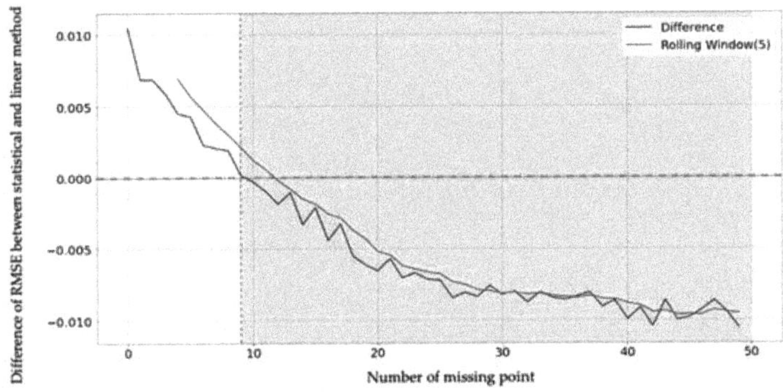

Figure 2. Difference of RMSE between statistical and linear imputation as the number of missing values increases.

In Figure 2, the difference between the linear and statistical interpolation RMSE is negative during the grayed-out intervals, indicating that statistical imputation method to be advantageous in periods with missing intervals of 10 points or more. Therefore, in this study, for missing intervals with a length equal to or less than 9 points, the analysis is performed by interpolating linearly and for missing intervals with a length greater than 9 points, the statistical imputation method is used.

Based on the grouping results, the energy consumption of home customers is divided into five patterns. Figure 3 shows the optimal number of groups using the elbow method. In Figure 3, the slope is reduced at point which the sum of the total SSE of five cluster. In Figure 4, 1133 customers are classified into Day (133 households), M-pattern (326 households), Stair (278 households), Owl (108 households), and Evening types (288 households) based on their energy consumption patterns.

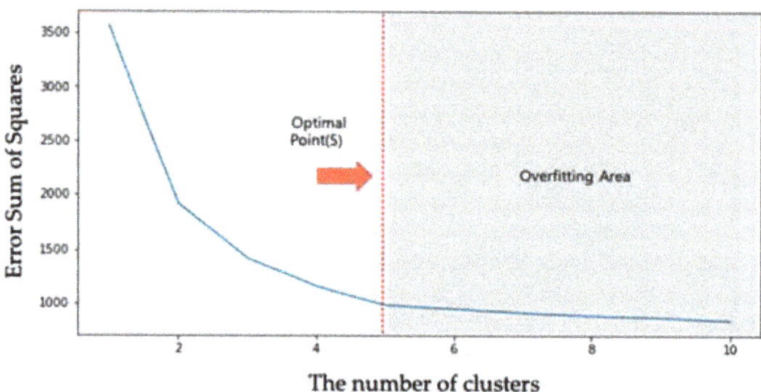

Figure 3. Sum of squares error (SSE) of residential customers clustered using K-means and elbow method.

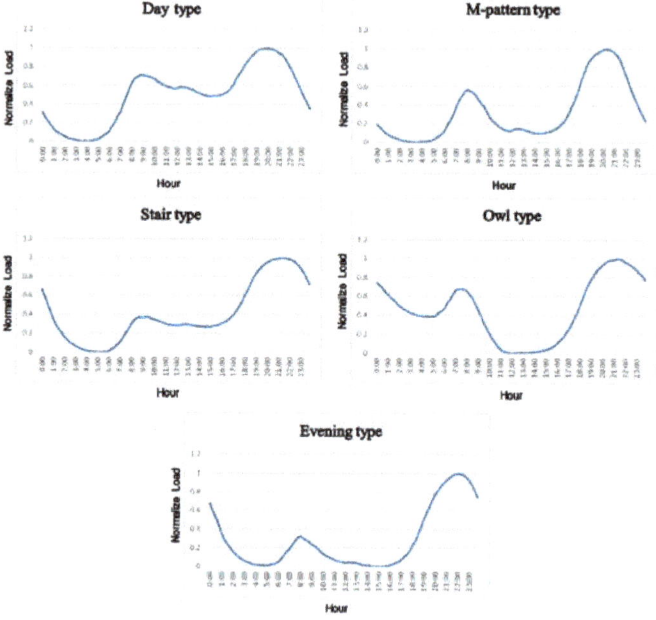

Figure 4. Results of five residential customer clustering utilizing K-means.

The day type shows energy use during the day which means that someone uses appliances in the house at lunchtime. The M-pattern type shows that there is a temporary increase in energy consumption during the morning period for going to work and no electricity usage until the evening before energy usage increased during the evening hours after work. It is assumed in households in this pattern that all family members have a job. The stair type shows a usage pattern where electricity consumption increases in the morning, is constant in the afternoon, and increases in the evening. The owl type shows high energy consumption during the night and maybe the typical energy use pattern of the single-person residential customer. The evening type shows a group whose consumption increases briefly in morning hours, then is rarely used during the day, before increasing again in the evening. The evening type is the typical energy use pattern of customers not being home in the afternoon and beginning their evening routine after leaving the office.

4.2. Empirical Results for Impact of Social Distancing in Korea

First, the impact of COVID-19 on household energy consumption is analyzed by using Equation (1) during January–April 2020, which is the period included in social distancing enforced by COVID-19. To check the robustness of estimating the significance of the parameter at different sets of household and time dummies, the individual fixed effect is introduced with three sets of dummies. The household level means estimating each coefficient for the entire date period for each household. For the household month level, each coefficient is estimated by clustering by month for each household. Finally, at the household-month-weekday level, each coefficient is estimated by clustering the weekdays and weekends of the month for each household. Therefore, the robustness check of the proposed household energy consumption model was performed by different individual fixed effects, as shown in Table 4.

Table 4. Models to estimate the impact of COVID-19 on household energy consumption.

	Dependent Variable: ln (Use)		
	M1	**M2**	**M3**
R_j	0.004 ***	0.003 ***	0.003 ***
	(0.0003)	(0.0003)	(0.0003)
temp	−0.002 ***	−0.0004 **	−0.001 ***
	(0.0002)	(0.0002)	(0.0002)
$temp^2$	0.00001	0.00002	0.00002 **
	(0.00001)	(0.00001)	(0.00001)
humid	0.001 ***	0.001 ***	0.001 ***
	(0.00003)	(0.00003)	(0.00003)
holiday	0.008 **	0.022 ***	0.023 ***
	(0.004)	(0.003)	(0.003)
price	0.0005 ***	−0.001 ***	−0.001 ***
	(0.0001)	(0.00005)	(0.00004)
Individual fixed effect	household-level	household-month level	household-month-weekday level
Time fixed effect	weekday-hour	weekday-hour	weekday-hour
Observations	3,289,907	3,289,907	3,289,907
R-squared	0.453	0.495	0.503
Adjusted R-squared	0.453	0.495	0.501

*,**,***, significance at 10%, 5%, and 1%.

The inspection of robustness for the interested coefficient of reproduction number of COVID-19 shows statistical significance in each model although the individual effect is changed. The time-fixed effect has not changed because it is obvious that household energy consumption has different patterns on weekends and weekdays. Therefore, the significant positive impact of the time-varying reproduction number on household energy consumption is shown. Given all the models presented, the estimated impact is robust and statistically significant at the level of 0.01. The house energy consumption models presented in Table 4 show that an average increase of 0.3% in the hourly average energy usage is caused by a unit increase in the time-varying reproduction number of COVID-19. Other statistical control variables such as temperature, humidity, hourly electricity price, and holiday dummy are also statistically significant. The result of the temperature factor presents that the relationship between electricity consumption and temperature shows a U-shape. This shape means that if the temperature rises to a mild level, electricity use decreases. If the temperature rises to a level beyond mild, the electricity use increases. A similar result has also been presented in previous impact of climate studies [21,34].

Second, since the M3 model has the highest adjusted R-squared among the models in Table 4, the M3 model is performed to analyze each target group such as income status, family member, and housing area. Figure 5 shows whether the impact of the time-varying reproduction number of COVID-19 on household energy consumption in response to demographic characteristics for each group. The income group is divided into three groups: low-income, middle-income, and high-income. The family group is divided into four groups including under or two, three, four, and five group. Moreover, the area group consists of four groups, which are very small, small, medium, and large. Detailed statistical information for each group is given in Table 3.

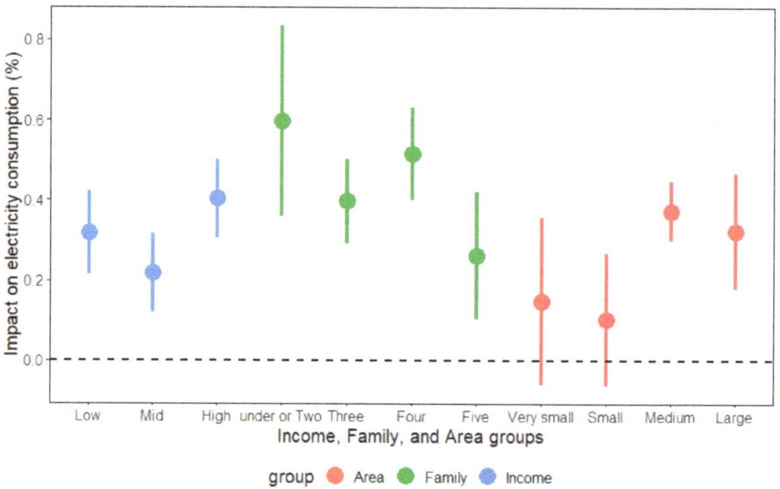

Figure 5. Change in each groups' household energy consumption due to increase in the time-varying reproduction number of COVID-19.

Because The M3 model has most high Adjusted R-squared between Models in Table 4, the M3 model is performed to conduct regression analysis for each target groups, which are income status, family member, and house area. In Figure 5, the solid dots represent the coefficient converted into an impact percentage that measures the impact on household energy consumption when the time-varying reproduction number of COVID-19 increases by one unit, and the vertical lines represent 95% confidence levels. Looking at income group, low income tends to live in house that are not energy efficient and to have appliances with low energy efficiency, as shown in existing studies, which can cause a high increase in electricity consumption of low-income households compared to mid-income households. Furthermore, mid-income groups show less impact on energy consumption compared to both low-income and high-income group, which means mid-income groups have energy-efficient homes and appliances, compared to low-income and high-income groups who might need more energy due to large areas occupied and the number of family members [20,35,36]. According to the results of the number of family members, groups of two household members or less shows the largest increase in household electricity consumption in response to a unit increase in the time-varying reproduction number. Because single-member and two-group consumers have little activity at home during the pre-social distancing period, working from home during the social distancing period may increase in electricity consumption. As the number of family members increases, the change in electricity consumption affected by social distancing tends to decrease, the large number groups might use more electricity before social distancing. For area groups, large area consumers increase household energy consumption more than other area groups. The empirical results imply that if the area of the house is large, there might be many family members and the income might be high, so the household energy consumption increases the most during the social distancing period.

Finally, in Figure 6, using Equation (3) with the load shape group for household consumers, the results show the impact of the time-varying reproduction number of COVID-19 on the hourly household energy consumption during the first week of social distancing. The shaded areas, which represent 95% confidence intervals, and solid lines are generated by the bootstrapping procedure using Equations (2) and (3) [21]. In daytime hours, all groups' changes in energy consumption have a positive impact on the household energy usage. This reflects that people tend to have curtailed outdoor activities during daytime hours because of social distancing. During night-time hours, the M-pattern type group was most affected by social distancing, because it usually had the least usage at night-time compared to other groups as shown in Figure 4. For the opposite reason, the groups that used electricity at night-time during pre-social distancing period was less affected than the M-pattern. This means that there is less opportunity to increase indoor activities at home in the night-time. In morning hours, all groups except for owl and M type group show the small negative change in energy consumption due to work from home and remote classes. The reason is that people slept more during morning hours because there was no need to prepare to go out.

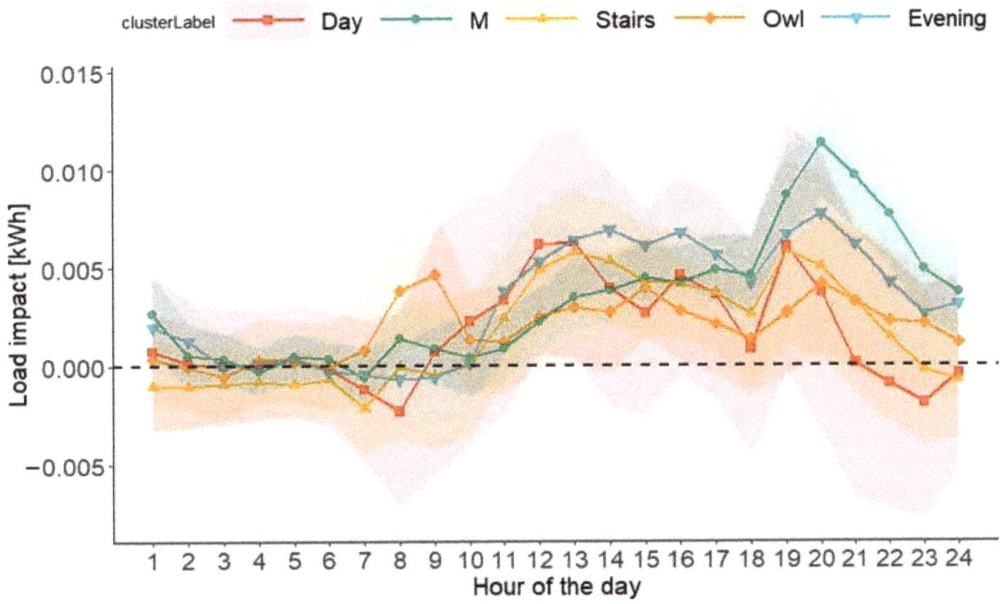

Figure 6. Change in load shape groups' household hourly energy consumption due to increase in the time-varying reproduction number of COVID-19.

5. Discussion

In this study, statistical and linear interpolation methods are applied by creating a scenario in which missing data occurred using the Monte Carlo technique. If the missing data are short, the linear interpolation method performs better. The statistical interpolation method has better results if the missing data is longer. K-means clustering is performed using the representative load pattern. The elbow method is used to optimize the grouping of K-means clustering. As a result, 1133 households are classified into five groups, which are Day type, M-pattern type, Stair type, Owl type, and Evening type.

In the results of household energy consumption model, this study shows three key implications. First, a numerical time-varying reproduction number of COVID-19 is a useful factor for reflecting the impact of the COVID-19 pandemic on household energy consumption. In order to reflect the intensity of the COVID-19 spread situation, a numerical time-varying reproduction number is used instead of a dummy variable. As a result, an average increase of 0.3% in the hourly average energy consumption is influenced by a unit increase in the time-varying reproduction number of COVID-19. This result might seem rather low compared to the increase rate presented in previous studies [8,14–16]. However, unlike other countries, Korea has responded to the spread of COVID-19 through social distancing rather than lock-down [37]. Therefore, the increase in household electricity consumption would be relatively small as Koreans could move more freely during the COVID-19 pandemic than people from other countries. Furthermore, whereas previous studies showed a large percentage of change by estimated in specific industries, regions, and national sectors [8,14–16,37–39], this study estimated the COVID-19 impact on the hourly electricity usage of each household during the social distancing period.

Second, the results of this study remind policymakers of the problem that COVID-19 is more impact on low-income families than others. Many countries imposed mandatory stay-at-home orders and take various actions to discount electricity bills. Therefore, the low-income families should be more supported by the government if a pandemic is repeated like in Indonesia. In Indonesia, poor people could use free electricity from 24 April 2020 [40].

Last, The COVID-19 impact on each load shape is related to their energy consumption patterns. In addition, there are the same patterns as previous studies. As like [16,39], In morning hours, patterns are shown the negative change in energy consumption due to work from home and remote classes. However, previous studies showed the greatest impact in the daytime, otherwise in this paper, when subdivided by pattern and analyzed, some groups were more affected by social distancing during the night-time hours. This result indicates that if some households did not used electricity in night-time hours before the pandemic, their electricity usage could increase significantly in the daytime as well as night-time.

6. Conclusions

The household energy consumption model is used to analyze the impact of COVID-19 reproduction number on household energy consumption. The impact of COVID-19 reproduction numbers is significantly positive; they increase household energy consumption by 0.3%. As a result, if the COVID-19 reproduction number increases due to increasing local infection, most households' energy consumption will increase. Most people do not have to physically go to work and school because they can opt for work from home and remote classes. This could be seen as a decrease in energy consumption in the early morning. In addition, the energy consumption from daytime to night-time hours increases. An analysis of the impact on each demographic characteristic group by using a robust model is conducted. For income groups, the impact of social distancing on household energy consumption is small to mid-income groups' energy consumption. However, low-income groups' energy consumption is more affected by social distancing than the mid-income group. For family member groups, the smaller the number of family members, the greater the impact of COVID-19 on household electricity consumption. For house area groups, the change enforced by social distancing in electricity consumption increases as the house area increases. The extended household energy consumption model is used to investigate how the different impacts of COVID-19 reproduction numbers are analyzed by hours depending on each load shape. According to each load shape, the effect of electricity consumption is different in the morning, daytime, and night-time hours.

However, there is the limitation of the availability of the time-varying reproduction number of COVID-19. Because it is not possible to count the number of COVID-19 infected people in real-time, calculating and adopting the time-varying reproduction number of COVID-19 in an hour could be physically difficult.

The empirical results show that the method presented with time-varying reproduction numbers has been able to reach the best estimator of the impact of social distancing. Thus, the results of this study have implications for electrical utility power systems in response to the COVID-19 pandemic.

Author Contributions: M.J. performed the research and wrote the manuscript. H.C.J. helped with the clustering algorithm and analysis. T.K. helped with the original draft and revised the manuscript. D.H.S. helped design the model and performed the numerical simulations. S.-K.J. provided guidance for the research and revised the manuscript. All authors have read and agreed to the published version of the manuscript.

Funding: This work was supported by the Korea Institute of Energy Technology Evaluation and Planning (KETEP) and the Ministry of Trade, Industry & Energy (MOTIE) of the Republic of Korea (No. 20181210301430). This research was supported by the Basic Research Program through the National Research Foundation of Korea (NRF) funded by the MSIT (No. 2020R1A4A1019405).

Institutional Review Board Statement: Not applicable.

Informed Consent Statement: Not applicable.

Data Availability Statement: Not applicable.

Acknowledgments: This work was supported by the Korea Institute of Energy Technology Evaluation and Planning (KETEP) and the Ministry of Trade, Industry & Energy (MOTIE) of the Republic of Korea (No. 20181210301430). This research was supported by the Basic Research Program through the National Research Foundation of Korea (NRF) funded by the MSIT (No. 2020R1A4A1019405).

Conflicts of Interest: The authors declare no conflict of interest.

References

1. WHO Website. 2021. Available online: https://covid19.who.int/ (accessed on 1 January 2021).
2. Pillay, A. COVID-19 Crisis: Lockdown Knocks Power Use down 20–40% in Some Cities. The Business Standard. 19 June 2020. Available online: https://www.business-standard.com/article/economy-policy/lockdown-knocked-power-consumption-down-by-up-to-40-in-large-cities-120061800951_1.html (accessed on 21 January 2021).
3. IEA. International Energy Agency, COVID-19 Impact on Electricity. 2021. Available online: https://www.iea.org/reports/covid-19-impact-on-electricity (accessed on 21 January 2021).
4. ETEnergyWorld. Worldwide Lockdowns to Reduce Nuclear Power Output by 3 Per Cent in 2020. 1 May 2020. Available online: https://energy.economictimes.indiatimes.com/news/power/worldwide-lockdowns-to-reduce-nuclear-power-output-by-3-per-cent-in-2020/75493688 (accessed on 9 January 2021).
5. Meinrenken, C.J.; Modi, V.; Mckeown, K.R.; Culligan, P.J. New Data Suggest COVID-19 Is Shifting the Burden of Energy Costs to Households. 21 April 2020. Available online: https://news.climate.columbia.edu/2020/04/21/covid-19-energy-costs-households/ (accessed on 1 February 2021).
6. Times, T.E. India's Power Consumption Shrinks 9.24% at 100.13 BU in March 2020. 2020. Available online: https://economictimes.indiatimes.com/industry/energy/power/indias-power-consumption-shrinks-9-24-at-100-13-bu-in-march/re_show/74933363.cms (accessed on 1 February 2021).
7. IESO. Reliability Outlook: An Adequacy Assessment of Ontario's Electricity System from October 2020 to March 2022. September 2020. Available online: https://www.ieso.ca/en/Sector-Participants/Planning-and-Forecasting/Reliability-Outlook (accessed on 17 February 2021).
8. Aruga, K.; Islam, M.; Jannat, A. Effects of COVID-19 on Indian energy consumption. *Sustainability* **2020**, *12*, 5616. [CrossRef]
9. Pesaran, M.H.; Shin, Y.; Smith, R.J. Bounds testing approaches to the analysis of level relationships. *J. Appl. Econom.* **2001**, *16*, 289–326. [CrossRef]
10. Chhetri, R. Effects of COVID-19 pandemic on household energy consumption at College of Science and Technology. *Energies* **2020**, *3*, 1383–1387.

11. Carvalho, M.; Bandeira de Mello Delgado, D.; de Lima, K.M.; de Camargo Cancela, M.; dos Siqueira, C.A.; de Souza, D.L.B. Effects of the COVID-19 pandemic on the Brazilian electricity consumption patterns. *Int. J. Energy Res.* **2021**, *45*, 3358–3364. [CrossRef]

12. Lusis, P.; Khalilpour, K.R.; Andrew, L.; Liebman, A. Short-term residential load forecasting: Impact of calendar effects and forecast granularity. *Appl. Energy* **2017**, *205*, 654–669. [CrossRef]

13. Fan, S.; Methaprayoon, K.; Lee, W.-J. Multiregion load forecasting for system with large geographical area. *IEEE Trans. Ind. Appl.* **2009**, *45*, 1452–1459. [CrossRef]

14. Krarti, M.; Aldubyan, M. Review analysis of COVID-19 impact on electricity demand for residential buildings. *Renew. Sustain. Energy Rev.* **2021**, *143*, 110888. [CrossRef]

15. Snow, S.; Bean, R.; Glencross, M.; Horrocks, N. Drivers behind residential electricity demand fluctuations due to COVID-19 restrictions. *Energies* **2020**, *13*, 5738. [CrossRef]

16. Bielecki, S.; Skoczkowski, T.; Sobczak, L.; Buchoski, J.; Maciag, Ł.; Dukat, P. Impact of the Lockdown during the COVID-19 Pandemic on Electricity Use by Residential Users. *Energies* **2021**, *14*, 980. [CrossRef]

17. Gostic, K.M.; McGough, L.; Baskerville, E.B.; Abbott, S.; Joshi, K.; Tedijanto, C.; Kahn, R.; Niehus, R.; Hay, J.A.; De Salazar, P.M.; et al. Practical considerations for measuring the effective reproductive number, Rt. *PLoS Comput. Biol.* **2020**, *16*, e1008409. [CrossRef]

18. Yoo, M.S.; Kim, Y.J.; Baek, S.J.; Kwon, D.H. The concept of reproduction number and changes according to government response policies. *Wkly Health Dis.* **2020**, *14*, 282–289.

19. Liu, S.; Kong, G.; Kong, D. Effects of the COVID-19 on air quality: Human mobility, spillover effects, and city connections. *Environ. Resour. Econ.* **2020**, *76*, 635–653. [CrossRef] [PubMed]

20. He, P.; Liang, J.; Qiu, Y.L.; Li, Q.; Xing, B. Increase in domestic electricity consumption from particulate air pollution. *Nat. Energy* **2020**, *5*, 985–995. [CrossRef]

21. Eom, J.; Hyun, M.; Lee, J.; Lee, H. Increase in household energy consumption due to ambient air pollution. *Nat. Energy* **2020**, *5*, 976–984. [CrossRef]

22. Li, Y.; Pizer, W.A.; Wu, L. Climate change and residential electricity consumption in the Yangtze River Delta, China. *Proc. Natl. Acad. Sci. USA* **2019**, *116*, 472–477. [CrossRef] [PubMed]

23. Ito, K. Do consumers respond to marginal or average price? Evidence from nonlinear electricity pricing. *Am. Econ. Rev.* **2014**, *104*, 537–563. [CrossRef]

24. Peppanen, J.; Zhang, X.; Grijalva, S.; Reno, M.J. Handling bad or missing smart meter data through advanced data imputation. In Proceedings of the 2016 IEEE Power & Energy Society Innovative Smart Grid Technologies Conference (ISGT), Minneapolis, MN, USA, 6–9 September 2016.

25. Borges, C.E.; Kamara-Esteban, O.; Castillo-Calzadilla, T.; Andonegui, C.M.; Alonso-Vicario, A. Enhancing the missing data imputation of primary substation load demand records. *Sustain. Energy Grids Netw.* **2020**, *23*, 100369. [CrossRef]

26. Ferrenberg, A.M.; Swendsen, R.H. Optimized Monte Carlo data analysis. *Comput. Phys.* **1989**, *3*, 101–104. [CrossRef]

27. Jang, M.; Jeong, H.C.; Kim, T.; Joo, S. Load Profile-Based Residential Customer Segmentation for Analyzing Customer Preferred Time-of-Use (TOU) Tariffs. *Energies* **2021**, *14*, 6130. [CrossRef]

28. Tran, D.H.; Luong, D.-L.; Chou, J.-S. Nature-inspired metaheuristic ensemble model for forecasting energy consumption in residential buildings. *Energy* **2020**, *191*, 116552. [CrossRef]

29. Chicco, G.; Napoli, R.; Piglione, F. Comparisons among clustering techniques for electricity customer classification. *IEEE Trans. Power Syst.* **2006**, *21*, 933–940. [CrossRef]

30. Zhang, Y.; Chen, W.; Xu, R.; Black, J. A cluster-based method for calculating baselines for residential loads. *IEEE Trans. Smart Grid* **2015**, *7*, 2368–2377. [CrossRef]

31. Saputra, D.M.; Saputra, D.; Oswari, L.D. Effect of distance metrics in determining k-value in k-means clustering using elbow and silhouette method. In *Sriwijaya International Conference on Information Technology and Its Applications; SICONIAN.*; Atlantis Press: Padang, Indonesia, 2020; pp. 341–346. Available online: https://www.atlantis-press.com/proceedings/siconian-19/125939938 (accessed on 1 November 2021).

32. Zeng, J.; Wang, J.; Guo, L.; Fan, G.; Zhang, K.; Gui, G. Cell scene division and visualization based on autoencoder and k-means algorithm. *IEEE Access* **2019**, *7*, 165217–165225. [CrossRef]

33. Jeong, H.C.; Jang, M.; Kim, T.; Joo, S.-K. Clustering of load profiles of residential customers using extreme points and demographic characteristics. *Electronics* **2021**, *10*, 290. [CrossRef]

34. Davis, L.W.; Gertler, P.J. Contribution of air conditioning adoption to future energy use under global warming. *Proc. Natl. Acad. Sci. USA* **2015**, *112*, 5962–5967. [CrossRef] [PubMed]

35. Nair, G.; Gustavsson, L.; Mahapatra, K. Factors influencing energy efficiency investments in existing Swedish residential buildings. *Energy Policy* **2010**, *38*, 2956–2963. [CrossRef]

36. Liang, J.; Liu, P.; Qiu, Y.; Wang, Y.D.; Xing, B. Time-of-use electricity pricing and residential low-carbon energy technology adoption. *Energy J.* **2020**, *41*. Available online: https://www.iaee.org/energyjournal/article/3500 (accessed on 21 February 2021). [CrossRef]

37. Kang, H.; An, J.; Kim, H.; Ji, C.; Hong, T.; Lee, S. Changes in energy consumption according to building use type under COVID-19 pandemic in South Korea. *Renew. Sustain. Energy* **2021**, *148*, 111294. [CrossRef]

38. Cvetković, D.; Nešović, A.; Terzić, I. Impact of people's behavior on the energy sustainability of the residential sector in emergency situations caused by COVID-19. *Energy Build.* **2021**, *230*, 110532. [CrossRef] [PubMed]
39. Rouleau, J.; Gosselin, L. Impacts of the COVID-19 lockdown on energy consumption in a Canadian social housing building. *Appl. Energy* **2021**, *287*, 116565. [CrossRef] [PubMed]
40. Qarnain, S.S.; Muthuvel, S.; Bathrinath, S. Review on government action plans to reduce energy consumption in buildings amid COVID-19 pandemic outbreak. *Mater. Today Proc.* **2021**, *45*, 1264–1268. [CrossRef] [PubMed]

Article

Isolated Areas Consumption Short-Term Forecasting Method

Guillaume Guerard *, Hugo Pousseur † and Ihab Taleb

Research Center, Leonard de Vinci Pole Universitaire, 92916 Paris La Défense, France;
hugo.pousseur@edu.devinci.fr (H.P.); ihab.taleb@devinci.fr (I.T.)
* Correspondence: guillaume.guerard@devinci.fr
† Student in Master Thesis.

Abstract: Forecasting consumption in isolated areas represents a challenging problem typically resolved using deep learning or huge mathematical models with various dimensions. Those models require expertise in metering and algorithms and the equipment needs to be frequently maintained. In the context of the MAESHA H2020 project, most of the consumers and producers are isolated. Forecasting becomes more difficult due to the lack of external data and the significant impact of human behaviors on those small systems. The proposed approach is based on data sequencing, sequential mining, and pattern mining to infer the results into a Hidden Markov Model. It only needs the consumption and production curve as a time series and adapts itself to provide the forecast. Our method gives a better forecast than other prediction machines and deep-learning methods used in literature review.

Keywords: time series; Hidden Markov Model; short-term forecast

Citation: Guerard, G.; Pousseur, H.; Taleb, I. Isolated Areas Consumption Short-Term Forecasting Method. *Energies* **2021**, *14*, 7914. https://doi.org/10.3390/en14237914

Academic Editors: Zbigniew Leonowicz and Michał Jasinski

Received: 25 October 2021
Accepted: 19 November 2021
Published: 25 November 2021

Publisher's Note: MDPI stays neutral with regard to jurisdictional claims in published maps and institutional affiliations.

1. Introduction

The power grid became a smart grid [1]. In fact, from the first electric transmission line installed by Thomas Edison in 1882 to nowadays super-large electric grids, power systems evolved with the advance and progress of human industrial civilization. The power grid has evolved in how electricity is generated and distributed.

Over the decade, electricity forecasting (or energy forecasting) focuses not only on the grid side of power grids but also on the client-side to balance load and demand. Short-term consumption forecasting is used to manage, plan, and schedule energy use in various structures, from small-scale houses to vast complexes as factories, malls, or hospitals. It makes it possible to monitor energy consumption, reduce losses, minimize risks, maintain reliability for uninterrupted operation. As a decision-making tool, consumption and production forecasting are used in simulation to plan future investments and where to invest, to schedule maintenance, to manage plants use and battery use.

To optimize and regulate energy flows and prices, the electric system requires the knowledge of future short-term consumption. This means knowing the global consumption and any local consumption, i.e., scaling from a country to a house, to be able to forecast their future. Estimating energy consumption is one of the critical challenges of our time and yet, the interest in this field is still increasing. Besides the energy flow optimization, predicting energy consumption allows us to acquire a more thorough understanding of our modes of consumption.

There is currently no standard defining the types of forecasts. *Hong and Fan* have classified very short term, short term, medium term, and long term forecasts into cut-off horizons of 1 day, 2 weeks, and 3 years [2]. In this paper, we consider only very short-term forecasting with cut-off horizons of 0.5 to 4 h.

Our study's context concerns the European islands, more especially French Mayotte island through the H2020 MAESHA project (https://www.maesha.eu/, accessed on 15 June 2021). MAESHA is the acronym for "deMonstration of smArt and flexible solutions for a decarboniSed energy future in Mayotte and other European islAnds", but also means "future" in Shimaore, the local dialect of the French island of Mayotte that will host the demonstration site. Islands use mainly thermal energies to produce electricity, for example, Mayotte has sixteen diesel engines producing 95% of its energy in 2018 (Source: French energies regulator CRE 2020's report.). Followers include Futuna islands, Saint Barthélémy, Canary islands, Favignana and Gozo.

The goal of the H2020 MAESHA project is to simulate the grid to improve flexibility and include renewable energies and demand-side strategies. Indeed, islands are disconnected from the mainland grid and are primarily dependent on coal. Since the grid is impacted by the geography and activities on the island, coal offers the most flexible and instant source of power to maintain the offer according to the demand. However, the islands tend to cut off the use of coal to include more renewable energies and batteries.

Islands contain many houses, factories, warehouses, farms only connected to the grid by one point. One key challenge is to be able to predict the consumption and production of those isolated areas to manage the offer and demand as well as an appropriate distribution of energy across the islands. To predict isolated areas, we formulate the hypothesis to be able to predict the consumption or production of each device of this area. It will provide a more accurate forecast than a prediction of the whole system.

We define three principal applications of our forecasting tool after being deployed at Mayotte islands:

1. CONSUMPTION FORECAST: The sum of the forecast of each device produces a prediction of an isolated system.
2. CONSUMPTION SCHEDULING: Since each appliance produces its prediction, one can schedule their consumption over time with smart appliances or smart house control.
3. DEMAND-SIDE MANAGEMENT: One may determine the effect of demand-side management strategies on their consumption. It is therefore possible to generate strategies to reduce consumption, switch off certain devices or reduce consumption of appliances, or to plan their renewable energies and batteries with the market.

Concerning the proposed method, it is based on various processes to optimize the accuracy of the forecast and to be adapted to any consumption or production curves. Here are the insights of our paper:

*   We provide a method to transform a time series into sequences. Those sequences represent the cycles of consumption/production of devices.
*   The sequences are reworked to limit the noise while keeping recurrent motifs as well as outliers. This processing produces a more understandable and practical set of sequences.
*   To be able to perform forecasts, we build through ALERGIA algorithm a Hidden Markov Model for each device based on their reworked sequences. A Hidden Markov Model is flexible and can be adapted and updated with new data.
*   In our paper, we provide a step-by-step analysis of our method. Moreover, we compare our results with well-known methods used in forecasting.

After presenting the literature review in the following Section 2, we present the materials and the method in the Section 3. The rest paper is organized as follows:

1. DATASET: first of all, time series are decomposed into sequences (See Section 4).
2. MOTIFS DISCOVERY: to increase the quality of the forecast, a motif discovery algorithm is applied to the sequence (See Section 5).
3. GRAMMATICAL INFERENCE: this method builds a sequence prediction machine as a Hidden Markov Model (See Section 6).
4. VALIDATION: we also implemented famous forecasting methods to compare with our method (See Section 7).
5. EXPERIMENTS: the Sections 8 and 9 show a pedagogical instance and the general results with comparison and step by step analysis.

We conclude the paper in Section 10.

2. Literature Review

Since energy consumption is a set of ordered values representing an evolution of a quantity over time, it is handled as a time series forecasting problem. Current contributions in this field use data-driven models to predict consumption over time. In the following literature review, we cite popular articles, based on high-rank journals and conferences (CORE rank). We also privilege current trends, without being exhaustive, by citing less than five years old articles if possible.

2.1. Context and Surveys

Many surveys list forecasting methods. Zhao et al. [3], Daut et al. [4], Wang et al. [5] categorize those methods in three categories: mathematical and statistical methods; Artificial Intelligence (AI) based methods; and hybrid methods, combining the two mentioned. They also provide useful references and compare those methods. About the two last categories, Raza et al. [6] references various AI-based and hybrid methods at various short-term forecasting scales, from building to a country.

Predicting the energy consumption is frequently done in the short term, almost 60% of the studies employing data-driven models make hourly predictions [7]. It is equally interesting to see that most of the studies focus mainly on the Cooling/Heating (HVAC) consumption estimation. Indeed, HVAC represents between 40 and 50% of the overall consumption of vast buildings like offices, faculties or hotels [8]. Forecasting this consumption is often easier since the HVAC is fairly continuous over time and depends on external parameters like local weather or time of the year [9]. Note that no study considers all sockets/appliances of a household to predict the whole system.

Since energy consumption prediction is a time series forecasting problem, it's not surprising to find familiar methods like AutoRegressive Integrated Moving Average (ARIMA), applied to this field. As it is shown in this research work [10], the most used techniques are ARIMA, Support Vector Machines (SVM), and Artificial Neural Networks (ANN). Some models combine neural networks with more classic models of time series forecasting. This combination aims to strengthen predictions and increase performance. They are designated as hybrid models.

2.2. Most-Used Methods

Although getting old, ARIMA is still being used for its simplicity, its accuracy improved by moving average and its confidence interval. ARIMA is unsuited for long-term prediction and cannot deal with nonlinear relationships. Despite more robust models being used, making ARIMA less privileged nowadays, this method can be found combined with other techniques as a hybrid model. Making a hybrid model with ARIMA allows to combine it with techniques that can support non-linear relationships like SVM or ANN [11–13].

ANNs are the most used method in forecasting building energy consumption. They are widely employed in this field for their numerous advantages. Indeed, the complexity of this task is considerable due to several factors/parameters which are a well-suited problem for ANNs and their capacity to deal with non-linear relationships. ANNs are extremely robust and flexible, especially MultiLayer Perceptron (MLP). They are easy to implement but require some specialized knowledge to configure them. They can be employed alone [14], they can also be combined with other models to get a hybrid model [11,13]. One of the disadvantages is the massive amount of data required to train the network. If there aren't enough data to train the ANN, it will have difficulties to generalize and risks to overfit.

A type of neural network widely used in the field of time series forecasting is the Long Short-Term Memory (LSTM). It's an artificial recurrent neural network architecture that can process entire sequences of data, making it a privileged model for handwriting recognition, speech recognition, and time-series data. It can be implemented alone [15,16] or with a convolutional layer for better results [17].

2.3. Forecasting Isolated Areas

One problem is frequently overlooked or ignored: how to predict the energy consumption of an isolated house, with only its consumption data. Indeed, a single household is strongly dependant on the inhabitants. Building a good forecasting model is hard due to the human factor, the lack of external context (no sensors), and the small or null amount of energy consumed most of the day.

Our study is based on sequence prediction machines. They build a prediction or decision tree from the time series. Those methods are mostly ignored in energy forecasting because of the massive number of heterogeneous data. However, those methods provide outstanding results with a homogeneous dataset like Deoxyribonucleic acid (DNA) forecasting. The key challenge of our method is to build an understandable and useful dataset to be used by sequence prediction machines.

3. Materials and Methods

3.1. Materials

Mayotte island contains various cities and villages close to the coast and rivers as shown in the density map Figure 1. The mainland is essentially mountains, quickly going up to 900 m above sea level. The economy of Mayotte is essentially based on agriculture, fishing, and aquaculture.

The island has only one high voltage line, energy circulates across the island through low voltage. The network forms the shape of a bubble, but inland areas are mainly served by linear lines which made the network unstable. Moreover, the two diesel plants are in the north of the island; in the south many brownouts and blackouts happen every day (Figure 2). Isolated areas frequently cause discrepancies between demand and distribution, triggering local outages or even cascading over part of the network. Our study is a necessity for the network, before considering building new plants and new batteries.

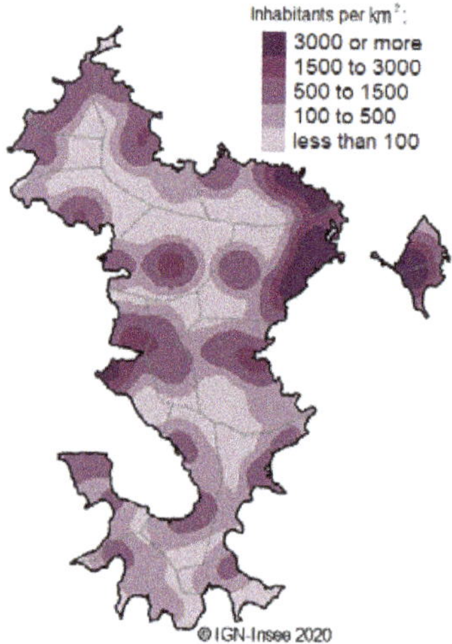

Figure 1. Density map of Mayotte from IGN-Ensee 2020©.

Figure 2. Electric low voltage network of Mayotte from EDM Open Data 2021©.

Since there are no current data for isolated areas in the Mayotte island, the proposed method has been evaluated on the SMART* (http://traces.cs.umass.edu/index. php/Smart/Smart, (accessed on 8 May 2021)) databases. This dataset is composed of seven instrumented and isolated smart homes gathering real-life data each minute for more than two years. Each device, appliance, or socket gets its time series. Those data are consolidated, i.e., they do not have missing data or abnormal values. Since we aim to forecast isolated areas in many European islands, this dataset perfectly suits our goal. The Table 1 presents a dataset overview. The columns refer respectively to the home ID, the number of meters, the percent of total energy from renewable energy.

Table 1. Analysis of the dataset.

HomeID	# Meters	%RE
A	31	0
B	26	0
C	20	10
D	70	39
F	22	73
G	45	0
H	18	0

3.2. Methods

The goal of our method is to build a prediction machine from a time series. Indeed, in isolated areas, it's easy to add meters to any socket, but it becomes harder and more expansive to instrumentalize the area and locally compute a deep-learning method. After preprocessing the time series into a set of sequences, we implement clustering methods, sequence mining, and motif mining (common subsequences) to transform those sequences into meaningful sequences. They are inferred into a Hidden Markov Model, a data structure well-known to generate the next items given a sequence. This flexible prediction machine can learn if the forecast encounters new patterns.

The proposed method contains various steps explained in the following sections. To be more understandable, we provide in this section an overview of the method with the goals of each part. Figure 3 presents the steps of the proposed method. We briefly explain each part with its purpose:

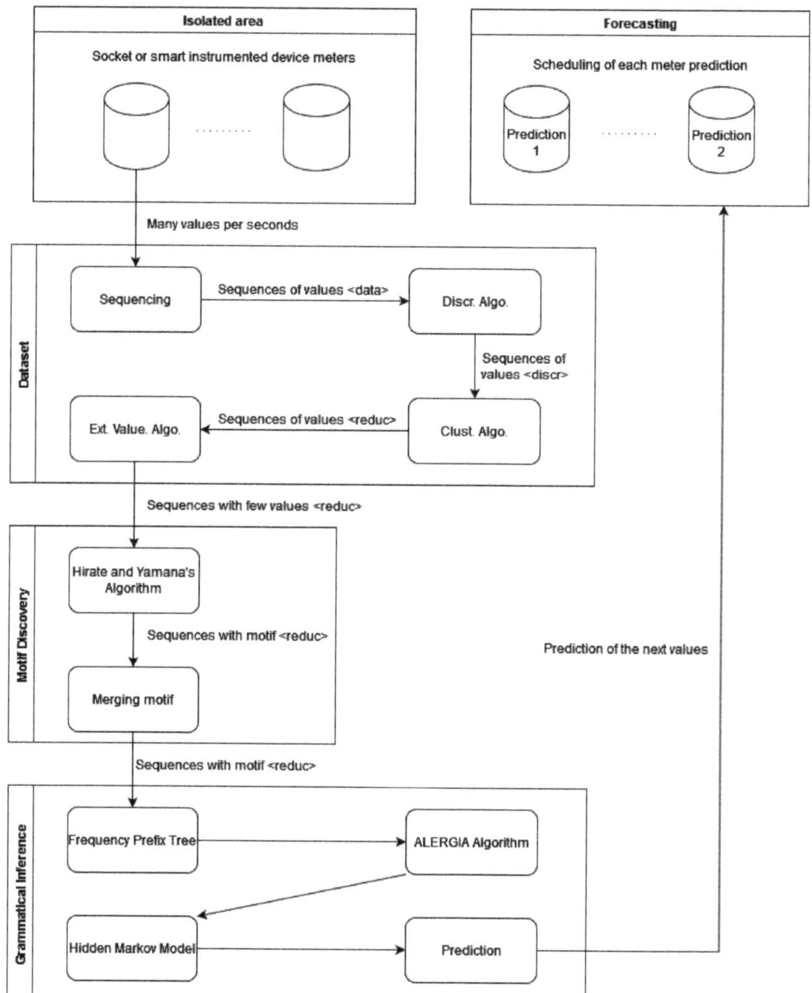

Figure 3. Overview of the method.

1. METERING: Some meters report the consumption or production (we use consumption for the following explanation) of devices or sockets in the isolated area. Each meter provides a time series with a timestep in milliseconds.
2. DATASET: Since meters give several values per second, data are preprocessed. The first step is to cut the time series into sequences of consumption. Then, values are summed together at a common timestep. Values of consumption are replaced by the

slope between two values to better reflect the behavior of the device. To limit the number of unique values, we use two clustering algorithms and a new algorithm to retrieve extreme values hidden in the clusters. This step produces a meaningful dataset for any prediction algorithm, especially those based on sequence mining.

3. MOTIF DISCOVERY: To increase the knowledge contained in the sequences, we implement two algorithms to determine recurrent motifs of consumption inside the sequences. This step increases the range of prediction. Indeed, since a motif starts to be predicted, the whole motif of consumption is already known.

4. GRAMMATICAL INFERENCE: Most of the sequence prediction machines can only predict known sequences or cannot predict further than one recognized sequence. To build a more accurate and complete prediction machine, we transform the sequence into a Hidden Markov Model, a combination of Markov Process and Automaton. This model possesses an algorithm to update the probability and to predict future values.

5. FORECASTING: Since we can forecast the consumption for each meter, we can schedule or sum the future consumption and production of the isolated area.

4. Dataset's Building

To know the consumption of an appliance, a meter is plugged. Each appliance produces a curve of consumption. This one saves a time series where each tuple delivers the following information: ID_device, consumption (in Wh), and date. Each time series (one per appliance) is cut into various sequences as described below. Those sequences form the dataset used by our method to produce a forecast.

4.1. Sequence

A sequence of consumption starts from a zero value of consumption to a non-zero value of consumption on two consecutive time steps. As soon as there is a zero-value of consumption at the following step, the sequence stops. Some devices may have programmed cycles of consumption with small break times. The following method merges two sequences of consumption if it is considered on the same cycle:

$$\Delta B \leqslant \Delta T_i \text{ and } \Delta B \leqslant \Delta T_j$$

with the following variables: ΔB the break duration; ΔT_i the ith sequence of consumption duration; ΔT_j: jth sequence of consumption duration. The Figure 4 presents an example of merging multiple sequences (separated by vertical lines) to constitute a unique cycle of consumption.

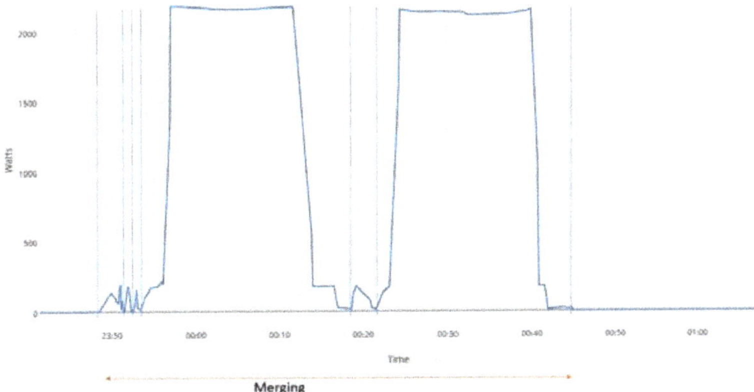

Figure 4. A washing machine consumption's cycle.

Since most of the appliance consumes a little amount of energy in idle mode, the time series must be studied to determine how to set a "zero" value, below this threshold, all values are considered as zero. The consumption curve is studied to set the "zero" value thanks to some mathematical tools as histogram, distribution function before the method, and square error after appreciating the results. They provide an overview of the data to find a good threshold for the "zero" value.

Given the enormous number of values that all the sequences of an appliance may produce, it is relevant to smooth the data by the following methods: a data discretizing named *Discr. Algo.*, a data clustering named *Clust. Algo.*, an algorithm to retrieve extreme values named *Ext. Value. Algo.*

4.2. Data Discretizing

Since most meters provide several values per second, we need to provide a suitable structure for any time series. *Discr. Algo.* provide a discretized dataset at a constant timer which starts at the value of 5 min. Put differently, we sum the instant consumption of the device for 5 min. Thus, the curve of consumption only shows the consumption value every 5 min.

Once all the consumption's values are computed in the following dataset *discretized data*, the integral of the *data* curve is compared to the integral of the *discretized data* curve. The ratio between those two curves provides an estimation of the loss of information. This ratio must not exceed a value α.

As long as this ratio is strictly less than α, it is possible to consider the data with a greater timer between two values. On the contrary, if the ratio is greater than α, one needs to shorten the timer between two values. The algorithm stops when the ratio is close to α. The process is described in Algorithm 1.

Algorithm 1: *Discr. Algo.*

    Input   : dataset of consumption named $\langle data \rangle$
    Output: a discretised dataset of consumption named $\langle discr \rangle$
    $\alpha \in [0, 1]$
    $\epsilon \ll \alpha$
    $timer \leftarrow 5\ min$
    Function *Discretizing(data α)* **is**
    $\langle discr \rangle \leftarrow empty$
    $Integral\_data \leftarrow$ integral of the curve generated by $\langle data \rangle$
    $test \leftarrow true$
    while *test* **do**
        $Integral\_discr \leftarrow 0$
        $x_i \leftarrow$ first value of $\langle data \rangle$
        insert x_i in $\langle discr \rangle$
        foreach x_j *from* $\langle data \rangle$ *at a timer ahead* x_i **do**
            insert x_j in $\langle discr \rangle$
            $Integral\_discr{+}= (\frac{(x_j - x_i)}{2} + min(x_i, x_j)) * timer$
            $x_i \leftarrow x_j$
        incr is the incremental difference between $\langle discr \rangle$ and $\langle data \rangle$
        if $\frac{incr}{Integral\_data} < \alpha - \epsilon$ **then**
            increase the value of *timer*
            reset $\langle discr \rangle$
        if $\frac{incr}{Integral\_data} > \alpha + \epsilon$ **then**
            decrease the value of *timer*
            reset $\langle discr \rangle$
        else
            $test \leftarrow false$

4.3. Slope between Two Values

We transform the value in each sequence. We don't consider the value of consumption but compute the slope between two values. The slope provides a more thorough understanding of the device's behavior than its consumption in Wh. Indeed, some smart appliances have similar cycles of consumption but at various scales of consumption.

Each data of a sequence is referenced with: its value, its sequence ID, its position in the sequence, and from the second position the value refers to the slope between two consecutive values.

4.4. Data Clustering

The data may contain millions of distinctive slope values. Reducing the noise in each sequence is useless since sequences have less than one hundred values. In consequence, we need to reduce the noise in the whole dataset by a clustering method.

Clustering represents the process of determining typical groups, called clusters, in a dataset. The objective is to find homogeneous clusters that are as distinct as possible from other clusters. More formally, the grouping should maximize inter-cluster variance while minimizing intra-cluster variance.

A classification of clustering methods for various static data is proposed in [18]. In the context of our dataset, agglomerative hierarchical-based clustering is suitable because it considers the distance between each value. It runs the distance matrix and plots a dendrogram that displays a hierarchical relationship among the data based on the nearest neighbors for each value. The metric is the Euclidian distance between each value.

To determine the number of cuts, the Elbow method is employed. The total Within-cluster Sum of Squares (WSS) in the dataset is computed for various cuts (i.e., in the function of the number of the cluster). The WSS curve forms an elbow, the number of cuts is equal to the value where the WSS starts to be stabilized. Then, two hierarchical algorithms are implemented:

1. DIvise ANAlysis clustering: DIANA is completely described in Kaufmann and Rousseeuw's book [19]. At each iteration, the cluster with the largest dissimilarity between any two of its observations is selected. The algorithm selects its most disparate observation. Those values initiate the splintering process, and the algorithm reassigns closer observations to both splinters.
2. AGglomerative NESting: AGNES is thoroughly described in Kaufmann and Rousseeuw's book [19]. It constructs a hierarchy of clusterings. Initially, each observation remains a small cluster by itself. Clusters are merged until only one large cluster remains which contains all the observations. At each stage, the two nearest clusters are combined to form one larger cluster.

Various methods are implemented as Ward, Complete linkage, or Flexible linkage to offer various hierarchies to maximize the Silhouette and minimize the WSS.

The Silhouette refers to a method of interpretation and validation of consistency in data clustering and provides a graphical representation of how well each point has been classified.

Let $a(i)$ be the average distance between a point i to all other data points in the same cluster; $b(i)$ be the smallest average distance of point i to any other cluster of which i is not in the set. The Silhouette of point i is defined as follows:

$$s(i) = \begin{cases} 1 - a(i)/b(i), & \text{if } a(i) < b(i) \\ 0, & \text{if } a(i) = b(i) \\ b(i)/a(i) - 1, & \text{if } a(i) > b(i) \end{cases}$$

the Silhouette provides a result between -1 and 1. An $s(i)$ close to one means that the data is appropriately clustered. A value of -1 means the data is in the wrong cluster. In other words, the Silhouette qualifies if the data fit in its cluster or not. A mean Silhouette close to 1 induced the lowest WSS.

After the hierarchical cut, the data are assimilated to their respective nearest cluster center (as the mean of its data). The data with a small Silhouette or with a negative Silhouette means the value will be replaced with an inappropriate one. To decrease the loss of information due to a small Silhouette's value, a K-means algorithm adjusts the observations to the good cluster. Note that the mean of each cluster and the affectation of the observations may differ from the hierarchical clustering.

Since the clustering methods computed the distance between data, in consequence, the dataset must be sampled to avoid memory shortage. The sample must be representative of the diversity of the values. A Monte-Carlo procedure is employed to re-sample a given observation at each level of sampling. The expectation is the mean of the re-sampling will approach Shannon's diversity index at that sample level.

Since we minimize the WSS, each data is now as close to its cluster center as possible. At this step, all values of the sequence are now changed to their respective cluster means. The number of unique values is drastically reduced.

4.5. Extreme Points Retrieval

Once the K-means converges, the Silhouette is computed again. If the Silhouette of some points is not higher than a threshold $0 < silh < 1$, then those points are considered as *noise*. Indeed, the more the Silhouette is close to zero, the more the WSS increase. Put differently, the values with a Silhouette under the threshold *silh* decrease the quality of the clustering, they better fit with a new cluster. A new cluster is produced with each *noise* point.

Otherwise, if the Silhouette provides a negative value, the point is conceded to the corresponding cluster. Then, the Silhouette are updated. This process stops when there are no more *noise* points. Note that through iterations, some *noise* points may be grouped.

This process is useful to keep extreme and isolated values that have been erased through the clustering process. Indeed, those values refer to extreme slope values (mostly very high values) and we need those to accurately predict the consumption.

The proposed method is linear on numbers of clusters and numbers of iteration, and log-linear on the number of points. The Algorithm 2 describes how to retrieve the extreme points *Ext. Value. Algo.* Following this process, each value is changed to their cluster mean.

Algorithm 2: *Ext. Value. Algo.*

Input : Kmeans results $\langle reduc \rangle$, a Silhouette threshold *silh*.
Output: $\langle reduc \rangle$ with extremal values.
Function *Retrieval($\langle reduc \rangle$, silh)* **is**

  while *A cluster has a point with a Silhouette $<$ silh* **do**
    foreach *Points of this cluster* **do**
      if *The silhouette is negative* **then**
        ⌊ Give the point to the corresponding cluster
      if *The silhouette is $<$ silh* **then**
        ⌊ Put the point in a new cluster C
    Compute Silhouette

---

5. Motifs Discovery

At this point, the sequences are meaningful but can include various values. We formulate the hypothesis that some subsequences appear multiple times in the dataset. A device may have consumption patterns/schemes which happen in several sequences. To limit the length of the sequences and to improve the effectiveness of the prediction, those schemes should be considered as a unique symbol, i.e., once the pattern is detected or suggested, the prediction will provide all the concerned motifs.

This section is composed of *Disco. Motif. Algo.*, an algorithm designed to discover a frequent motif inside the dataset.

Sequential pattern mining is an important data mining method that can extract similar subsequences while maintaining their order. However, it is critical to identify data intervals of sequential patterns extracted by sequential pattern mining. *Disco. Motif. Algo.* is based on *Hirate and Yamana* [20] generalized sequential pattern mining with data interval.

Those intervals can be a data gap and/or a time interval. The motif on the consumption scheme must be continuous and contiguous. The generalized sequential pattern mining is set on contiguous data with no time interval.

Besides the common subsequence, its Support is indicated. The Support provides the frequency of this subsequence among the dataset but does not give if the motif is recurrent inside the same sequence. After finding the motifs, only the ones with the most significant length and Support are considered without any conflict on data.

To resume, the *Disco. Motif. Algo.* is as follows:

1. Implement *Hirate and Yamana*'s Algorithm.
2. Use the algorithm on the dataset $\langle reduc \rangle$ from Section 4.4.
3. Return a list of motifs and their support.
4. Merge the symbol of the motif in the dataset from the top motif's length and Support to the lowest. Do not merge if data is already in a better motif.

Let's observe the process through an example. We consider three sequences where each letter refers to a cluster's mean: *aaabacc*, *bacaaa*, *ccbacaaa*. *Hirate and Yamana*'s Algorithm provides, for example, subsequences *bac*, *aaa*, *ba*, *aa* with a support of 100% because those sequences are observed in the three sequences. For example, the subsequence *cc* has support of 66% because of its presence in the first and third sequence. In conclusion, we rewrite the three sequences considering in braces the motif: $(aaa)(bac)c$, $(bac)(aaa)$ and $cc(bac)(aaa)$.

Time series motifs are pairs of individual sequences, or subsequences of a longer sequence, which are extremely similar to each other. One hypothesis is made after *Disco. Motif. Algo.*: some recurrent motifs may differ by at least one symbol.

The goal of *Ext. Motif. Algo.* is to discover extended versions of the motifs in the sequences. For example, the sequences *aaaeaaa* and *aaafaaa* only differ about one symbol. If the symbol e and f have close values, the two symbols are considered equal with a new symbol $g = \frac{e+f}{2}$.

The Figure 5 represents the sequences in Figure 4 after *Ext. Motif. Algo.* Each motif is identified by a number and a color (from 1 to 4).

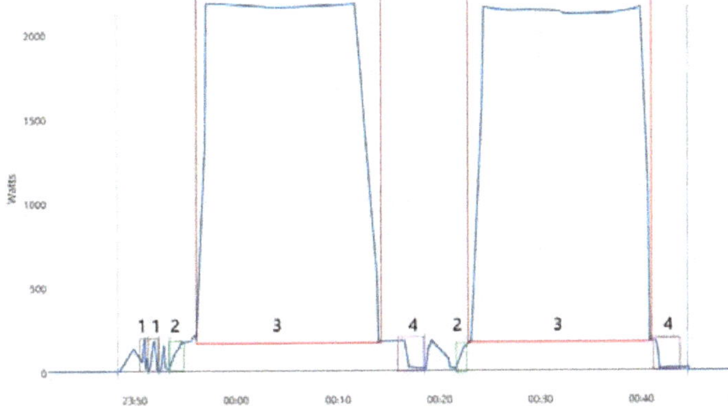

Figure 5. A washing machine consumption's curve after motif discovery.

6. Grammatical Inference

Following the previous steps, the cycles of consumption are now composed of cluster's means and motifs. The goal of this section is to build a prediction machine from those sequences. First of all, the sequences are displayed into a frequency prefix tree, also known as a trie.

Since the trie is PAC-learnable, for Probably Approximately Correct, it is presently possible to make grammatical inference on it [21]. The grammatical inference has been significantly developed by *Colin de la Higuera* whose works are presented in these books [22,23]. Grammatical inference represents the process of learning a formal grammar of a set of data. Note that the trie is pre-parsed, the grammatical inference algorithms will be faster and more efficient.

Many methods of grammatical inference exist, often competing in the PAutomaC [24] contest, a competition of PAC-learning. Among the algorithms, ALERGIA and MDI are non-deterministic algorithms allowing the reduction of a probabilistic deterministic automaton.

6.1. ALERGIA

The ALERGIA algorithm is detailed in the following articles [25,26]. ALERGIA has been applied to information extraction from text or structured documents and speech-language modeling. ALERGIA follows the same process as the well-known RPNI algorithm: some sequences, a compatibility test, and a merging operation.

The compatibility test is based on Hoeffding bounds. In the trie, let n_i be the number of subsequences arriving at node i; f denotes the count of the transitions: $f_i(a)$ be the number of subsequences following arc with the value a; $f_i(.)$ be the number of sequences terminating at node i. Calculate the following probabilities: $p_i(a) = \frac{f_i(a)}{n_i}$ for each transition.

Two nodes i and j are compatible if for each transition (including termination) and their corresponding children are going to be checked recursively:

$$|p_i(a) - p_j(a)| < \sqrt{0.5 \ln \frac{2}{\epsilon}} \left(\frac{1}{\sqrt{n_i}} + \frac{1}{\sqrt{n_j}} \right)$$

where ϵ is called the acceptance range.

The ALERGIA algorithm checks all pairs of nodes, if no pairs are compatible then the algorithm stops. ALERGIA produces a deterministic probabilistic finite automaton. As a compromise between the automaton reduction and the Kullback-Leiber divergence [27], a value $\epsilon = \frac{1}{N^3}$ is taken with N the number of states.

6.2. Prediction and Updating

To be able to use efficiently the probabilistic automaton, we transform it into a Hidden Markov Model (HMM).

The process to transform a stochastic automaton into an HMM is described and proved by Dupont et al. [28] and Harbrard et al. [29]. The steps of the process are not discussed in our paper. This process has been developed in previous works and published in Demessance et al. [30]. The algorithm has been adapted to our context. Let us describe the structure of an HMM and how to perform a prediction.

An HMM is a graph where nodes and arcs possess probabilities. A set of couples [*item*, *probability*] are referenced on each node where *item* is a cluster's means, an outlier, or a motif; and *probability* refers to its probability to be generated on this node. The item # refers to the end of a sequence. By convention, all items are represented in each node, even if its probability is equal to zero.

The root nodes of an HMM are characterized by an ongoing arc without a start node. The root nodes are nodes where we can start to read the sequence. They are computed through the ALERGIA algorithm. Last but not least, arcs possess a *probability* to go from a node to another node.

To perform a prediction, we introduce two operations on an HMM: *jump* and *observation*. The jump process is similar to a random walk in a Markov chain. Then, one can generate an *item* on the last node visited. The combination of those two processes is called observation. The probability to observe an item is equal to the product of the jump and item generation.

Since we need to predict an item from a given sequence, the first step is to start the observation from the right node. Indeed, we need to read the given sequence in the HMM before performing a new observation. This process is done thanks to the *Viterbi* algorithm [31]. Given an input sequence, this algorithm computes the most probable sequence of observations.

Finally, the prediction process occurs at the node of the last observation of this sequence. A set of all predicted observations are generated from this node. This means all jumps from the start node to another node and all generated items at these nodes are computed.

Each observation produces a couple of *(suffix, probability)* where *suffix* refers to the generated item with its *probability* to be observed. Some couple has the same *suffix*, thus its probability is summed.

To observe at a range of 2, we produce a new observation starting at the last visited node of each *suffix*. The new *suffix* is the concatenation of the two observed items. And so on for a larger range of predictions.

In our context, we want to predict the next slope value (consumption or production) of a device knowing its past. Following the prediction process, the forecast sequence is equal to the most probable *suffix*.

Let us show the prediction process with an example (see Figure 6). The HMM has four nodes named 11, 12, 21, 22. The item and probability in brackets are items for each node. The weighted arcs between nodes represent the probability to jump. For this example, we start at node 11 and build the *suffixes* at two steps forward.

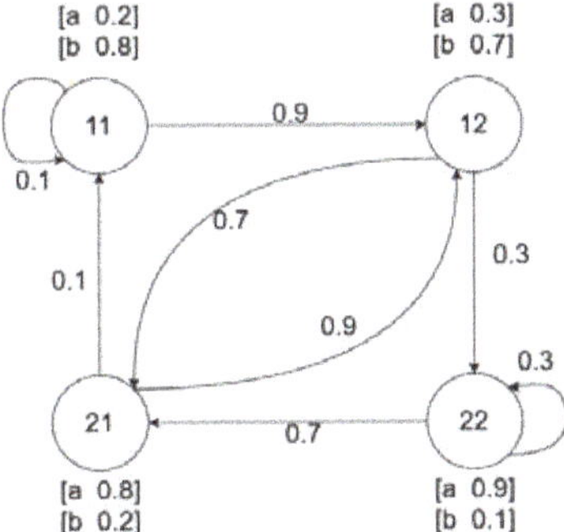

Figure 6. Example of Hidden Markov Model.

Concerning the prediction of the next item, there are two possibilities to jump: to 12 with a probability 0.9 and to 11 with a probability 0.1. To perform an observation, each of these probabilities are multiplied by the probability of each item at nodes. On node 11, we generate item a at 0.2 and b at 0.8. The probabilities of observations from node 11 to node 11 are: a with a probability $0.1 * 0.2 = 0.02$ and b with a probability $0.1 * 0.8 = 0.08$. For the observation from node 11 to node 12, we have: a with a probability $0.9 * 0.3 = 0.27$ and b with a probability $0.9 * 07 = 0.63$. Since both a and b are common *suffixes*, we add their probability to be observed: a with a probability $0.02 + 0.27 = 0.29$ and b with a probability $0.08 + 0.63 = 0.71$ (note the sum of probabilities of *suffixes* is always equal to 1).

To perform a prediction of two items, we need to compute the next observation considering we start at node 11 with a probability 0.1 and at node 12 with a probability 0.9. The predicted item is the *suffix* with the highest probability.

In our model, the predicted item can be a single value or a motif. If we consider a timestamp of 5 min, i.e., 5 min between each value, an observation represents a forecast of at least 5 min. Indeed, predicting a motif extends the range of the forecast. In our model, we consider a forecast of at least 20 minutes, i.e., at most a prediction of 4 items.

6.3. Update

If the HMM provides not good enough results, or when we face new behaviors, we must update the HMM.

The update process adapts the probabilities to jump and the probabilities to generate items for a sequence set. To update the HMM, we use the *Baum-Welch algorithm* [31,32]. In our method, we implement this algorithm without any modification, so it isn't discussed in this paper.

7. Validation and Comparison

To validate our approach, we compare our results with famous sequence prediction machines and deep-learning methods.

7.1. Sequence Prediction Machines

The task of sequence prediction consists of predicting the following data of a sequence based on the previously observed data. Most of the sequence prediction machines are based on Markov properties as Prediction by Partial Matching (PPM) [33], Dependancy-Graph (DG) [34], All-K-Order-Markov (AKOM) [35], Transition Directed Acyclic Graph (TDAG) [36], Probabilistic Suffix Tree (PST) [37], Context Tree Weighting (CTW) [37].

The drawback of these algorithms is the increasing of temporal and spatial complexity when one wants to increase their accuracy. If the Markovian hypothesis doesn't hold, the accuracy severely decrease. Some models are based on compression such as LZ78 [38], Active Lezi [39] and more recently CPT [40] and CPT+ [41].

To validate our model, we implemented the following methods with the same sequence preprocessing: TDAG, CPT+, AKOM. Note that those sequence prediction machines do not possess an update method.

7.2. Deep-Learning Methods

To validate our method, we implement four deep-learning methods, trained and tested directly on the time series of the whole house after discretizing. The four methods are the most known concerning consumption forecasting as seen in the literature review: CNN-LSTM, CNN, LSTM, and MLP [42].

8. Pedagogical Instance

The data processing (sequencing, evolution of consumption) and the hierarchical clustering are done with the R Studio software. The sequence prediction machines are implemented with JAVA using the SPMF Library developed by Fournier-Viger et al. [43]. The deep-learning methods, ALERGIA algorithm, and HMM methods are implemented in Python language. A Python main script regroups all the processes.

Our method takes one minute to run from the data gathering to the prediction. Tests are executed on a personal computer running on Windows 10 with a processor Intel Core i5-6400 CPU at 2.70 Ghz with 8Go DDR RAM, with 250k-1M observations per appliance. The method requires a similar time with any dataset since it's sampled.

In pedagogical terms, we consider the dishwasher from Home C, Meter 1.

8.1. Data Analysis

First of all, one must analyze the general trends of the time series shown in Figure 7. The consumption values present 503,910 observations from 0.0001 Wh to 1 4018 Wh. Over 98% of the values are under 0.5 Wh and represent the idle mode of the dishwasher. In consequence, a threshold of 0.5 Wh is used as "zero". Close to 90% of the total consumption is kept.

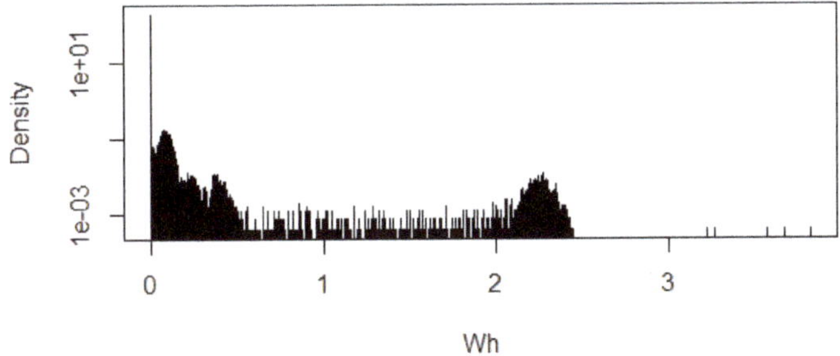

Figure 7. Density of the data.

After sequencing the time series, the dataset provides 748 sequences with an average of 10.6 values of slopes per sequence for a total of 10,834 values where 92.7% are singletons. The slope's percentiles are shown in Figure 8. A representative sample of 1000 slope's values is taken from the dataset and is used to reduce the noise through clustering.

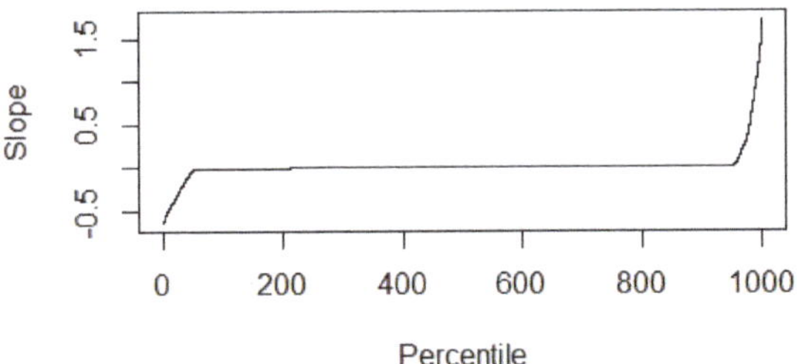

Figure 8. Percentile of the slope.

8.2. Method Step by Step Analysis

The first step of clustering is to determine the optimal number of clusters employing the Elbow method to the sample. The Elbow method executed on the sample gives an optimal number of clusters of 4 clusters and an estimated WSS at 35.8 by extrapolating to the whole dataset.

The Figure 9 presents the inertia of AGNES-flexible for 1 to 20 clusters. The inertia tells how far away the points within a cluster are. As seen with the inertia of the hierarchical clustering, 4 clusters are a good choice. Indeed, the inertia jumps from a maximal value of 1.8 to 3.8. With 5 clusters, the inertia will be at 1.6 which has no change compared to 4 clusters.

Subsequently, the method determines the best hierarchical clustering among DIANA and AGNES algorithms with several metrics. Table 2 presents the average Silhouette, total WSS, and WSS for each cluster for all the hierarchal clustering algorithms. The suffix of AGNES denotes the method: Complete, Ward, Flexible. In this example, the highest average silhouette is given by AGNES with the flexible method.

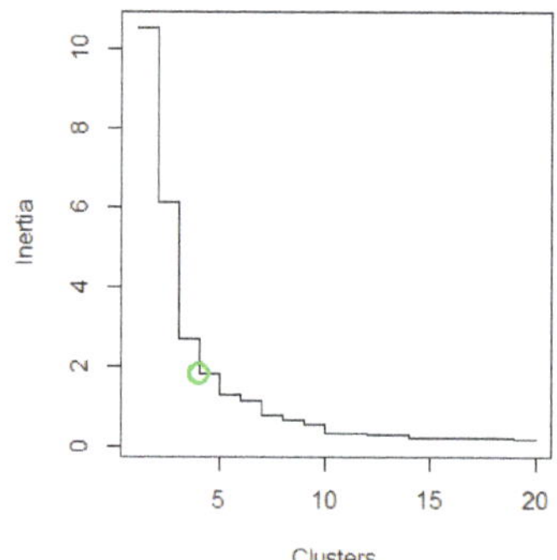

Figure 9. Inertia of AGNES-flexible on the sample.

Table 2. Comparison of WSS for each hierarchical clustering.

Algo	Silh	TWSS	WSS
DIANA	0.937	12.8	2.85 2.36 3.46 4.16
AGNES-c	0.94	12.4	1.90 3.01 3.34 4.16
AGNES-w	0.945	12.1	1.64 2.29 3.34 4.83
AGNES-f	0.945	12.1	1.64 2.29 3.34 4.83

The Figure 10 shows the Silhouette after merging the dataset. The average Silhouette before merging is equal to 0.95. The difference between the Silhouette before and after merging is -0.01. It may vary depending on the sample. The more the dataset contains a vast diversity of values, the more the Silhouette will decrease after merging the dataset. Indeed, the sample is representative of the diversity, not of the distribution.

Figure 10. Silhouette after merging the dataset.

Finally, the K-means algorithm adjusts the data in the right cluster to maximize the silhouette. The Figure 11 presents the Silhouette of the dataset after implementing the K-means algorithm. The average Silhouette is close to the sample. Among the dataset, 586 different data (5.5% of the dataset) have changed their cluster.

Table 3 compares the WSS at each step of our method: after the "zero" threshold (*Zero*), after the hierarchical clustering (*Hcl*), after the K-means (*End*). The total WSS (TWSS) slightly decreases between the hierarchic clustering results and the K-means algorithm results. The WSS is similar for each cluster, and the mean is more adequate to the range of values.

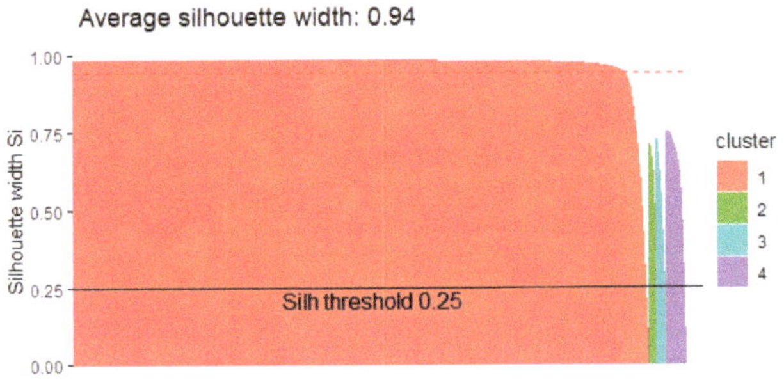

Figure 11. Silhouette of the dataset after K-means algorithm with 4 clusters.

Table 3. Step by step comparison of the proposed method.

	Zero	Hcl	End
TWSS	35.8	37.7	35.4
WSS	none	4.02	8.42
		9.45	8.92
		14.11	8.64
		10.11	9.45
Elements	10,834	10,075	10,179
		405	348
		232	173
		122	134
Mean	0.012	0.001	0.001
		−0.358	−0.39
		0.057	0.551
		1.256	1.225
Min/max	−0.631/1.711	−0.19/0.22	−0.19/0.27
		−0.63/−0.13	−0.63/−0.20
		0.22/0.92	0.28/0.89
		0.93/1.71	0.89/1.71

At this point, we retrieve the extreme points with a *Silh* threshold at 0.25 as shown in Figure 11. The number of outliers is 11 (0.10% of the dataset), each outlier has its cluster. Thus, in the beginning, we got 10,834 different values. They are transformed into 15 values, 4 from the means of the clusters, and 11 outliers. We reduce the number of different values by 99.89%.

Concerning the motif discovery, we start with 748 sequences with an average of 10.6 values. We discover several motifs, from length 16 to length 2. The average number of values drops to 3.4 values per sequence, reducing the length by 67.9%.

Finally, we transform the dataset into a stochastic automaton thanks to ALERGIA algorithm. We do not provide the graph of the HMM since it's understandable for a human's eye (18 nodes, an average of 3 arcs per node, 15 items per node).

We perform some metrics to compare the true consumption to the predicted consumption. First of all, we transform the predicted sequence into a curve that constitutes the predicted consumption. The HMM prediction has an accuracy (the probability to predict the right value) of 99.8%. The Mean Square Error (MSE) is 0.0183 Wh and the Mean Absolute Error (MAE) is 0.0956 Wh which is low compared to the data values: the minimal consumption is 0.5016 Wk; the maximal consumption is 1.4018Wh; the mean is 1.3047 Wh. The Mean Absolute Percentage Error (MAPE) is 7.4%. The MAPE is relevant since it quickly increases with small values.

8.3. Impact of Zero Threshold

The accuracy with a value's threshold at 0.5 Wh is 99.8 with a MAPE at 7.4%. To determine how the Zero value impacts the results, we perform the same test with a Zero threshold at 0.1 Wh. Our model's accuracy drops to 78.8 with a MAPE at 12.8%. As shown, the threshold must be representative of a device in operation to provide relevant results.

8.4. Impact of the Number of Clusters

The Table 4 presents the accuracy (Acc.), average Silhouette (Sil.), MSE (in 10^{-3} Wh), MAE (in 10^{-3} Wh) and MAPE (in %) in the function of the number of clusters, we don't count outliers as clusters. Note that MSE, MAE, and MAPE are computed considering only true predictions. The average Silhouette drops after 10 clusters because their center becomes too close to each other. The MSE, MAE, and MAPE decrease with the number of clusters, but the accuracy becomes quickly low.

Table 4. Prediction efficiency in function of the number of clusters.

# Cluster	Acc.	Sil.	MSE	MAE	MAPE
4	99.8	0.94	18	96	7.4
5	92.6	0.93	17	96	7.3
6	85.4	0.94	8	64	4.9
7	73.5	0.94	5	50	3.9
10	40.8	0.92	3	42	3.2
15	30.1	0.78	1	31	2.4
20	18.7	0.76	1>	21	1.16

9. Overall Results

This subsection presents the results for various households.

Table 5 presents the results for a sample of 10 appliances taken randomly. The table shows in order the number of optimal clusters, the accuracy, the average Silhouette, and the MAPE in percent. Some devices are more suitable for our method than others. The appliances with the highest accuracy are the most used and those with distinct cycles of consumption (dishwasher as an example). The appliances with the lowest accuracy are the less used and those which depend on external parameters or human interaction (ovens for example).

Table 5. Results for a sample of 10 appliances.

Test	# Cluster	Acc	MAPE %
1	4	99.8	7.4
2	5	96.1	4.7
3	5	98.3	1.8
4	7	99.9	6
5	9	95.5	1.5
6	3	99.9	5.2
7	5	98.9	0.9
8	4	98.1	1.3
9	4	99.2	4.2
10	8	96.8	1.1

9.1. Comparison with Deep-Learning Methods

To validate our method, we implement four deep-learning methods, trained and tested on the whole Home C discretized time series. The four methods are CNN-LSTM, CNN, LSTM, and MLP. Their architecture is shown in Table 6. The window size is optimized to keep the lowest error. All models are trained to give a 10 min forecast, similar to our method with a range of 2.

Table 6. Models parameters & architecture.

Models	CNN_LSTM	CNN	LSTM	MLP
Architecture	1x Conv1D 2x LSTM 2x Dense Output	1x Conv1D 1x Flatten 2x Dense Output	2x LSTM 1x Dense Output	1x Flatten 2x Dense Output
Epochs	17	25	30	50
Optimizer	Adam	Adam	Adam lr = 0.003	Adam

The MAPE results are: 40% for the CNN-LSTM; 47% for CNN; 54% for LSTM and 48% for MLP. As said in the introduction, deep-learning methods fail to forecast isolated areas due to the low values and the huge impact of human behavior on the system.

9.2. Comparison with Sequence Prediction Machines

We compare our results with sequence prediction machines: TDAG at order 4, CPT+, AKOM at order 4. They show similar results as our method with an equal or less MAPE with no more than 10 points less due to lower accuracy than our method. Note that the coverage is at 100%, i.e., all the sequences are used to train the model.

In this case, all prediction machines provide outstanding accuracy due to the quality of the sequence's processing we propose. Note those prediction machines can only predict already known sequences, and cannot predict when there are at the end of a sequence. Videlicet, those prediction machines are useless if they encounter a new behavior and for a long forecast.

9.3. Algorithms and Best Results

We calculate the number of times an algorithm in the method provides the best result. Indeed, depending on the distribution function of the values, the clustering methods (Dividing, agglomerative, Ward, Complete linkage, Flexible linkage) provide different dendrograms. Thus, the distribution and clusters change, which implies a variation in WSS and Silhouette and the remaining methods. About the prediction machines, they implement various algorithms to reduce the decision tree. Depending on the trie in input, they may provide another prediction from each other.

We compute the percentage of presence of each clustering method and prediction machines to obtain the best prediction based on 232 tests in the Table 7. We don't include deep-learning methods since they reveal their limits for forecasting isolated areas.

When the values follow a Gaussian distribution, AGNES-ward and AGNES-flexible are the most used; else DIANA or AGNES-complete give a better clustering. Concerning the prediction, our method based on HMM gives 91% of time the best accuracy. Sequence machines have most of the time a lesser or equal accuracy than the HMM due to the outliers. Those results validate our approach.

Table 7. Used of each algorithms for an optimal solution.

Algo	% Use	Algo	% Use
DIANA	9	TDAG	>1
AGNES-c	11	CPT+	1
AGNES-w	41	AKOM	8
AGNES-f	39	HMM	91

9.4. Forecasting of an Isolated Area

Another application of our method is the consumption schedule. One identifies the sequences associated with the use of their appliances or the recurrent consumption cycles. Next, one manipulates them for a job scheduling with knapsack constraints [44].

Since the time series is decomposed into sequences, one can analyze the use of devices to compute a frequency or recurrence of sequences along days. More information is added to each sequence: the time when the sequence starts with its frequency as a stochastic process. Thanks to this, we can schedule the sequences of each device having a recurrent behavior over the day. The Figure 12 presents the common patterns of consumption of appliances in Home F during 500 min where each color refers to one appliance. The user can add more appliances consumption cycle according to its need or according to the most probable use time.

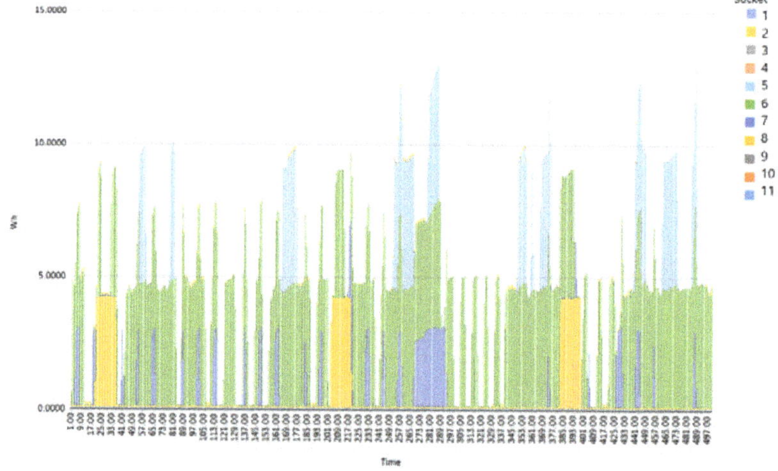

Figure 12. Manual consumption scheduling of recurrent consumption cycle of 11 appliances.

9.5. Discussion

Our method works for any time series, without time, season, weather contexts. Those external factors influence human behavior. Since we determine sequences of consumption reflecting those behaviors, our method is relevant in any context.

The drawbacks are the lack of flexibility of a prediction machine when it encounters unknown schemes of consumption and fails to forecast the good motif. The ability to adapt the HMM thanks to *Baum-Welch*'s algorithm makes our model more flexible to new behaviors.

Note that our model performs better than the most used methods in the literature for isolated areas. This statement becomes false when we need to forecast for a large area as a region. In this case, deep-learning methods with external factors are the most effective. Indeed, our method needs a huge amount of knowledge for each place and cannot compete with some deep-learning methods. Those run with the time series and the weather and provide results with less than 1% of MAPE over a region.

10. Conclusions

Consumption's prediction of isolated households represents a complex problem. Some deep learning methods or mathematical models generate substantial results but are unsuitable to isolated small prosumers.

We adopt a method principally employed to predict DNA to the energy forecasting problem. Our method works efficiently on any device (household appliances, industrial appliances, industrial motors). Moreover, the method is not a black or grey box and it can be embedded easily in any smart building by anyone.

The result of the sequence processing is used to build an HMM. The latter can provide a short-term forecast for the device. The sum of the forecast forms the forecast of the area. The sequences are also used by the owner to schedule its consumption and to be able to adapt its scheduling according to the demand-side management program.

Our method will be used in Mayotte Island, through the H2020 MAESHA's project. It will be included in a multi-agent system to simulate the behavior of the whole island. This sandbox is useful to test new strategies, renewable plants, and flexible markets on the island before considering real-world changes.

Author Contributions: Conceptualization, G.G. and H.P.; methodology, G.G.; software, H.P. and I.T.; validation, G.G., H.P. and I.T.; resources, I.T.; writing—original draft preparation, G.G. and H.P.; writing—review and editing, G.G.; visualization, G.G.; supervision, G.G.; project administration, G.G. All authors have read and agreed to the published version of the manuscript.

Funding: This project has received funding from the European Union's Horizon 2020 research and innovation programme under grant agreement No. 957843.

Institutional Review Board Statement: Not applicable.

Informed Consent Statement: Not applicable.

Data Availability Statement: Not applicable

Conflicts of Interest: The authors declare no conflict of interest.

References

1. Amin, M. The smart-grid solution. *Nature* **2013**, *499*, 145–147. [CrossRef]
2. Hong, T.; Fan, S. Probabilistic electric load forecasting: A tutorial review. *Int. J. Forecast.* **2016**, *32*, 914–938. [CrossRef]
3. Zhao, H.X.; Magoulès, F. A review on the prediction of building energy consumption. *Renew. Sustain. Energy Rev.* **2012**, *16*, 3586–3592. [CrossRef]
4. Daut, M.A.M.; Hassan, M.Y.; Abdullah, H.; Rahman, H.A.; Abdullah, M.P.; Hussin, F. Building electrical energy consumption forecasting analysis using conventional and artificial intelligence methods: A review. *Renew. Sustain. Energy Rev.* **2017**, *70*, 1108–1118. [CrossRef]
5. Wang, Z.; Srinivasan, R.S. A review of artificial intelligence based building energy use prediction: Contrasting the capabilities of single and ensemble prediction models. *Renew. Sustain. Energy Rev.* **2017**, *75*, 796–808. [CrossRef]

6. Raza, M.Q.; Khosravi, A. A review on artificial intelligence based load demand forecasting techniques for smart grid and buildings. *Renew. Sustain. Energy Rev.* **2015**, *50*, 1352–1372. [CrossRef]
7. Amasyali, K.; El-Gohary, N.M. A review of data-driven building energy consumption prediction studies. *Renew. Sustain. Energy Rev.* **2018**, *81*, 1192–1205. [CrossRef]
8. Pérez-Lombard, L.; Ortiz, J.; Pout, C. A review on buildings energy consumption information. *Energy Build.* **2008**, *40*, 394–398. [CrossRef]
9. Liu, T.; Tan, Z.; Xu, C.; Chen, H.; Li, Z. Study on deep reinforcement learning techniques for building energy consumption forecasting. *Energy Build.* **2020**, *208*, 109675. [CrossRef]
10. Deb, C.; Zhang, F.; Yang, J.; Lee, S.E.; Shah, K.W. A review on time series forecasting techniques for building energy consumption. *Renew. Sustain. Energy Rev.* **2017**, *74*, 902–924. [CrossRef]
11. Zhang, G. Time series forecasting using a hybrid ARIMA and neural network model. *Neurocomputing* **2003**, *50*, 159–175. [CrossRef]
12. Nie, H.; Liu, G.; Liu, X.; Wang, Y. Hybrid of ARIMA and SVMs for Short-Term Load Forecasting. *Energy Procedia* **2012**, *16*, 1455–1460. [CrossRef]
13. Wang, X.; Meng, M. A Hybrid Neural Network and ARIMA Model for Energy Consumption Forcasting. *J. Comput.* **2012**, *7*. [CrossRef]
14. Platon, R.; Dehkordi, V.R.; Martel, J. Hourly prediction of a building's electricity consumption using case-based reasoning, artificial neural networks and principal component analysis. *Energy Build.* **2015**, *92*, 10–18. [CrossRef]
15. Karevan, Z.; Suykens, J.A. Transductive LSTM for time-series prediction: An application to weather forecasting. *Neural Netw.* **2020**, *125*, 1–9. [CrossRef]
16. Wang, J.Q.; Du, Y.; Wang, J. LSTM based long-term energy consumption prediction with periodicity. *Energy* **2020**, *197*, 117197. [CrossRef]
17. Kim, T.Y.; Cho, S.B. Predicting residential energy consumption using CNN-LSTM neural networks. *Energy* **2019**, *182*, 72–81. [CrossRef]
18. Jaiwei, H.; Kamber, M. *Data Mining: Concepts and Techniques*; Morgan Kaufmann: San Francisco, CA, USA, 2006.
19. Rousseeuw, P.J.; Kaufman, L. *Finding Groups in Data*; Wiley Online Library: Hoboken, NJ, USA, 1990.
20. Hirate, Y.; Yamana, H. Generalized Sequential Pattern Mining with Item Intervals. *JCP* **2006**, *1*, 51–60. [CrossRef]
21. Haussler, D. *Probably Approximately Correct Learning*; University of California, Santa Cruz, Computer Research Laboratory: Santa Cruz, CA, USA, 1990.
22. De la Higuera, C. *Grammatical Inference: Learning Automata and Grammars*; Cambridge University Press: Cambridge, UK, 2010.
23. Eyraud, R.; De La Higuera, C.; Kanazawa, M.; Yoshinaka, R. Introduction to the Grammatical Inference Special Issue of Fundamenta Informaticae. 2016. Available online: https://hal.archives-ouvertes.fr/hal-01399434/document (accessed on 20 July 2021).
24. Verwer, S.; Eyraud, R.; De La Higuera, C. PAutomaC: A probabilistic automata and hidden Markov models learning competition. *Mach. Learn.* **2014**, *96*, 129–154. [CrossRef]
25. Carrasco, R.C.; Oncina, J. Learning stochastic regular grammars by means of a state merging method. In *International Colloquium on Grammatical Inference*; Springer: Berlin/Heidelberg, Germany, 1994; pp. 139–152.
26. Thollard, F.; Dupont, P.; De La Higuera, C. Probabilistic DFA inference using Kullback-Leibler divergence and minimality. In Proceedings of the Seventeenth International Conference on Machine Learning (ICML), Standord, CA, USA, 29 June–2 July 2000; pp. 975–982.
27. Kullback, S. *Information Theory and Statistics*; Courier Corporation: North Chelmsford, MA, USA, 1997.
28. Dupont, P.; Denis, F.; Esposito, Y. Links between probabilistic automata and hidden Markov models: Probability distributions, learning models and induction algorithms. *Pattern Recognit.* **2005**, *38*, 1349–1371. [CrossRef]
29. Habrard, A.; Denis, F.; Esposito, Y. Using pseudo-stochastic rational languages in probabilistic grammatical inference. In *International Colloquium on Grammatical Inference*; Springer: Berlin/Heidelberg, Germany, 2006; pp. 112–124.
30. Demessance, T.; Bi, C.; Djebali, S.; Guérard, G. Hidden Markov Model to Predict Tourists Visited Places. In Proceedings of the 2021 22nd IEEE International Conference on Mobile Data Management (MDM), Toronto, ON, Canada, 15–18 June 2021; pp. 209–216.
31. Kriouile, A.; Mari, J.F.; Haon, J.P. Some improvements in speech recognition algorithms based on HMM. In Proceedings of the International Conference on Acoustics, Speech, and Signal Processing, Albuquerque, NM, USA, 3–6 April 1990; pp. 545–548.
32. Baggenstoss, P.M. A modified Baum-Welch algorithm for hidden Markov models with multiple observation spaces. *IEEE Trans. Speech Audio Process.* **2001**, *9*, 411–416. [CrossRef]
33. Cleary, J.; Witten, I. Data compression using adaptive coding and partial string matching. *IEEE Trans. Commun.* **1984**, *32*, 396–402. [CrossRef]
34. Padmanabhan, V.; Mogul, J. Using prefetching to improve world wide web latency. *Comput. Commun.* **1998**, *16*, 358–368. [CrossRef]
35. Pitkow, J.; Pirolli, P. Mininglongestrepeatin g subsequencestopredict worldwidewebsurfing. In Proceedings of the UsENIX Symposium on Internet Technologies and Systems, Boulder, CO, Canada, 11–14 October 1999; p. 1.
36. Laird, P.; Saul, R. Discrete sequence prediction and its applications. *Mach. Learn.* **1994**, *15*, 43–68. [CrossRef]

37. Begleiter, R.; El-Yaniv, R.; Yona, G. On prediction using variable order Markov models. *J. Artif. Intell. Res.* **2004**, *22*, 385–421. [CrossRef]
38. Ziv, J.; Lempel, A. Compression of individual sequences via variable-rate coding. *IEEE Trans. Inf. Theory* **1978**, *24*, 530–536. [CrossRef]
39. Gopalratnam, K.; Cook, D.J. Online sequential prediction via incremental parsing: The active lezi algorithm. *IEEE Intell. Syst.* **2007**, *22*, 52–58. [CrossRef]
40. Gueniche, T.; Fournier-Viger, P.; Tseng, V.S. Compact prediction tree: A lossless model for accurate sequence prediction. In Proceedings of the International Conference on Advanced Data Mining and Applications, Hangzhou, China, 14–16 December 2013; pp. 177–188.
41. Gueniche, T.; Fournier-Viger, P.; Raman, R.; Tseng, V.S. CPT+: Decreasing the time/space complexity of the Compact Prediction Tree. In Proceedings of the Pacific-Asia Conference on Knowledge Discovery and Data Mining, Ho Chi Minh City, Vietnam, 19–22 May 2015; pp. 625–636.
42. Lim, B.; Zohren, S. Time-series forecasting with deep learning: A survey. *Philos. Trans. R. Soc. A* **2021**, *379*, 20200209. [CrossRef]
43. Fournier-Viger, P.; Gomariz, A.; Gueniche, T.; Soltani, A.; Wu, C.W.; Tseng, V.S. SPMF: A Java Open-Source Pattern Mining Library. *J. Mach. Learn. Res.* **2014**, *15*, 3389–3393.
44. Charikar, M.; Khuller, S. A robust maximum completion time measure for scheduling. In Proceedings of the Seventeenth Annual ACM-SIAM Symposium on Discrete Algorithm, Miami, FL, USA, 22–26 January 2006; pp. 324–333.

Article

Sparse Signal Reconstruction on Fixed and Adaptive Supervised Dictionary Learning for Transient Stability Assessment

Raoult Teukam Dabou [1], Innocent Kamwa [1,*], Jacques Tagoudjeu [2] and Francis Chuma Mugombozi [3]

[1] Electrical and Computer Science Engineering Department, Laval University, Quebec, QC G1V 0A6, Canada; raoult.teukam-dabou.1@ulaval.ca

[2] Department of Mathematics and Physical Science, National Advanced School of Engineering of Yaoundé, University of Yaoundé I, Yaoundé P.O. Box 8390, Cameroon; jacques.tagoudjeu@univ-yaounde1.cm

[3] Department of Power Systems Simulation and Evolution, Research Institute of Hydro Québec/IREQ, Varennes, QC J3X 1S1, Canada; mugombozi.chumafrancis@hydroquebec.com

* Correspondence: innocent.kamwa.1@ulaval.ca

Abstract: Fixed and adaptive supervised dictionary learning (SDL) is proposed in this paper for wide-area stability assessment. Single and hybrid fixed structures are developed based on impulse dictionary (ID), discrete Haar transform (DHT), discrete cosine transform (DCT), discrete sine transform (DST), and discrete wavelet transform (DWT) for sparse features extraction and online transient stability prediction. The fixed structures performance is compared with that obtained from transient K-singular value decomposition (TK-SVD) implemented while adding a stability status term to the optimization problem. Stable and unstable dictionary learning are designed based on datasets recorded by simulating thousands of contingencies with varying faults, load, and generator switching on the IEEE 68-bus test system. This separate supervised learning of stable and unstable scenarios allows determining root mean square error (RMSE), useful for online stability status assessment of new scenarios. With respect to the RMSE performance metric in signal reconstruction-based stability prediction, the present analysis demonstrates that [DWT], [DHT | DWT] and [DST | DHT | DCT] are better stability descriptors compared to K-SVD, [DHT], [DCT], [DCT | DWT], [DHT | DCT], [ID | DCT | DST], and [DWT | DHT | DCT] on test datasets. However, the K-SVD approach is faster to execute in both off-line training and real-time playback while yielding satisfactory accuracy in transient stability prediction (i.e., 7.5-cycles decision window after fault-clearing).

Keywords: sparse signal decomposition; supervised dictionary learning; dictionary impulsion; singular value decomposition; discrete cosine transform; discrete Haar transform; discrete wavelet transform; transient stability assessment

Citation: Dabou, R.T.; Kamwa, I.; Tagoudjeu, J.; Mugombozi, F.C. Sparse Signal Reconstruction on Fixed and Adaptive Supervised Dictionary Learning for Transient Stability Assessment. *Energies* **2021**, *14*, 7995. https://doi.org/10.3390/en14237995

Academic Editors: Zbigniew Leonowicz and Michał Jasinski

Received: 26 October 2021
Accepted: 22 November 2021
Published: 30 November 2021

1. Introduction

Dynamic state estimation (DSE) is a fast-developing tool for stability monitoring and control [1]. According to the IEEE TF on DSE [2], enhanced system observability using DSE based internal angle and speed estimates will lead to several breakthroughs in Wide-Area System Integrity Protection Systems (WA-SIPS) functions, such as faster out-of-step detection, more realistic location of runaway generator and minimal amount of generation/load to be shed in order to preserve system integrity without knowing the topology accurately. However, instrumental in these achievements is the ability to predict fault-induced instability from DSE based information in real-time, reliably, securely, and fast enough to enable timely and effective countermeasures [3]. In the context of the integration of renewable energy, development of smart grids and artificial intelligence, the authors in [4] proposed a literature review of power system restoration to analyze dynamic decision-making. More generally, application of AI techniques including supervised machine learning in the context of power systems event detection and diagnosis is a field of growing interest [5].

1.1. Background on Sparse Dictionaries Learning

Sparse signals are characterized by a few non-zero coefficients in one of their transformation domains [6]. Therefore, sparse signal decomposition consists in modelling data vectors as sparse linear combinations of basis elements. The basis set used in the decomposition is also called a dictionary. Sparse dictionary learning has attracted widespread scientific attention in signal processing and power system areas. Sparse signal decomposition on hybrid dictionaries is used in [7] for detecting and classifying single and multi-power quality disturbances. In [8], discrete wavelet transform (DWT) and deep neural networks (DNN) are developed for detecting series arc faults (with accuracy recognition rate of 97.75%) in low-voltage residential power distribution. Alternatively, several approaches are proposed in [9] for sparse signal decomposition based on DCT+K-SVD, DCT+ method of optimal directions (MOD), DWT+K-SVD, and DWT+MOD. In [10], new automatic transient detection and localization approaches are proposed for power quality analysis based on an over-complete dictionary (OCD) and $\ell - 1$ norm minimization. Dictionary learning sparse decomposition is implemented in [11] for accurate and fast classification of power quality (PQs) disturbances generated from the IEEE Power Energy Society database. Damped-sinusoidal functions integrated in a redundant dictionary are used in [12] to detect and characterize power systems oscillatory transients in the portion of the time-domain signal based on matching pursuit. A Stockwell transform (ST) approach is proposed in [13] to extract significant disturbance features, used as the input dataset of K-nearest neighbor algorithms (K-NN), decision trees (DT) and support vector machines (SVM). A compressive-information sparse representation based on sparse recovery with a new high-dimensional convex hull is developed in [14] for fast and reliable PQs event classification due to the new challenge of smart grids. The sparse representation and fully connected neural networks based on sparse coding, intelligent features learning, and neural networks are developed in [15] for extracting and classifying idiographic signal residential fault. A fault diagnosis approach based on sparse representation and SVM is used in [16] for computing sub-dictionaries and representing the testing sample while in [17], K-SVD, compressive sampling matching pursuit, and Gauss random matrix are used for reconstructing power quality events. For a signal-to noise ratio of 30 dB, a K-SVD-CoSaMP algorithm reconstructs PQs better than the DCT algorithm.

According to [18], linear sensitivity distribution factor is performed based on injection shift factor of the line with respect to active power injection at all buses. The phasor data angle and DC linear power flow reformulation are used in [19] for identifying multiple line outages based on solving a sparse representation problem via coordinate descent iteration. The cluster-based sparse coding (CSC) algorithm is proposed in [20] for multiple events detection, recognition, and temporal localization in large-scale power systems while in [21], sparse representation classification and random dimensionality reduction projection are used to extract, reduce feature dimensionality and classify power system faults. The singular value decomposition (SVD) and total least square estimation via rotational invariance techniques are used in [22], to analyze and extract amplitude, frequency, attenuation coefficient, initial phase of combined PQs. Among all dictionaries-based learning schemes, which built on K-SVD and OMP algorithms, used e.g., in [23,24] for short-term prediction of solar power fluctuations, is often considered as the most powerful computationally and accuracy wise. In [25], the features extracted from active power are incorporated into a dictionary learning based defense scheme for understanding the cyberattack process in smart grids. As alternatives to dictionary learning, recurrent neural networks (RNN), long short-term memory (LSTM), gated recurrent unit (GRU) and convolutional neural network-LSTM (CNN-LSTM) have also been proposed for time series forecasting of solar irradiance and photovoltaic power [26]. In the above work, the LSTM achieved the best performance in terms of the root-mean square error evaluation error (RMSE). As a further application to time series, [27] used a nonlinear dictionary learning to decompose electricity signals and learn the most representative temporal features introduced by operating devices.

1.2. Contributions of the Present Paper

The previous literature review established clearly that dictionary-based learning applications in power systems have been focused hitherto on PQs analysis and time-series forecasting and characterization. A timely contribution of this paper is therefore to extend dictionary-based learning to transient stability assessment. Even though machine learning has been studied extensively on this problem [28], there is no reported application of K-SVD + OMP dictionary-based learning of post-disturbance time-responses for fast stability prediction in the context of WA-SIPS or contingency screening in control rooms. Furthermore, the paper investigates performance comparisons of single, hybrid, and adaptive dictionaries for wide-area stability assessment which has not yet been addressed in power system literature. In order to determine which dictionary learning approach is the best, according to metrics such as accuracy, reliability, and security, in addition to on-line and off-line computational requirements, this paper proposes to design twenty stable and unstable supervised dictionary learning schemes for sparse reconstruction and classification of stability signal responses based on generator time-series recorded by PMU in post-disturbance stage. Our findings highlight the reconstruction and classification performance of single, hybrid (with two and three sub-dictionaries) and learned dictionaries on the test dataset (rotor speed and corresponding stability status).

The rest of this paper is organized as follows. Section 2 presents design approaches of fixed and adaptive overcomplete dictionary learning for sparse signal decomposition. Section 3 presents performance measures of supervised dictionary learning. Section 4 presents a top-down approach of sparse signal decomposition on fixed and adaptive dictionary learning. Section 5 presents the experimental results of ten supervised dictionary learning algorithms for TSA. Finally, some conclusions and perspectives are given in Section 6.

2. Design of Fixed and Adaptive Supervised Dictionary Learning

Given $Y_{rs} = [y_1, \ldots, y_N] \in \mathbb{R}^{n \times N}$ the post-contingency rotor speed, which contains N input signals of n dimension. The sparse decomposition of Y_{rs} on dictionaries $D_{rs} \in \mathbb{R}^{n \times K}$ ($K \succ n$, over-complete dictionary) is computed by:

$$Y_{rs} = \sum_{i=1}^{K} D_{irs} X_{irs} + \varepsilon_{irs} \tag{1}$$

where: X_{rs} and ε_{irs} denote sparse coding and reconstruction errors of Y_{rs} sparse signal decomposition. The orthogonal matching pursuit (OMP) is the greedy method for sparse coding mostly used in supervised dictionary learning, for minimizing the response reconstruction over a finite number of iterations [29]. These dictionaries are used to evaluate the transient stability which is the ability of the power system to remain in synchronism immediately following a disturbance [30]. The architecture of singles, hybrids, and adaptive supervised dictionary learning for reconstructing and classifying the signals and rapidly predicting the transient stability status immediately after clearing a fault is presented in Figure 1.

2.1. Fixed Single Supervised Dictionary Learning

From the PQD and time-series forecasting experiences, fixed dictionaries can be used to reconstruct rotor speeds and predict the transient stability of generator responses. For this purpose, the cosine transform, sine transform, wavelet transform and Haar transform are simply applied to each signal to extract sparse and redundant features for stability of signals reconstruction.

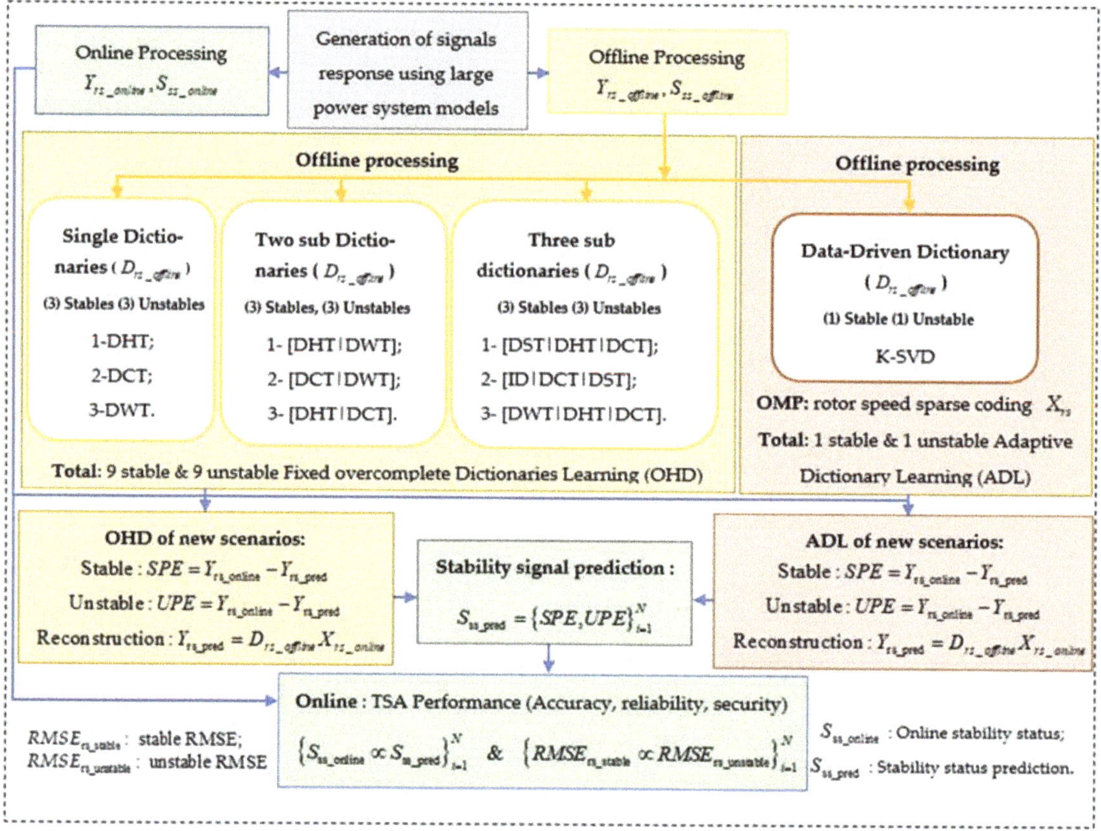

Figure 1. Proposed flowchart of sparse signal decomposition on supervised dictionaries for transient stability assessment.

2.1.1. Discrete Cosine Transform

The discrete cosine transform (DCT) expresses a time-series generator response in terms of a sum of cosine functions oscillating at different frequencies introduced by the fault occurrence. The DCT kernel projection is defined by:

$$\text{DCT} = D_{rs} = \sqrt{\frac{2}{n}} \left[\xi_i \cos\left(\frac{\pi(2j+1)i}{2n} \right) \right]$$

$$i, j = 0, 1, 2, 3, \ldots, K - 1$$

(2)

where $\xi_i = 1/\sqrt{2}$ for $i = 0$, otherwise $\xi_i = 1$. The DCT proposes optimal de-correlation coefficients of the stability signals, while grouping energy contained in the signals in low frequency coefficients [7].

2.1.2. Discrete Sine Transform

The discrete sine transform (DST) derived from the DCT, allows expressing the signal as a sum of sinusoids with different frequencies and amplitudes. The DST kernel projection is computed by:

$$\text{DST} = D_{rs} = \sqrt{\frac{2}{n}} \left[\xi_i \sin\left(\frac{\pi(2j+1)(i+1)}{2n} \right) \right]$$

$$i, j = 0, 1, 2, 3, \ldots, K - 1$$

(3)

where $\zeta_i = 1/\sqrt{2}$ for $i = n - 1$, otherwise $\zeta_i = 1$. The DCT and the DST are almost similar operations to the discrete Fourier transform (DFT). The only difference lies in the projection kernels, which give real coefficients for the DCT, imaginary coefficients for the DST and complex coefficients for the DFT [7].

2.1.3. Discrete Haar Transform

Discrete Haar transform (DHT) consists of orthogonal switched rectangular waveforms, which can take zero value and sample points in subintervals of $t \in [0, 1]$ and where the amplitude can differ from one function to another as follows [31]:

$$\text{DHT} \equiv D_{rs} = \frac{1}{\sqrt{N}} \begin{cases} 2^{m/2}, \frac{k-1}{2^m} \le t < \frac{k-(1/2)}{2^m} \\ -2^{m/2}, \frac{k-(1/2)}{2^m} \le t < \frac{k}{2^m} \\ 0, \text{ otherwise in } [0, 1) \end{cases} \tag{4}$$

$$r = 2^m + k - 1, \ t = n/N, \ 0 \le r, n \le N - 1$$

where: m and k represent the integer decomposition of the index r. The time-frequency Haar function is unitary and invariant under circulant shift.

2.1.4. Discrete Wavelet Transform

The discrete wavelet transform (DWT) is calculated by convolving the post-contingency signal with a mother wavelet transform $\Psi_{m,n}$ [32].

$$\text{DWT} = D_{rs} = \frac{1}{\sqrt{a_0^m}} \sum_k \Psi_{m,n} \left(\frac{k - n b_0 a_0^m}{a_0^m} \right) \tag{5}$$

where: n and m are positive integers. The DWT allows extracting information content at different positions and scales for subsequently reconstructing post-contingency stability signals.

2.1.5. Impulse Dictionary

The impulse dictionary (ID) allows expressing signal as a linear combination of the Dirac basis vectors representing the frequency response pic [7]. It is constructed as:

$$\text{ID} = D_{rs} = [I]_{K \times K} \tag{6}$$

2.2. Fixed Hybrids Supervised Dictionaries Learning

The fixed hybrid supervised dictionary learning is developed from the concatenation of DCT, DHT, DST, DWT, and ID, with the aim of improving online reconstruction and classification of the single dictionaries.

2.2.1. Two Sub-Dictionaries Concatenation

Three concatenations of two sub-dictionaries are constructed for sparse features extraction and classification. Equations (7)–(9) represent these strings of dictionaries joining single structure end-to end.

$$Y_{rs_{n \times N}} = D_{rs} X_{K \times N} = [DHT|DCT]_{n \times K} X_{K \times N} \tag{7}$$

$$Y_{rs_{n \times N}} = D_{rs} X_{K \times N} = [DWT|DHT]_{n \times K} X_{K \times N} \tag{8}$$

$$Y_{rs_{n \times N}} = D_{rs} X_{K \times N} = [DCT|DWT]_{n \times K} X_{K \times N} \tag{9}$$

2.2.2. Three Sub-Dictionaries Concatenation

To improve the singles and hybrids structures with two sub-dictionaries, three string concatenations are developed, see for instances Equations (10)–(12). The fixed predefined overcomplete dictionary learning enables online TSA.

$$Y_{rs_{n \times N}} = D_{rs} X_{K \times N} = [DST|DHT|DCT]_{n \times K} X_{K \times N} \tag{10}$$

$$Y_{rs_{n \times N}} = D_{rs} X_{K \times N} = [ID|DCT|DST]_{n \times K} X_{K \times N} \tag{11}$$

$$Y_{rs_{n \times N}} = D_{rs} X_{K \times N} = [DWT|DHT|DCT]_{n \times K} X_{K \times N} \tag{12}$$

2.3. Adaptive Supervised Dictionary Learning

The K-SVD is a generalization of a k-means clustering algorithm introduced by [24] for dictionary learning after estimation of the representation matrix. During the first learning step of K-SVD, dictionary $D_{rs} = [d_1 \ldots d_K] \in \mathbb{R}^{n \times K}$ ($K \succ n$, over-complete dictionary) is fixed, and the best sparse coding $X_{rs} = [x_1 \ldots x_N] \in \mathbb{R}^{K \times N}$ is found, under the sparsity constraint T. The K-SVD approach is evaluated based on solving an optimization problem defined in Equation (13).

$$\langle D_{rs}, X_{rs} \rangle = \underset{D_{rs}, X_{rs}}{\mathrm{argmin}} \|Y_{rs} - D_{rs} X_{rs}\|_2^2 \text{ subject to } \forall_i \|x_i\|_0 \leq T \tag{13}$$

The OMP is used for sparse coding fixed, hybrids, and adaptive overcomplete dictionary learning (ODL). Given D_{rs}, the sparse coding $X_{rs} \in \mathbb{R}^{K \times N}$, is evaluated using the ℓ_0 norm by:

$$x_{irs} = \underset{x_{irs}}{\mathrm{argmin}} \|Y_{rs} - D_{rs} X_{rs}\|_2^2 \text{ subject to } \forall i, \|x_i\|_0 \leq T \tag{14}$$

where: $D_{rs} \in \mathbb{R}^{n \times K}$ and T denote overcomplete dictionaries and sparsity constraint respectively. The update of each d_i atom necessarily leads to that of the non-zero entries in a row vector $x_i^T \neq 0$. The dictionary is learned by solving to following minimization problem:

$$\underset{d_i, x_i}{\min} \|E_i - d_i x_i^T\|_2^2 \text{ subject to } \|d\|_2^2 = 1 \tag{15}$$

where: $E_i = Y_{rs} - \sum_{j \neq i} d_j x_j^T$ denotes the approximation error matrix, which can be easily decomposed into $U \Delta V^T$. The solution d_j is a column of U and the coefficient vector x_j^T is a column of $V \times \Delta(1, 1)$ [33].

3. Performance Measures of Supervised Dictionary Learning

The confusion matrix is used to evaluate the classification performance of the dictionary learning algorithms on the testing database. Three main metrics are calculated from Table 1.

Table 1. Confusion matrix for TSA classification.

Observation		Predicted Class	
		Positive (Secure)	Negative (Insecure)
Actual	Positive	*TP*	*FN*
Class	Negative	*FP*	*TN*

The accuracy metric defines the ability of learned dictionaries for correctly classifying stable and unstable cases in the testing database:

$$Accuracy\ (\%) = \frac{TP + TN}{TP + TN + FN + FP} \tag{16}$$

The reliability metric in Equation (17) allows for evaluating the performance of learned dictionaries on the unstable cases:

$$Reliability\ (\%) = \frac{TN - FN}{TN} \tag{17}$$

The security metric in (18) makes it possible to define the capacity of learned dictionaries to predict the stable case:

$$Security\ (\%) = \frac{TP - FP}{TP} \tag{18}$$

The RMSE defined in (19) is used to evaluate the difference between the predicted Y_{rs_pred} and observed Y_{rs_online} signal responses:

$$\mathrm{RMSE} = \sqrt{\frac{1}{n}\sum_{i=1}^{n}\left(Y_{irs\_online} - Y_{irs\_pred}\right)^2} \tag{19}$$

4. Top-Down Approach of Sparse Signal Decomposition on Fixed and Adaptive Dictionaries Learning

The simulation of a hundred faults on the line and generator buses of the IEEE test system allowed generating generator rotor-speed responses and their stability status. Figure 2 describes top-down architecture of the proposed system along with the details of the experiment for an even greater clarity of SLOD.

Seventy percent of signals are used for offline-supervised dictionary learning while 30% are kept aside for online TSA. For an offline learning approach, the training signals are labelled and separated into two databases: stable rotor speeds (Database 1) and unstable rotor speeds (Database 2). Database 1 is used as input data for implementing 9 fixed dictionaries D_{rs} for stable cases reconstruction using: DHT, DCT, DWT, [DHT | DWT], [DWT | DCT], [DHT | DCT], [DST | DHT | DCT], [ID | DCT | DST] [7], and [DWT | DHT | DCT] dictionaries. Similarly, Database 1 is used as input data for developing an adaptive stable dictionary D_{rs} based on K-SVD approach. The second Database 2 is used as input for establishing fixed dictionaries D_{rs} for unstable case reconstruction using: DHT, DCT, DWT, [DHT | DWT], [DWT | DCT], [DHT | DCT], [DST | DHT | DCT], [ID | DCT | DST] [7], and [DWT | DHT | DCT] dictionaries. Similarly, database 2 is used as input data for realizing an adaptive unstable dictionary D_{rs} based on K-SVD approach.

The orthogonal matching pursuit algorithm is developed to carry out sparse coding X_{rs} of rotor speeds from the test database (Database 3). The test database contains both stable and unstable rotor speed signals Y_{rs}. The stable and unstable sparse coding X_{rs} are individually projected onto the 20 learned dictionaries D_{rs} to determine playback degree of stability. According to the projection result, if RMSE index of the signal obtained from a stable dictionary is lower than RMSE index predicted by the unstable dictionary, the test signal is targeted as being stable (stability status = "0"). In the opposite case, the signal is labelled as being unstable (stability status = "1"). This analysis criterion is used for evaluating accuracy, reliability (false positive success rate), and security (true negative success rate) of each supervised dictionary-learning scheme on the testing database.

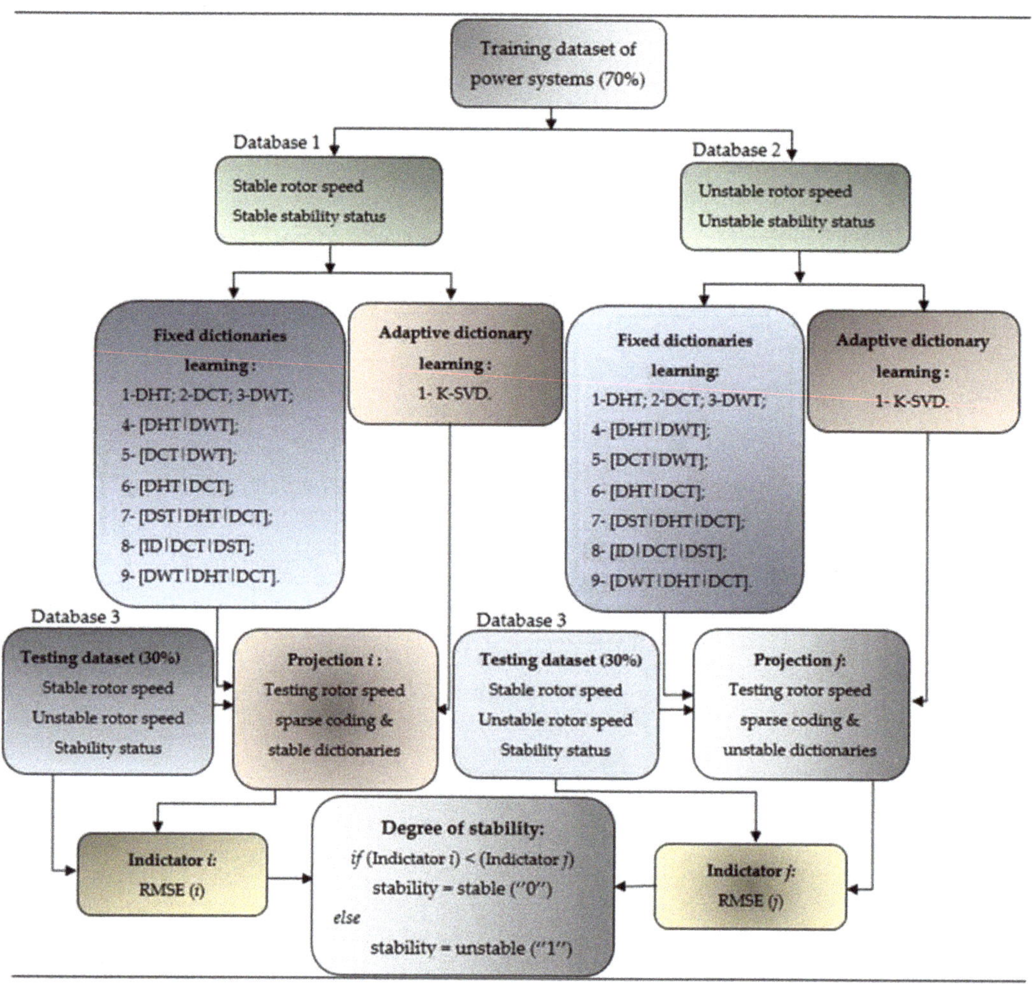

Figure 2. Top-down approaches of supervised dictionary learning for TSA.

5. Experimental Results

The analysis of supervised single ([DCT], [DHT], [DWT]) dictionaries, hybrids with two ([DHT I DCT], [DCT I DWT], [DHT I DWT]) and hybrids with three ([DST I DHT I DCT], [ID I DCT I DST], and [DWT I DHT I DCT]) sub-dictionaries for TSA was performed on the IEEE 68 -bus test system. The $75 \times 20{,}216$ (i.e., 75×8664 allocated for online testing) post-contingency speed signals used to train dictionary learning are generated by simulating various types of fault with wide-range of duration applied on each line and at each generator terminal bus of the test system under various power dispatch and topologies conditions. The generated generator responses used for learning are summarized in stable training set (i.e., $75 \times 19{,}477$) and unstable training set (i.e., 75×739). The unstable cases thus represent a mere 3.8% of the cases, which means that the dataset is highly skewed toward stable cases. These two dictionaries learned separately will allow classifying a joint database containing stable and unstable scenarios. Each of the signals will be projected on each of the two stable and unstable dictionaries. The learned stable or unstable dictionaries that give the lowest RMSE will define the stability status of the new scenario as either stable or unstable.

5.1. Description of Dataset for TSA

To generate a diversity of cases, fault duration is gradually increased (per step of 0.5 cycle) until critical loss-of synchronism limit is reached along each line and close to each generator bus of the test system. For each study case, the numerical simulation duration is set to 10 s and transient fault occurs at $t = 1$ s. However, for the unstable scenarios, numerical simulation duration varies according to the time instant of loss-of-synchronous. For each simulation, only 75 post-contingency samples (sampling rate: 600 samples/s) per rotor speed are used for dictionary learning which means a time-window of 125 ms for making decision (7.5 cycles of fundamental). The re-sampling start time of the generator signal responses varies according to the simulation fault clearing time instant.

In the presence of a disturbance, the differences in the absolute values of rotor angles of all the combinations (i, m) of the generators are calculated and compared to 180°. If the rotor angle does not exceed 180°, or if the disturbance causing the rotor swing is promptly removed, the generator may remain in synchronism with the system. Hence, during fault simulations, the stability of the power system can take two possible status after fault clearing, "$S_{ss} = 0$" corresponding to the stable status, or "$S_{ss} = 1$" corresponding to the unstable status [34,35]. Figure 3 presents an example of stable (a) and unstable (b) speed recorded during the search of critical fault clearing time on line 1 of the IEEE 68-bus test system.

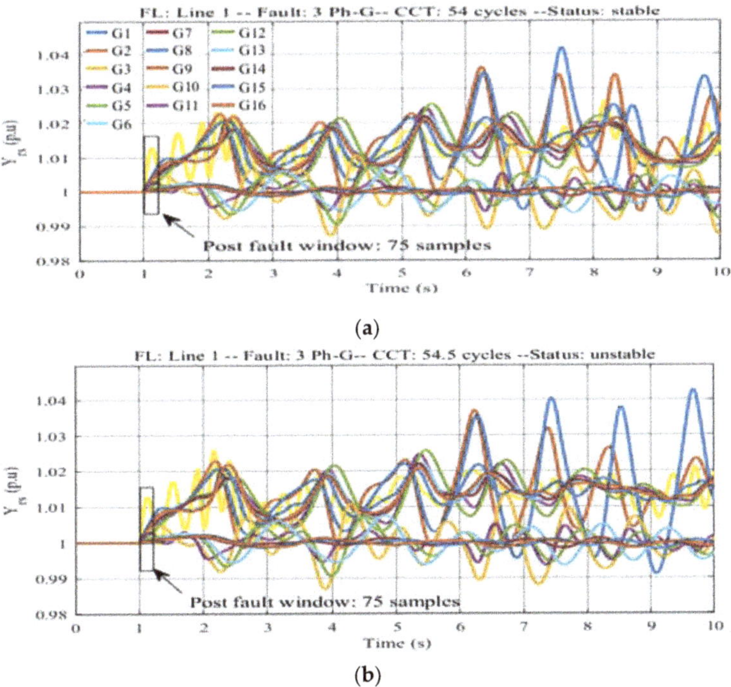

Figure 3. Stable (**a**) and unstable (**b**) rotor speed obtained by simulating three-phase ground fault.

Table 2 presents the dataset configuration used for reconstructing and classifying the unknown testing database (containing 8664 signals: 8341 stable scenarios and 323 unstable scenarios i.e., 3.72% of the cases) based on supervised learning of overcomplete dictionaries.

Table 2. Configuration of stable and unstable sparse signal decomposition for TSA.

Sparse Signal	Offline	Online
Decomposition	Features Extraction	Reconstruction and TSA
Adaptive Dictionary Learning (ADL)	Rotor speed	Prediction error
	(For D_{rs} and X_{rs} computed)	(For RMSE computed)
Fixed Dictionaries Learning (1,2,3H OHD)	Rotor speed	Prediction error
	(For D_{rs} and X_{rs} computed)	(For RMSE computed)

5.2. Fixed and Adaptive Sparse Features Extraction

The number of fixed and hybrid features (i.e., 25 indicators) extracted per signal depends on the number of iterations defined by the orthogonal matching pursuit [36]. However, the adaptive dictionary learning allows extraction of 75 features per generator response for online transient stability assessment. Table 3 summarizes the sparse signal decomposition developed off-line and used for online TSA.

Table 3. Parameters used to train single, hybrids and adaptive supervised dictionary learning.

Configuration	Learning Approaches	Dictionaries/ Feature Size	Sparse Expansion Coefficient Size
Adaptive dictionary (ADL)	K-SVD	$75 \times 20{,}216$	$8700 \times 20{,}216$
Single dictionary (SOD)	DHT		$75 \times 1{,}617{,}280$
	DCT		$75 \times 1{,}516{,}200$
	DWT		$75 \times 1{,}556{,}632$
Two sub-dictionaries (2H-OHD)	[DHT|DWT]	$25 \times 20{,}216$	$75 \times 3{,}133{,}480$
	[DCT|DWT]		$75 \times 3{,}153{,}696$
	[DHT|DCT]		$75 \times 2{,}304{,}624$
Three sub-dictionaries (3H-OHD)	[DST|DHT|DCT]		$75 \times 3{,}800{,}608$
	[ID|DCT|DST] [7]		$75 \times 4{,}589{,}032$
	[DWT|DHT|DCT]		$75 \times 3{,}820{,}824$

The stable K-SVD & OHD dictionaries are initialized and trained with stable signals as input, while unstable K-SVD & OHD dictionaries are trained with unstable responses only. For the fixed sparsity (i.e., matrix of numbers that includes many zeros or values that will not be significant) $T = 10$, the coefficients are computed using OMP and the maximum of number iterations in set as 90. K-means is the method used for defining the maximum iterations able to converge the learning algorithm. It allows defining the threshold beyond which no point changes during the sparse signal decomposition. This method supposes that the last iterations have the least contribution to the percent of correct representation [37]. From the online sparse coding extracted using OMP, stable, and unstable K-SVD & OHD dictionary learning will allow reconstructing and predicting transient stability signals. Sample K-SVD Figure 4a and OHD Figure 4b atoms extracted from the learning of rotor speed responses. Figure 4 presents some examples of adaptive K-SVD (a) and fixed OHD (b) atoms randomly extracted in the offline rotor speed Y_{rs}.

Adaptive K-SVD atoms well describe the non-stationary and transient response of generators signals immediately after clearing fault. However, usually sinusoidal fixed OHD atoms (except DHT) are not satisfactory for analysis of the transient behavior of generators. To illustrate the atoms performance extracted from the training database, we selected two signals from the testing database, which are respectively stable and unstable. Figure 5 presents sparse decomposition of testing signals ID 1 (in blue color) & ID 23 (in red color) based on a separate stable and unstable K-SVD dictionary learning. The K-SVD demonstrates here its ability for an almost perfect fit of the signals in each of the dictionary. This pattern of success is repeated for all signals in the training dataset.

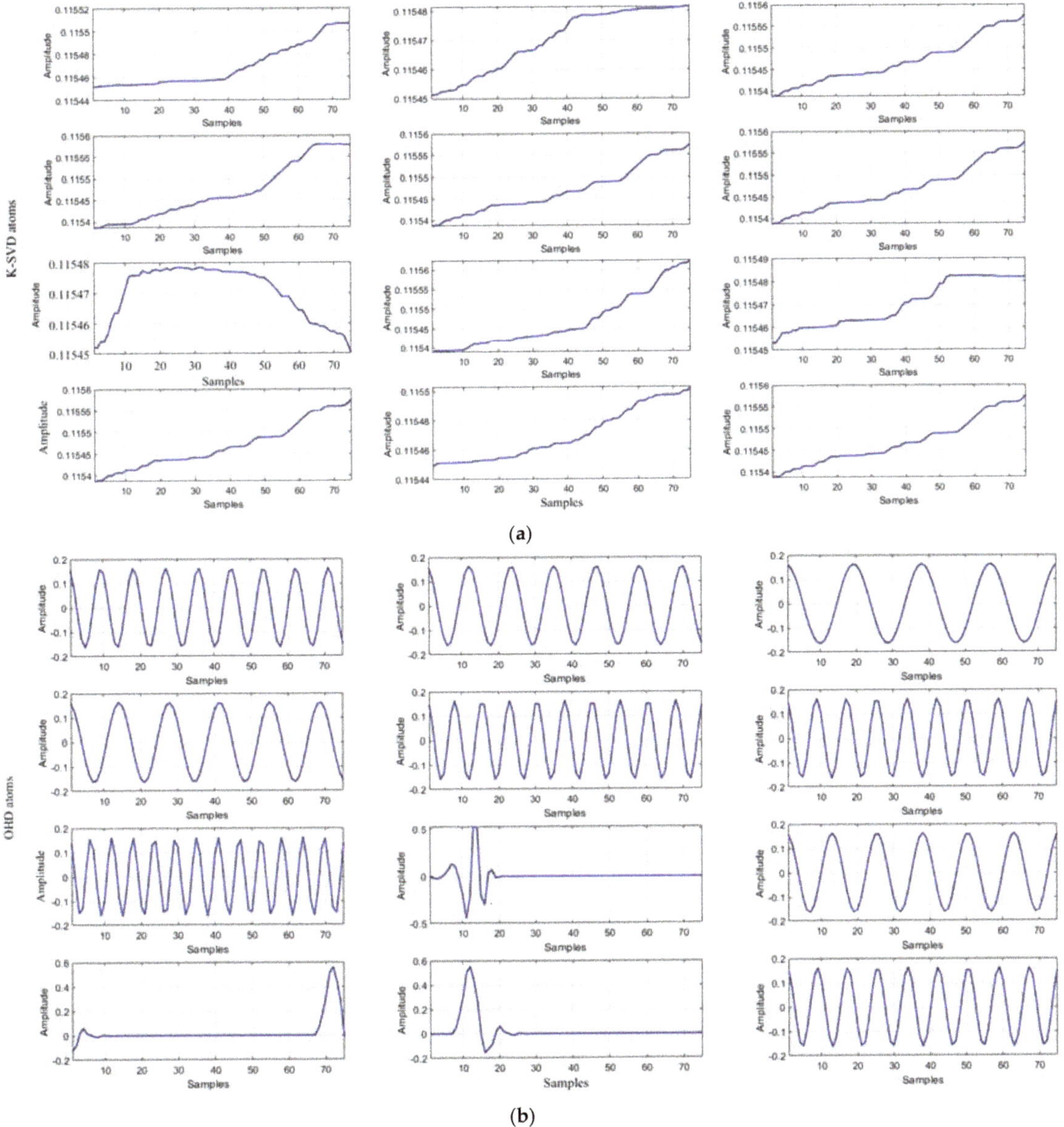

Figure 4. K SVD (**a**) and OHD (**b**) atoms used for TSA reconstruction and classification.

The OHD atoms, such as single (DCT, DWT, DHT) dictionaries, two ([DHT | DCT], [DCT | DWT], [DHT | DWT]) and three ([DST | DHT | DCT], [ID | DCT | DST], [DWT | DHT | DCT]) sub-dictionaries are next used for reconstructing stable (blue color) & unstable (red color) rotor speed and tracking online (black color) response, see Figure 6.

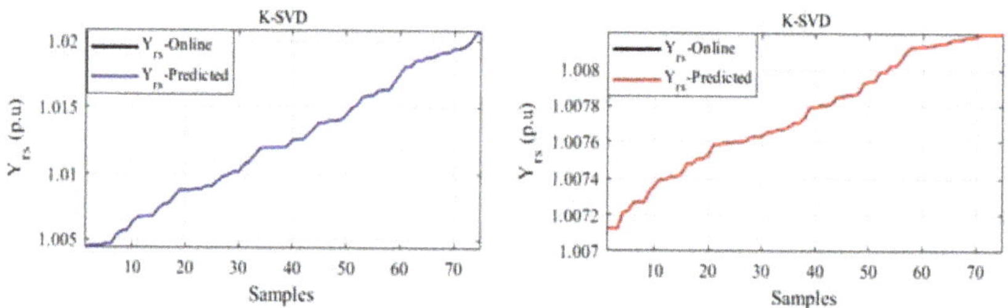

Figure 5. Online & K-SVD reconstruction of stable (blue) and unstable (red) of sample signal's response in the test dataset.

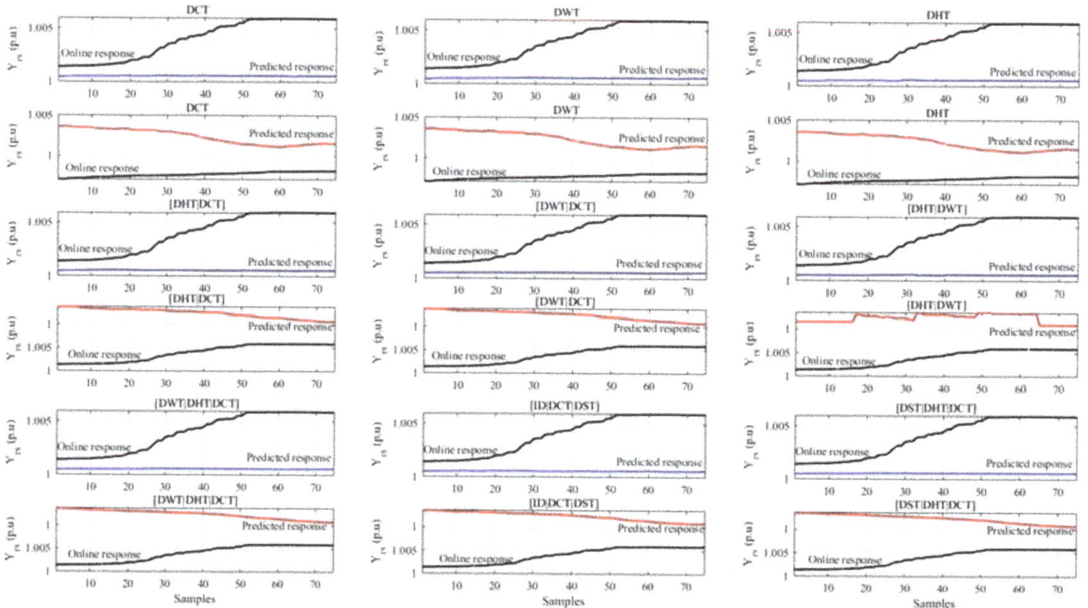

Figure 6. Online & OHD reconstruction of stable (blue) and unstable (red) of sample signal's response in the test dataset.

5.3. Online Sparse Features Classification Based Overcomplete Dictionaries Learning

The projection of signals ID 1 & ID 23, on stable and unstable dictionary learning (SOD, 2H-OHD, 3H-OHD & K-SVD), will allow evaluation of the online stability status according to the atoms structure which gives the lowest RMSE. Figure 7 presents RMSE prediction value obtained by projecting signals ID 1 (blue) and ID 23 (red) on stable and unstable SOD and K-SVD respectively.

According to SOD, the DWT is the best descriptor of stable rotor speed, while for the unstable scenarios prediction DHT is most indicated. The RMSE resulting from hybrids of [DHT|DCT], [DCT|DWT] & [DHT|DWT] are similarly used to perform fixed dictionary learning for online TSA. The projection of signals ID 1 (blue) & ID 23 (red) on two hybrids dictionaries learned separately, allows for a correct prediction of the online stability status based on the lowest RMSE, see Figure 8.

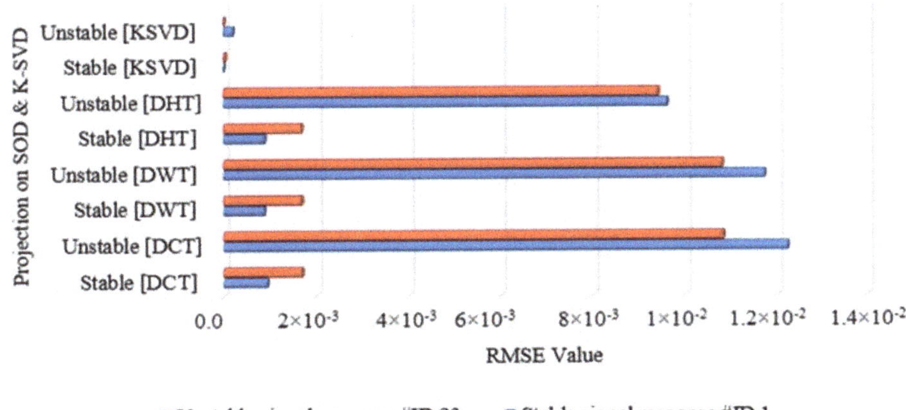

Figure 7. Projections of signals ID 1 and ID 23 on stable and unstable SOD and ADL.

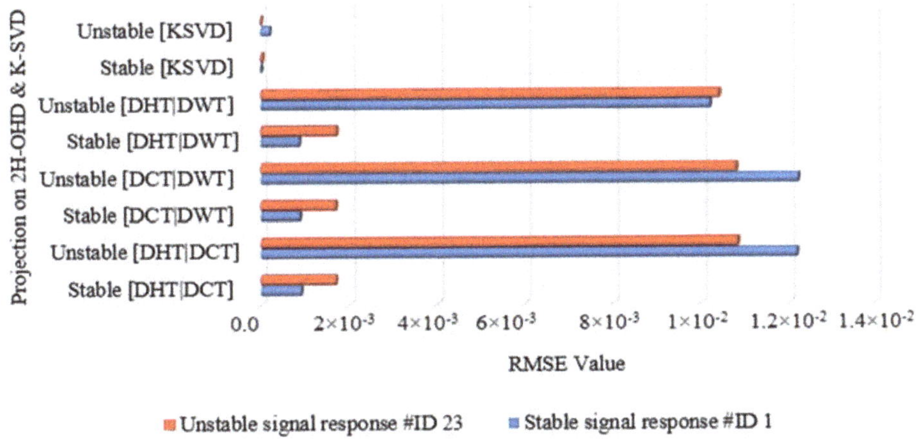

Figure 8. Projections of signals ID 1 and ID 23 on stable and unstable 2H-OHD and ADL.

Regarding 2H-OHD, the [DHT | DWT] has lower RMSE predictors for the stable and unstable stability status responses compared to [DHT | DCT], [DCT | DWT]. The projection of signals ID 1 (blue) & ID 23 (red) on three hybrids dictionary learning allows effective prediction of online TSA, with a slightly improved performance compared to single dictionaries & hybrids with two sub-dictionaries, see Figure 9. Regarding 3H-OHD, the [DST | DHT | DCT] has lower RMSE predictors for the stable and unstable stability status responses compared to [ID | DCT | DST] and [DWT | DHT | DCT], which means a crisper separation between stable and unstable cases.

Figure 10 presents the stability degree of signals ID 1 and ID 23, developed based on stable and unstable dictionary learning. By definition, it is computed as the ratio between RMSE of unstable dictionary divided by the RMSE of stable dictionary-based reconstruction. Therefore, a small RMSE results in a high degree of stability meaning a higher probability of the case being stable while a small value means that the case is likely classified as unstable. For the stable case ID 1, the K-SVD gives a very large stability degree compared to [DCT], [DHT], [DWT], [DHT | DCT], [DCT | DWT], [DWT | DHT], [DST | DHT | DCT],

[ID | DCT | DST], and [DWT | DHT | DCT]. Similarly, K-SVD gives a very low stability degree of ID 23, which confirms a good stability separability of K-SVD based dictionary.

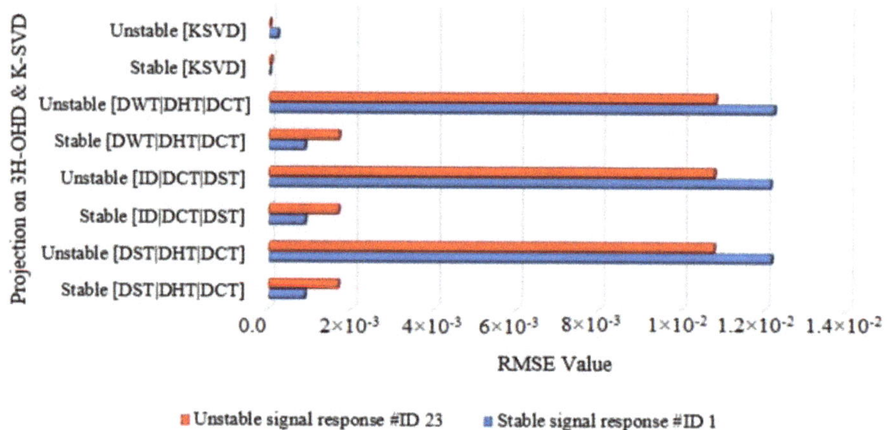

Figure 9. Projections of signals ID 1 and ID 23 on stable and unstable 3H-OHD and ADL.

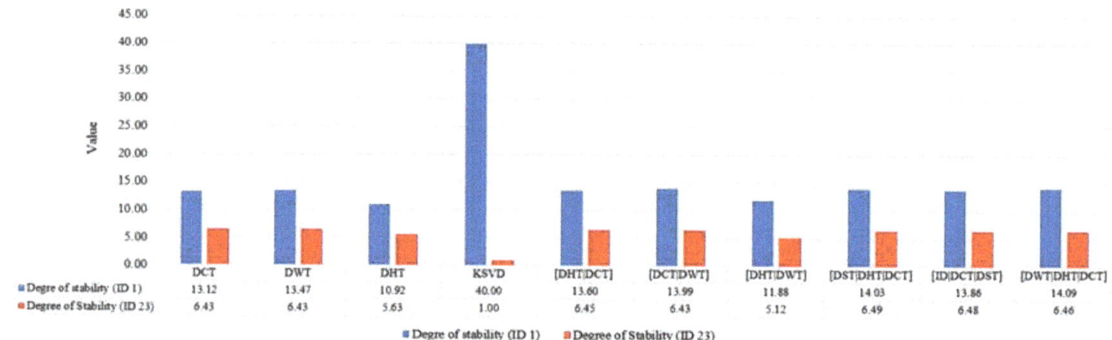

Figure 10. Stability degree of adaptive, single, hybrids with two and three overcomplete dictionary learning.

5.4. Performance Metrics of Sparse Decomposition on Fixed and Adaptive Overcomplete Dictionaries Learning

The metrics (namely: accuracy, reliability, and security) are used to evaluate the performance of fixed and adaptive overcomplete dictionary learning. The accuracy metric defines the ability of learned dictionaries for correctly classifying stable (i.e., 8341 scenarios) and unstable (i.e., 323 scenarios) in the testing database (i.e., 8664 scenarios). The reliability metric makes it possible to evaluate the performance of learned dictionaries to predict unstable scenarios. The security metric allows defining the capacity of learned dictionaries to predict the stable scenarios. They broadly succeed in projecting the new transient scenarios on each stable and unstable dictionary and evaluating the probability for each of them to have a low RMSE reconstruction. For each signal, the stability status classification is confirmed or not, according to the absolute values of generators rotor angles. Table 4 summarizes the online TSA effectiveness of each OHD and ADL developed based on separated training datasets. Moreover, the performance of OHD and K-SVD based stable/unstable dictionaries are compared to the supervised learning of overcomplete dictionaries (SLOD) developed in [34], based on both rotor speed and stability status taken as the training input and using a single dictionary containing both stable and unstable cases. It appears from

Table 4 that all fixed dictionaries provided a decent performance with right to all metrics using separate stable/unstable datasets in contrast to K-SVD dictionary whose reliability is pretty-weak.

Table 4. TSA metric comparisons of fixed and adaptive supervised dictionary learning.

Configuration	Learning Approaches	Accuracy	Reliability	Security
Adaptive dictionary (ADL) (with a separate training dataset)	K-SVD	93.42	81.68	93.35
SLOD [34] (with a joint training dataset)	TK-SVD	99.99	99.99	99.99
Single dictionary (SOD) (with a separate training dataset)	DCT	94.76	93.05	94.52
	DHT	95.79	94.10	95.66
	DWT	94.96	93.40	94.74
Two sub-dictionaries (2H-OHD) (with a separate training dataset)	[DHT I DWT]	95.41	93.75	95.24
	[DCT I DWT]	94.79	93.05	94.56
	[DHT I DCT]	94.82	93.05	94.59
Three sub-dictionaries (3H-OHD) (with a separate training dataset)	[DST I DHT I DCT]	94.88	93.05	94.66
	[ID I DCT I DST] [7]	94.88	93.05	94.66
	[DWT I DHT I DCT]	94.81	93.05	94.58

The single dictionary based on DHT seems in average to be the best post-disturbance stability predictor with a 94.1% reliability success rate and 95.79% overall accuracy. By contrast, the reliability of the ADL (K-SVD) algorithm is only 81.68%. This compared poorly with the SLOD (TK-SVD) algorithm proposed in [34], which provided 99.99% success rate across all performance metrics.

5.5. Computational Efficiency of Fixed and Adaptive Overcomplete Dictionaries Learning

The computational performance of fixed and adaptive dictionary learning was assessed on a DELL computer configured with the Intel processor i7-7700HQ 4-core running at 2.80 GHz with 16 GB of RAM. Although the CPU time is relatively significant, the actual code is in Matlab scripting language and can therefore be made faster using C-programming. Table 5 summarizes all CPU computational time for the offline learning and online playback for TSA. The proposed adaptive overcomplete dictionaries take more offline computation time (i.e., 298,475.89 s or about 83 h) for sparse features extraction. However, K-SVD sparse features enable a very quick online transient stability prediction (i.e., 4.45 s computation time) based on the generator signal responses (i.e., 8664 scenarios). Compared to the best dictionary learning algorithm, the SLOD (TK-SVD), which has a 409,744.06 s or about 114 h offline training time and 6.59 s online playback time, the ADL (K-SVD) is 37% and 48% faster respectively. Moreover, several paths for improving the performance metrics of the ADL (K-SVD) to fully benefit from this improved computational performance remains to be investigated, such as a longer data window for decision-making and enhanced data normalization before dictionary learning, i.e., by using for instance rotor speed deviation instead of the absolute speed.

Table 6 provides a summary of the advantages and disadvantages of supervised dictionary learning for transient stability assessment.

The different methods are further compared in Figure 11 using data from Table 4. The bottom line is that the reconstruction-based TSA is better done with fixed dictionaries. The DHT turned out to be the best fixed dictionary in terms of both reliability and security although the difference is not huge across all types of fixed dictionaries.

Table 5. CPU time (in sec.) of fixed and adaptive supervised dictionary learning.

Configuration	Learning Approaches	Offline Processing Time/s	Online Processing Time/s
Adaptive dictionary (ADL) (with a separate training dataset)	K-SVD	298,475.89	4.45
SLOD [34] (with a joint training dataset)	TK-SVD	409,744.06	6.59
Single dictionary (SOD) (with a separate training dataset)	DHT	16,518.44	6061.06
	DCT	42,290.29	17,547.32
	DWT	45,882.93	17,366.59
Two sub-dictionaries (2H-OHD) (with a separate training dataset)	[DHT ǀ DWT]	31,535.18	13,854.68
	[DCT ǀ DWT]	52,812.70	18,519.29
	[DHT ǀ DCT]	55,830.41	20,278.90
Three sub-dictionaries (3H-OHD) (with a separate training dataset)	[DST ǀ DHT ǀ DCT]	33,949.11	14,489.51
	[ID ǀ DCT ǀ DST] [7]	28,617.70	12,320.68
	[DWT ǀ DHT ǀ DCT]	40,062.59	16,521.75

Table 6. Advantages and disadvantages of fixed and adaptive dictionary learning for TSA.

Approaches	Advantage	Disadvantage
ADL	—Satisfactory accuracy and security; —Good online CPU time; —Implemented based on data-driven algorithm (K-SVD); —Used separate approach for ADL.	—Not satisfactory reliability. —Takes longer time for offline learning
SLOD [34]	—Good accuracy, reliability, security; —Satisfactory online CPU time; —Implemented based on data-driven algorithm (K-SVD); —Used joint approach for SLOD.	Takes longer time for offline learning.
SOD	—Satisfactory accuracy security and reliability; —Rectangular waveforms, which can take zero value and sample points in subintervals of $t \in [0, 1]$; —Grouping energy contained in the signals in low frequency coefficients; —Extracting information content at different positions and scales for subsequently reconstructing post-contingency; —Used separate approach for SOD.	—Not satisfactory online CPU time; —Not suitable for transient signal reconstruction.
2H-OHD	—Satisfactory accuracy security and reliability; —Allows extracting simultaneously the non-sinusoidal information, the frequencies of the oscillations and a representation in a unitary sub-interval, invariant by shifts. —Allows extracting simultaneously the real low frequency coefficients, the non-sinusoidal information and the oscillation frequencies; —Allows extracting simultaneously, a representation in a unit sub-interval, invariant by shifts and the low frequency real coefficients; —Used separate approach for SOD.	—Not satisfactory online CPU time; —Not suitable for transient signal reconstruction.
3H-OHD	—Satisfactory accuracy security and reliability; —Allows extracting simultaneously in the stability signals, the imaginary coefficients, the representations in a unitary sub-interval, invariant by shifts and the low frequency real coefficients; —Allows extracting frequency peaks, real low frequency coefficients and imaginary coefficients; —Allows extracting in a way, the real coefficients low frequency, the non-sinusoidal information, scales, representations in a sub-unit interval, invariant by shifts and the imaginary coefficients; —Used separate approach for 2H-OHD.	—Not satisfactory online CPU time; —Not suitable for transient signal reconstruction.

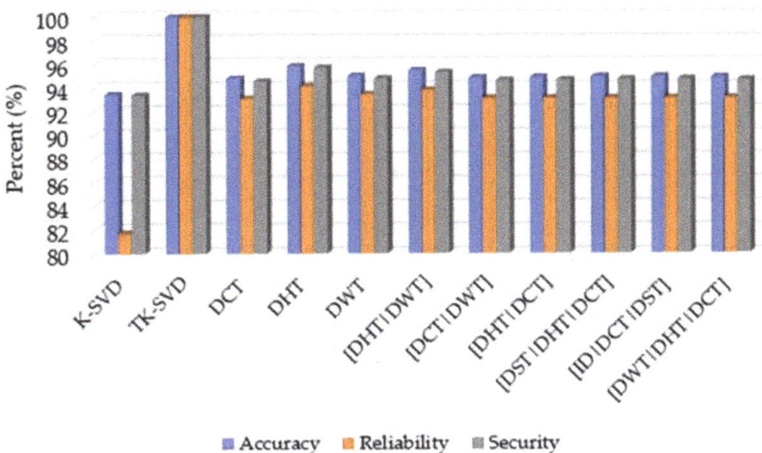

Figure 11. Comparison of adaptive, singles, and hybrids with two and three overcomplete dictionary learning.

Giving that DHT takes only about 50% of the computation time of the 3H-OHD from [7] in both supervised learning and online playback, it is definitely more promising and warrants further investigation in the TSA context. Additionally, given the short prediction window (75 ms) and the highly skewed two-data classes characterizing machine learning based TSA (with the unstable class typically representing only 3 to 10% of the total data set), the reconstruction-based classification using K-SVD appears unsuitable and rather, the TK-SVD based supervised learning algorithm of [34] should be further investigated and compared with the reconstruction-based classification using fixed DHT dictionary. These two options emerged as the best dictionary-based TSA prediction methods from the present study.

6. Conclusions

This work aims at reconstructing and predicting transient stability assessment based on fixed and adaptive overcomplete dictionary learning for online stability signal responses reconstruction. Practically speaking, this study develops and implements new arrangement of atoms able to improve wide-area stability monitoring and control using a dual-dictionary signal reconstruction approach. To this end, fixed and adaptive structures have been investigated based on DCT, DHT, and DWT dictionary learning for rotor speeds sparse reconstruction. Afterwards, stability status is determined based on the RMSE of rotor speed prediction using each stable/unstable dictionary to design the level of belonging to each class. The ratio of RMSE from these two separated predictions is proposed as measured of stability (or instability) of the given observed post-disturbance signal over a short 4.5-cyles data window.

Several concatenations of two (i.e., [DHT|DCT], [DCT|DWT], and [DHT|DWT]) and three (i.e., [DST|DHT|DCT], [ID|DCT|DST], and [DWT|DHT|DCT]) fixed sub-dictionaries are used to perform similar sparse decomposition on dual-dictionaries. Alternatively, the K-SVD adaptive approach is implemented using dual-dictionaries learned from the stable/unstable generator response datasets. While the adaptive K-SVD demonstrated a better reconstruction performance, in terms of prediction errors, its reliability performance was not satisfactory due to the highly skewed dataset derived from the IEEE 68-bus test system (3.8% of stable cases only) and short data frames (75 ms). Overall, this study proved that [DWT], [DHT|DWT], and [DST|DHT|DCT] fixed dictionaries are better stability indicators compared to adaptive K-SVD, [DHT], [DWT], [ID|DCT|DST], and [DWT|DHT|DCT] but none of these methods could match the good performance of the method reported in [35]. We will further evaluate the performance of the K-SVD and DHT based dictionary learning methods for TSA on the combined database of IEEE 39 and 68 -bus test systems in the presence of multi-faults and for various simultaneous disturbances. Moreover, to reduce the offline training time, we also plan to develop parallel computer-based dictionary learning of PMU based generator signal responses to enable online TSA on large-scale power systems.

Author Contributions: Conceptualization, R.T.D. and I.K.; methodology, I.K., J.T. and F.C.M.; software, I.K.; validation, I.K., J.T. and F.C.M.; formal analysis, R.T.D. and I.K.; investigation, R.T.D. and I.K.; resources, I.K.; data curation, R.T.D., I.K., J.T. and F.C.M.; writing—original draft preparation, R.T.D., I.K.; writing—review and editing, R.T.D. and I.K.; visualization, I.K.; supervision, I.K.; project administration, I.K.; funding acquisition, I.K. All authors have read and agreed to the published version of the manuscript.

Funding: Natural Sciences and Engineering Research Council of Canada (NSERC), Discovery Grant no. RGPIN-2021-02574, Innocent Kamwa, Laval University.

Institutional Review Board Statement: Not applicable.

Informed Consent Statement: Not applicable.

Data Availability Statement: The data are available on request to corresponding authors.

Conflicts of Interest: The authors declare no conflict of interest.

Abbreviations

K-SVD	K-singular value decomposition
TK-SVD	Transient K-singular value decomposition
DST	Discrete sine transform
DCT	Discrete cosine transform
ADL	Adaptive dictionary learning
DWT	Discrete wavelet transform
DHT	Discrete Haar transform
ID	Impulse dictionary
UPE	Unstable prediction error
SPE	Stable prediction error
OHD	Overcomplete hybrid dictionaries
SOD	Single overcomplete dictionaries
2H-OHD	Two overcomplete hybrid dictionaries
3H-OHD	Three overcomplete hybrid dictionaries
[DHT | DCT]	Concatenation of discrete Haar and cosine transforms
[DCT | DWT]	Concatenation of discrete wavelet and cosine transforms
[DHT | DWT]	Concatenation of discrete Haar and wavelet transforms
[DST | DHT | DCT]	Concatenation of discrete sine Haar and cosine transforms
[ID | DCT | DST]	Concatenation of discrete impulse, cosine and sine transforms
[DWT | DHT | DCT]	Concatenation of discrete wavelet, Haar and cosine transforms
Stable DHT	DHT dictionary learned from stable dataset
Unstable DHT	DHT dictionary learned from unstable dataset
Stable DWT	DWT dictionary learned from stable dataset
Unstable DWT	DWT dictionary learned from unstable dataset
Stable DCT	DCT dictionary learned from stable dataset
Unstable DCT	DCT dictionary learned from unstable dataset
Stable [DHT | DCT]	[DHT | DCT] dictionary learned from stable dataset
Unstable [DHT | DCT]	[DHT | DCT] dictionary learned from unstable dataset
Stable [DCT | DWT]	[DCT | DWT] dictionary learned from stable dataset
Unstable [DCT | DWT]	[DCT | DWT] dictionary learned from unstable dataset
Stable [DHT | DWT]	[DHT | DWT] dictionary learned from stable dataset
Unstable [DHT | DWT]	[DHT | DWT] dictionary learned from unstable dataset

Stable [DST | DHT | DCT]	[DST | DHT | DCT] dictionary learned from stable dataset
Unstable [DST | DHT | DCT]	[DST | DHT | DCT] dictionary learned from unstable dataset
Stable [ID | DCT | DST]	[ID | DCT | DST] dictionary learned from stable dataset
Unstable [ID | DCT | DST]	[ID | DCT | DST] dictionary learned from unstable dataset
Stable [DWT | DHT | DCT]	[DWT | DHT | DCT] dictionary learned from stable dataset
Unstable [DWT | DHT | DCT]	[DWT | DHT | DCT] dictionary learned from unstable dataset

References

1. Kamwa, I.; Jain, A.; Samantaray, S.R.; Geoffroy, L. Synchrophasors data analytics framework for power grid control and dynamic stability monitoring. *IET Eng. Technol. Ref.* **2016**, 1–22. [CrossRef]
2. Liu, Y.; Singh, A.K.; Zhao, J.; Meliopoulos, A.P.S.; Pal, B.; Ariff, M.A.B.M.; Van Cutsem, T.; Glavic, M.; Huang, Z.; Kamwa, I.; et al. Dynamic state estimation for power system control and protection. *IEEE Trans. Power Syst.* **2021**, *36*, 5909–5921. [CrossRef]
3. Paul, A.; Kamwa, I.; Joos, G. PMU Signals Responses-Based RAS for Instability Mitigation Through On-The Fly Identification and Shedding of the Run-Away Generators. *IEEE Trans. Power Syst.* **2020**, *35*, 1707–1717. [CrossRef]
4. Liu, Y.; Fan, R.; Terzija, V. Power system restoration: A literature review from 2006 to 2016. *J. Mod. Power Syst. Clean Energy* **2016**, *4*, 332–341. [CrossRef]
5. Yadav, R.; Pradhan, A.K.; Kamwa, I. Real-Time Multiple Event Detection and Classification in Power System Using Signal Energy Transformations. *IEEE Trans. Ind. Inform.* **2019**, *15*, 1521–1531. [CrossRef]
6. Stanković, L.; Sejdić, E.; Stanković, S.; Daković, M.; Orović, I. A Tutorial on Sparse Signal Reconstruction and Its Applications in Signal Processing. *Circuits. Syst. Signal Process.* **2019**, *38*, 1206–1263. [CrossRef]
7. Manikandan, M.S.; Samantaray, S.R.; Kamwa, I. Detection and Classification of Power Quality Disturbances Using Sparse Signal Decomposition on Hybrid Dictionaries. *IEEE Trans. Instrum. Meas.* **2015**, *64*, 27–38. [CrossRef]
8. Yu, Q.; Hu, Y.; Yang, Y. Identification Method for Series Arc Faults Based on Wavelet Transform and Deep Neural Network. *Energies* **2020**, *13*, 142. [CrossRef]
9. Qayyum, A.; Malik, A.S.; Naufal, M.; Saad, M.; Mazher, M.; Abdullah, F.; Abdullah, T.A.R.B.T. Designing of overcomplete dictionaries based on DCT and DWT. In Proceedings of the 2015 IEEE Student Symposium in Biomedical Engineering & Sciences (ISSBES), Shah Alam, Malaysia, 4 November 2015. [CrossRef]
10. Kathirvel, P.; Manikandan, M.S.; Maya, P.; Soman, K.P. Detection of power quality disturbances with overcomplete dictionary matrix and ℓ1-norm minimization. In Proceedings of the International Conference on Power and Energy Systems, Chennai, India, 22–24 December 2011; pp. 1–6.
11. Cai, D.; Li, K.; He, S.; Li, Y.; Luo, Y. A highly accurate and fast power quality disturbances classification based on dictionary learning sparse decomposition. *Trans. Inst. Meas. Control.* **2019**, *41*, 145–155. [CrossRef]
12. Zhu, T.X. Detection and Characterization of Oscillatory Transients Using Matching Pursuits With a Damped Sinusoidal Dictionary. *IEEE Trans. Power Deliv.* **2007**, *22*, 1093–1099. [CrossRef]
13. Bravo-Rodríguez, J.C.; Torres, F.J.; Borrás, M.D. Hybrid Machine Learning Models for Classifying Power Quality Disturbances: A Comparative Study. *Energies* **2020**, *13*, 2761. [CrossRef]
14. Babakmehr, M.; Sartipizadeh, H.; Simoes, M.G. Compressive Informative Sparse Representation-Based Power Quality Events Classification. *IEEE Trans. Ind. Inform.* **2020**, *16*, 909–921. [CrossRef]
15. Wang, Y.; Zhang, F.; Zhang, S. A New Methodology for Identifying Arc Fault by Sparse Representation and Neural Network. *IEEE Trans. Instrum. Meas.* **2018**, *67*, 2526–2537. [CrossRef]
16. Ren, L.; Lv, W.; Jiang, S.; Xiao, Y. Fault Diagnosis Using a Joint Model Based on Sparse Representation and SVM. *IEEE Trans. Instrum. Meas.* **2016**, *65*, 2313–2320. [CrossRef]
17. Liu, C.; Liu, J. Research on power quality signals reconstruction method based on K-SVD dictionary learning. In Proceedings of the Chinese Control Conference, Shenyang, China, 27–29 July 2020; pp. 2930–2934.
18. Chen, Y.C.; Dominguez-Garcia, A.D.; Sauer, P.W. A Sparse Representation Approach to Online Estimation of Power System Distribution Factors. *IEEE Trans. Power Syst.* **2015**, *30*, 1727–1738. [CrossRef]
19. Zhu, H.; Giannakis, G. Sparse Overcomplete Representations for Efficient Identification of Power Line Outages. *IEEE Trans. Power Syst.* **2012**, *27*, 2215–2224. [CrossRef]
20. Song, Y.; Wang, W.; Zhang, Z.; Qi, H.; Liu, Y. Multiple Event Detection and Recognition for Large-Scale Power Systems Through Cluster-Based Sparse Coding. *IEEE Trans. Power Syst.* **2017**, *32*, 4199–4210. [CrossRef]
21. Cheng, L.; Wang, L.; Gao, F. Power system fault classification method based on sparse representation and random dimensionality reduction projection. In Proceedings of the IEEE Power & Energy Society General Meeting, Denver, CO, USA, 26–30 July 2015; pp. 1–5.
22. Xiao, H.; Wei, J.; Li, Q. Identification of Combined Power Quality Disturbances Using Singular Value Decomposition (SVD) and Total Least Squares-Estimation of Signal Parameters via Rotational Invariance Techniques (TLS-ESPRIT). *Energies* **2017**, *10*, 1809. [CrossRef]
23. Chatterjee, A.; Yuen, P.W.T. Rapid estimation of orthogonal matching pursuit representation. In Proceedings of the IEEE International Geoscience and Remote Sensing Symposium, Waikoloa, HI, USA, 26–28 October 2020.
24. Aharon, M.; Elad, M.; Bruckstein, A. K-SVD: An Algorithm for Designing Overcomplete Dictionaries for Sparse Representation. *IEEE Trans. Signal Process.* **2006**, *54*, 4311–4322. [CrossRef]

25. Xiang, Z.; Huang, K.; Deng, W.; Yang, C. Blind topology identification for smart grid based on dictionary learning. In Proceedings of the IEEE Symposium Series on Computational Intelligence, Xiamen, China, 6–9 December 2019; pp. 1319–1326.
26. Rajagukguk, R.A.; Ramadhan, R.A.; Lee, H.-J. A Review on Deep Learning Models for Forecasting Time Series Data of Solar Irradiance and Photovoltaic Power. *Energies* **2020**, *13*, 6623. [CrossRef]
27. Khodayar, M.; Wang, J.; Wang, Z. Energy Disaggregation via Deep Temporal Dictionary Learning. *IEEE Trans. Neural Netw. Learn. Syst.* **2020**, *31*, 1696–1709. [CrossRef] [PubMed]
28. Kamwa, I.; Samantaray, S.R.; Joos, G. On the Accuracy Versus Transparency Trade-Off of Data-Mining Models for Fast-Response PMU-Based Catastrophe Predictors. *IEEE Trans. Smart Grid* **2012**, *3*, 152–161. [CrossRef]
29. Rubinstein, R.; Bruckstein, A.M.; Elad, M. Dictionaries for Sparse Representation Modeling. *Proc. IEEE* **2010**, *98*, 1045–1057. [CrossRef]
30. Kundur, P. *Power System Stability and Control*; McGraw-Hill: New York, NY, USA, 1994.
31. Akansu, A.; Haddad, R. *Multiresolution Signal Decomposition*, 2nd ed.; Academic Press: Cambridge, MA, USA, 2000; pp. 1–499.
32. Ma, B.; Zou, X. Discrete wavelet transform for signal processing in weight-in-motion system. In Proceedings of the International Conference on Electrical and Control Engineering, Washington, DC, USA, 25–27 June 2010; pp. 4668–4671.
33. Elad, M. *Sparse and Redundant Representations: From Theory to Application in Signal and Image Processing*; Springer: New York, NY, USA, 2010; pp. 1–376.
34. Pavella, M.; Ernst, D.; Ruiz-Vega, D. *Transient Stability of Power Systems: A Unified Approach to Assessment and Control*; Springer: New York, NY, USA, 2000; pp. 1–256.
35. Dabou, R.T.; Kamwa, I. Rapid design method for generating power system stability databases in SPS for machine learning. In Proceedings of the IEEE Canadian Conference on Electrical and Computer Engineering, London, ON, Canada, 30 August–2 September 2020; pp. 1–6.
36. Dabou, R.T.; Kamwa, I.; Tagoudjeu, J.; Mugombozi, C.F. Supervised learning of overcomplete dictionaries for on-line response based transient stability prediction. *IEEE Trans. Power Syst.* **2021**. submitted.
37. Lachiheb, O.; Gouider, M.S.; Said, L.B. An Improved MapReduce design of Kmeans with iteration reducing for clustering stock exchange very large datasets. In Proceedings of the 2015 11th International Conference on Semantics, Knowledge and Grids (SKG), Beijing, China, 19–21 August 2015; pp. 252–255.

Article

Intelligent Scheduling of Smart Home Appliances Based on Demand Response Considering the Cost and Peak-to-Average Ratio in Residential Homes

Nedim Tutkun [1], Alessandro Burgio [2], Michal Jasinski [3,*], Zbigniew Leonowicz [3] and Elzbieta Jasinska [4]

[1] Department of Electrical & Electronics Engineering, Istanbul Ticaret University, Istanbul 34840, Turkey; ntutkun@ticaret.edu.tr
[2] Independent Researcher, 87036 Rende, Italy; alessandro.burgio.phd@gmail.com
[3] Faculty of Electrical Engineering, Wroclaw University of Science and Technology, 50-370 Wroclaw, Poland; zbigniew.leonowicz@pwr.edu.pl
[4] Department of Operations Research and Business Intelligence, Wroclaw University of Science and Technology, 50-370 Wroclaw, Poland; elzbieta.jasinska@pwr.edu.pl
*   Correspondence: michal.jasinski@pwr.edu.pl

Citation: Tutkun, N.; Burgio, A.; Jasinski, M.; Leonowicz, Z.; Jasinska, E. Intelligent Scheduling of Smart Home Appliances Based on Demand Response Considering the Cost and Peak-to-Average Ratio in Residential Homes. *Energies* **2021**, *14*, 8510. https://doi.org/10.3390/en14248510

Academic Editor: Surender Reddy Salkuti

Received: 29 September 2021
Accepted: 13 December 2021
Published: 17 December 2021

Publisher's Note: MDPI stays neutral with regard to jurisdictional claims in published maps and institutional affiliations.

Abstract: With recent developments, smart grids assured for residential customers the opportunity to schedule smart home appliances' operation times to simultaneously reduce both the electricity bill and the PAR based on demand response, as well as increasing user comfort. It is clear that the multi-objective combinatorial optimization problem involves constraints and the consumer's preferences, and the solution to the problem is a difficult task. There have been a limited number of investigations carried out so far to solve the indicated problems using metaheuristic techniques like particle swarm optimization, mixed-integer linear programming, and the grey wolf and crow search optimization algorithms, etc. Due to the on/off control of smart home appliances, binary-coded genetic algorithms seem to be a well-fitted approach to obtain an optimal solution. It can be said that the novelty of this work is to represent the on/off state of the smart home appliance with a binary string which undergoes crossover and mutation operations during the genetic process. Because special binary numbers represent interruptible and uninterruptible smart home appliances, new types of crossover and mutation were developed to find the most convenient solutions to the problem. Although there are a few works which were carried out using the genetic algorithms, the proposed approach is rather distinct from those employed in their work. The designed genetic software runs at least ten times, and the most fitting result is taken as the optimal solution to the indicated problem; in order to ensure the optimal result, the fitness against the generation is plotted in each run, whether it is converged or not. The simulation results are significantly encouraging and meaningful to residential customers and utilities for the achievement of the goal, and they are feasible for a wide-range applications of home energy management systems.

Keywords: home energy management; binary-coded genetic algorithms; optimal power scheduling; demand response

1. Introduction

The gradual increase in electricity unit prices, for various reasons, brings an additional burden, especially for households. Meanwhile, technological developments in this field have brought new features to both the distribution of electrical energy and the use of electrical household appliances. This has made it possible to control electrical household appliances over the internet and to change the unit price of electricity for every hour, or even for every 15 minutes by means of a smart grid. With these developments, taking into account the usage characteristics of electrical household appliances and compromising daily usage habits, a significant reduction in daily energy costs can be achieved under the predefined constraints and conditions. This can be achieved by shifting the operating times of electrical household appliances to the optimal time intervals during the day, based on electricity tariffs and the avoidance of overload. There are a limited number of studies that have been conducted on this subject, and some of these studies are listed in the subsection below.

1.1. Related Works

Setlhaolo and Xia carried out an investigation considering the combined demand side management strategy for a limited number of houses from two aspects. The first one was the energy management system (EMS). The EMS considers demand side management strategies based on the minimization of consumer cost and the reduction of consumption from the power system, as well as CO_2 emission issues through the developed model [1]. Another investigation was carried out by Shakouri and Kazemi in [2]. In the paper, in order to obtain for the household the lower energy costs, multi-objective mixed integer linear programming (MILIP) was proposed. Additionally, the validation of the proposed model based on some scenarios indicated that the reduction of daily energy costs could be on the acceptable level [2].

The authors in [3] performed the study on home energy management for residential consumers. The aim of the investigation was to decrease both their electricity bills and the peak load demand of utility companies. It was assumed that it can be achieved with a smart energy storage system (ESS) [3]. Zhao et al. were other investigators studying home EMS to schedule power consumption in households in order to obtain a reduction of electricity bills and improve the peak-to-average ratio [4]. Paterakis et al. conducted another study on the home EMS to define the optimal schedule of appliances for the day of a smart household under hourly pricing, and to reduce peak power based on demand response (DR) strategies [5]. The authors in [6] proposed a new procedure to propose the schedule of power consumption in homes equipped with ESS. In this article, the aim of power scheduling was to reduce electricity bills and improve the peak-to-average ratio. In the investigation, the comfort of the residents was considered [6]. The authors of [7] investigated the process of the scheduling of the loads in a home EMS. This approach considers the multi-objective demand response optimization model. The indicated approach determines the scheduling of home appliances (HA) considering the non-dominated sorted genetic algorithm (GA). The authors in [8] presented a new home EMS controller. The proposed device is based on the genetic harmony search algorithm. The aim of the controller is to reduce electricity bills and the peak-to-average ratio, and to maximize user comfort. The investigated range considers a single home and multiple homes. In the investigation, real-time electricity pricing and critical peak pricing tariffs were applied. The author in [9] studied interruptible appliances (e.g., electric water heaters), although some researchers consider the problem of optimal scheduling for non-interruptible appliances; this problem was commonly formulated with integer decision variables. Zhu et al. in [10] carried out an investigation to consider the efficient solution of a complex combinatorial problem. The aim of the investigation was to define a schedule of household appliances in multiple smart homes. In the investigation, the improved cooperative heuristic approach was considered. The presented results confirmed that the proposed algorithm worked correctly. The article [11] proposed a satisfaction model for different types of household appliances. The aim of the optimization was the minimization of the energy costs. Different demand response strategies were considered e.g., demand limit based or injection limit. The article [12] presented an approach based on the fusion of the grey wolf and crow search optimization algorithm. In the investigation, the cost of electricity decrease, the extension of the users' comfort, and the minimization of the peak-to-average ratio for HA in the presence of real-time price signals using the indicated technique were taken into account. Adika and Wang proposed the new approach to residential customer-based demand-side management for smart charging and appliance scheduling in order to avoid overloading the power consumption, and to minimize the energy cost [11]. El-Baz and Tzscheutschler presented an algorithm that makes a simple, efficient, day-ahead electrical load prediction for any energy management system without requiring statistical or historical data, or using any kind of measurement sensors [12]. The authors in [13] performed an experimental study of the problem of scheduling HA from realistic points of view. Additionally, the problem of scheduling residential loads on the basis of the minimization of the cost of electricity was solved for consumer preferences [13]. Bouakkaz et al. presented a work that aims to propose the optimal strategy to schedule energy consumption to support homeowners in the reduction of energy costs, as well as saving energy in a residential home. The investigated objects were connected to a microgrid composed of a power grid system and photovoltaic and battery ESS [14]. Merdanoglu et al. investigated a model that solves the optimal scheduling problem embedded in a home EMS, enabling users to overcome the major obstacles in the implementation of DR programs. The investigation proposed the method of finding the minimum energy cost based on time-varying prices, the generation from RES, the usage demand for each HA, the battery ESS capacity and the grid constraints [15]. The authors in [16] proposed a new home energy management architecture with a renewable energy source and ESS, as well as the power grid, and in this home energy management, some mathematical models for the energy cost and peak-to-average ratio should be minimized. The applied solution is based on particle swarm optimization (PSO) and binary PSO during the day [16]. The results obtained from [16] clearly show that the proposed scheme worked well with the daily total electricity cost and the peak-to-average ratio, in comparison to those of similar works. Ahmad et al. worked to minimize the electricity bill using the scheduling of the HA and ESS in response to the dynamic pricing of the electricity market. In this study, different metaheuristic algorithms were applied, e.g., GA, binary PSO, wind-driven optimization, bacterial foraging optimization and hybrid PSO algorithms. The results corresponding to each algorithm were compared to each other through the proposed scheme [17]. It was seen that [17] is an exemplary study to test the validity of the approach and the scheme we propose compared to the studies given above; the results were produced considering the data used here, and necessary comparisons were made to obtain a better improvement in the problem of home energy management. The summary in terms of objectives, tariffs, and methods of investigated literature is given in Table 1.

Table 1. Summary in terms of objectives, tariffs, and methods of investigated literature.

Objective	Tariff	Algorithm	References
Cost	TOU, RTEP	MILP	[15]
cost & PAR	TOU, RTEP	MILP, GA, NSGA-II, GHSA, GWCSO	[2–8,12,16,17]
cost & comfort	TOU, RTEP	DA, NSGA-II, QBPSO, PSO, MILP	[5,9–11]
cost & consumption	TOU, RTEP	MINLP, MILP, PSO	[1,13,14]

1.2. The Original Contribution

The aim of this study is to perform a minimization of the daily total electricity bill and the peak-to-average ratio by optimizing the start times of the shiftable smart HA on the basis of hourly day-ahead real-time electricity pricing using binary-coded genetic algorithms. The designed home energy management system includes smart HA, a power grid, a solar PV generator, and ESS; in this system, surplus energy is sold to a power grid when it is most expensive, and deficit energy is purchased from the power grid at time slots when it is least expensive if possible. There are also a few constraints in the energy storage system, in which amounts of charging and discharging are not allowed to be more than 0.3 kWh in a single time slot. Thus, the optimization problem under consideration was solved for the optimal start times of the shiftable appliances in the home EMS using the binary-coded genetic algorithm approach. The results are compared to those obtained from few metaheuristic algorithms, and it appears that there is a considerable improvement, reducing both the electricity bill and peak-to-average ratio under similar conditions. This may be explained because, unlike the other meta-heuristic algorithms, the binary-coded genetic algorithms have more powerful genetic operators—such as selection, crossover, mutation, and elitism—to produce a variety of possible solutions and then refine them to reach the best possible solution over generations. It can be said that this work has contributed to this field of study, showing that binary-coded genetic algorithms with powerful genetic operators are highly efficient in improving the daily total electricity bill and peak-to-average ratio, as well as the comfort of the home.

2. The System Architecture, Problem Definition and Applied Methods

In this part of article, the system architecture, problem definition, and the applied approach to solve the defined optimization are detailed based on the real-time pricing scheme.

2.1. System Architecture

2.1.1. Home EMS

In general, a basic home EMS is composed of the metering infrastructure, smart metering, a home gateway, an energy management controller, smart HA, ESS, and a solar PV generator, as shown in Figure 1. In this architecture, the smart metering infrastructure covers a major role in supplying two-way communication between the smart meter and the utility for real-time electricity pricing. The smart meter installed between the infrastructure metering and the energy management controller is used to read energy consumption data, process it, and send it to the energy management controller for further handling. In this investigation, the smart household appliances are mainly divided into two groups: shiftable and non-shiftable. For instance, a washing machine or dishwasher is a shiftable home appliance, as it is operated in any time slot in the day. However, a refrigerator or lighting is a non-shiftable home appliance, because it must be operated in certain time slots during the day. Once the optimization procedure is complete, the optimal operating times of the switchable appliances are transmitted to the energy management controller for scheduling at the beginning of the day. The smart meter and the home gateway communicate with each other through Wi-Fi in the home local area network. The energy management controller in the home gateway schedules the operation time of the appliances after receiving a real-time electricity price signal from the utility.

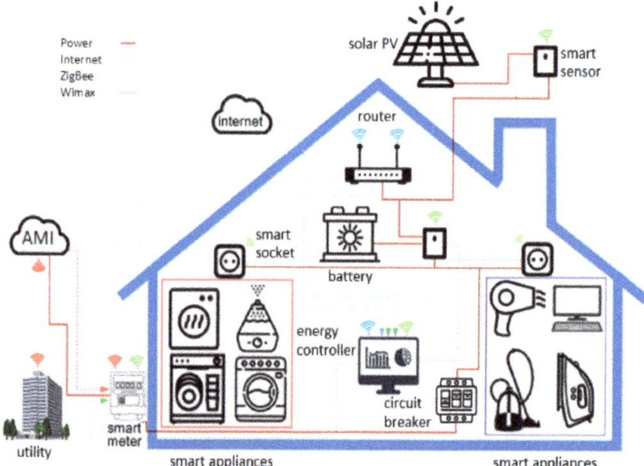

Figure 1. Basic architecture of the HEMS with a wireless home area network.

Because the first objective aims to reduce the daily electricity bill, shiftable HA—such as a washing machine, a dishwasher or a dryer for clothes—are expected to operate in the time slots when the electricity price is lowest; hence, this may lead to the overload of distribution lines and an increase in power losses beyond the typical values. However, householders usually prefer to operate them in the most convenient times for themselves, regardless of the maximum daily total electricity cost. For instance, the householders may intend to use the dishwasher after lunch, the washing machine in the morning, or the clothes dryer in the evening. It is obvious that there is a trade-off between the electricity cost, the householder's comfort, and the peak-to-average ratio. It is seen here that there are three objectives to be optimized, but among them the daily total electricity cost and a peak-to average ratio are more influential. Thus, it is essential to focus reasonably on these two objectives to be optimized, both in single and multiple cases. Table 2 shows the smart household appliances and their possible start- and end-time slots [17].

Table 2. Start, end and operation times of the appliances in use.

Appliance	Type	Power (kW)	Length *	Start *	End *
Washing machine		0.80	5	1	24
Air conditioner		1.30	10	1	24
Dryer for cloathes	Shiftable	0.70	4	1	24
Water heater		1.00	8	1	24
Dishwasher		0.20	3	1	24
Personnel computers		0.20	18	7	24
Security cameras		0.10	24	1	24
Microwave oven		0.50	7	14	20
Refrigerator	Non-Shiftable	0.90	20	3	22
TV		0.20	8	15	22
Lighting		0.10	6	17	22

* LENGHT is the duration of the work cycle, which is the number of time slots that must be scheduled for each appliance; these slots are consecutive only if the appliance is an uninterruptible load. START is the first time slot in which the work cycle of an appliance can start; it is not possible to schedule the start before START. END is the last time slot in which the work cycle of an appliance can end; it is not possible to schedule the end after END.

2.1.2. Rooftop PV Generator

The generation of electricity by a photovoltaic panel is highly dependent on the irradiance around the site where the system was built. The maximum power that can be extracted from a PV panel is usually given by the manufacturer's datasheet. However, in some cases it can be changed on the basis of the ambient conditions, e.g., the irradiance level and temperature, etc. The estimated power generation by a PV system in any time instant can be determined by the I-V and P-V characteristics after a manipulation. As can be seen from Figure 1, the rooftop photovoltaic panels generate a maximum of 3 kW power at certain times in a day. The maximum power output of the PV is defined on the basis of the maximum power point tracking system at any time instant. The power generated by the panels is first used to partly or fully supply smart home appliances, and then to charge the ESS if there is surplus power. The battery bank stores both the surplus energy from the rooftop PV system and the cheapest energy from the utility, if it is not fully charged at the slot.

2.1.3. ESS

The ESS has a capacity of 3 kWh, and it is charged from the photovoltaic system when surplus energy is available, and the grid when the electricity price is the lowest in a day. The energy storage system is preferred for use when the electricity price is the most expensive, and when it is urgently needed. However, the disadvantage of the photovoltaic-energy storage system is that it has a limited lifetime and requires periodic maintenance. In this investigation, the energy storage system can be formulated as follows:

Let B_s , $B_c^{(t)}$, $B_d^{(t)}$, η_c, η_d, B_c^{max} and B_d^{min} be the storage capacity of the battery bank, the charging energy during the tth slot, the discharging energy during the tth slot, the charging efficiency, the discharging efficiency, the maximum energy charging level and the minimum energy discharging level, respectively. The state of charge of the energy storage system during the tth slot can be expressed as

$$B_b^{(t)} = B_b^{(t-1)} + \eta_c B_c^{(t)} - \frac{B_d^{(t)}}{\eta_d} \tag{1}$$

where $B_b^{(t)} \leq B_s$. It should be noted that the battery cannot be charged or discharged at the same slot, hence $B_c^{(t)} > 0$, $B_d^{(t)} = 0$ or $B_c^{(t)} = 0$, $B_d^{(t)} > 0$. Table 3 illustrates the characteristics of the energy storage system.

Table 3. The characteristics of the ESS in use [17].

Parameter	Value
Roundtrip Efficiency	0.95
Charging Efficiency	0.95
Discharging Efficiency	0.95
Maximum Energy Capacity	3 kWh
Minimum Energy Capacity	0.5 kWh
Initially Stored Energy	0.5 kWh
Max. Charging/Discharging Energy	0.3 kWh

2.2. Problem Definition

In order to formulate the optimization problem, one day is divided into 24 equally spaced time slots, and the minimum time resolution is an hour. Let T denote a set of time slots; it can be defined by

$$T = \{1, 2, 3, \ldots, 120\} \forall t \in T \tag{2}$$

Let A denote a set of the interruptible, uninterruptible and baseline household appliances:

$$A = \{a_1, a_2, a_3, \ldots, a_i, \ldots, a_{16}\} \tag{3}$$

where a_i is the ith appliance of the set.

The power consumption vector for appliance a_i can be defined as

$$P_{a_i} = \left[p_{a_i}^{(1)}, p_{a_i}^{(2)}, \ldots, p_{a_i}^{(t)}, \ldots, p_{a_i}^{(120)} \right] \tag{4}$$

where $p_{a_i}^{(t)}$ is the average power consumption for appliance a_i during the tth interval.

Let $\alpha_{a_i} \in T$ and $\beta_{a_i} \in T$ be the start and end times of the operation interval respectively, and $\alpha_{a_i} < \beta_{a_i}$ for appliance a_i. Similarly, let l_{a_i} be the operation length of appliance a_i; thus, l_{a_i} must satisfy the constraint $\beta_{a_i} - \alpha_{a_i} \leq l_{a_i}$. It should be emphasized that the larger $\beta_{a_i} - \alpha_{a_i}$, the more possible solutions to the problem. Now, we need to assume that appliance a_i operates during l_{a_i} without interruption. Let s_{a_i} and e_{a_i} be the start and end times of appliance a_i, respectively; hence $e_{a_i} = s_{a_i} + l_{a_i}$ and $\alpha_{a_i} \leq s_{a_i} \leq \beta_{a_i} - l_{a_i}$.

Now, the power consumption for appliance a_i during the tth interval can be expressed as

$$P_{a_i}^{(t)} = \begin{cases} P_{ra_i}, & t \in [s_{a_i}, s_{a_i} + l_{a_i}] \\ 0, & t \notin [s_{a_i}, s_{a_i} + l_{a_i}] \end{cases} \tag{5}$$

where P_{a_i} is the average power consumption in kW.

The total daily power consumption can be calculated by

$$P_{total} = \sum_{t=1}^{120} \left(\left(\sum_{i=1}^{13} p_{a_i}^{(t)} \right) + \left(\sum_{j=1}^{5} p_{b_j}^{(t)} \right) \right) \tag{6}$$

where P_{b_j} is the average power consumption for the jth fixed time operated appliance in kW.

Although there was a considerable improvement in the reduction of the daily cost, householder comfort inevitably decreased. Therefore, it is necessary to reduce the delay time in the scheduled operation as much as possible using the multi-objective optimization approach. The delay time for appliance a_i can be calculated by

$$D_{a_i} = (s_{a_i})_{sch} - (s_{a_i})_{unsch} \tag{7}$$

where $(s_{a_i})_{sch}$ and $(s_{a_i})_{unsch}$ are the start times for appliance a_i in the scheduled and unscheduled operations, respectively.

If the electricity cost is the only objective, the fitness function can be expressed as

$$f_{cost} = Min\left(\sum_{t=1}^{24}\left(\left(\sum_{i=1}^{5} p_{a_i}^{(t)}\right) + \left(\sum_{j=1}^{6} p_{b_j}^{(t)}\right)\right)\right) \tag{8}$$

If the delay ratio is only objective, the fitness function can be expressed as

$$f_{PAR} = Min\left(\frac{Max\left(\left(\sum_{i=1}^{5} p_{a_i}^{(t)}\right) + \left(\sum_{j=1}^{6} p_{b_j}^{(t)}\right)\right)}{\frac{1}{24}\sum_{t=1}^{24}\left(\left(\sum_{i=1}^{5} p_{a_i}^{(t)}\right) + \left(\sum_{j=1}^{6} p_{b_j}^{(t)}\right)\right)}\right) \tag{9}$$

If the cost and the peak-to-average-ratio are the only two objectives, the fitness function can be expressed as

$$f = Min(\omega_1 f_{cost} + \omega_2 f_{PAR}) \tag{10}$$

where $\omega_1 = 1/3$ and $\omega_2 = 2/3$.

2.3. Proposed Method

The BCGA is highly fitted for solving the single objective and multi-objective optimization problems for the optimal start times of the shiftable appliances. The flowchart of BCGA method is presented in Figure 2. First, the start times of the shiftable appliances were generated by random binary strings, and their fitness values were calculated by the fitness function. The most fitted individuals were selected through the tournament selection mechanism, and the selected individuals were copied to the mating pool. A single point crossover was implemented on the selected population with a probability of 0.8, and the inverse mutation operation was applied to the current population, with a probability of 0.1. The elitist strategy was implemented on the mutated population to keep the best solution through the generations. The indicated process is repeated by the time the number of generation reaches the maximum number of generation, and the naturally possible solutions are improved through the successive generations, as expected. During the genetic process, the population size and the maximum number of generation were 200 and 300, respectively.

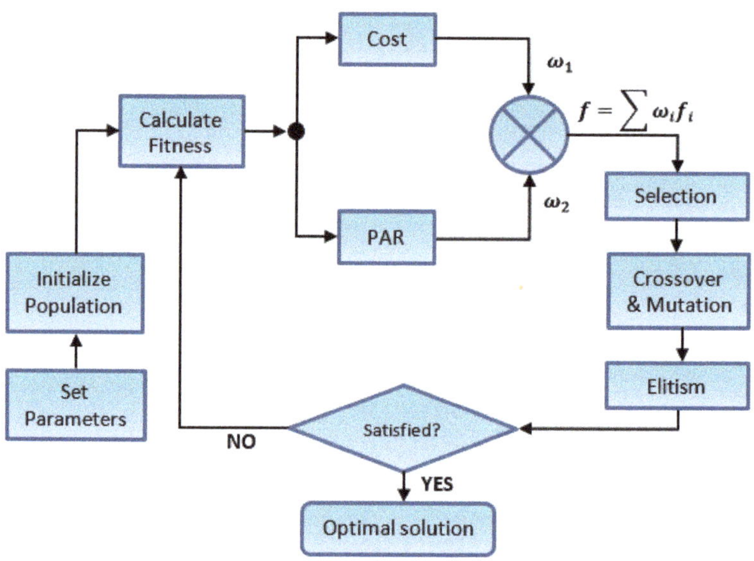

Figure 2. Flowchart of the proposed BCGA optimization.

3. Results and Discussion

In the above section, the simulation results obtained by the applied method are presented. This simulation includes daily energy cost minimization based on the electricity prices shown in Figure 3, after the use of electrical appliances in a typical house during the day. Divided into 24 equal time slots, the first time slot of the day indicates the 00–01-h interval, the second time slot shows the 01–02-hour interval, and so on. In this study, the optimization process includes: (I) minimizing the daily total energy cost, and (II) minimizing both the daily total energy cost and the PAR. Hence, the optimization problem was solved for an optimal solution for a single objective and two objectives within the framework of the specified conditions. As in the previous study [17], we have a 3 kWh-energy storage system and a 2 kW-solar PV system connected to the main grid. The power demand for the smart home appliances was first met from the solar PV system; in case it was insufficient, it was met from the energy storage system. If these two are in short supply, the remaining amount is supplied from the main grid. In the case of surplus power generated by the solar PV, it is sold to the grid at the grid purchase price. The optimization problem was solved by a desktop computer with Intel (R) Core (TM) i5-10210U CPU@1.60 GHz 16.0 GB RAM in five minutes. The results obtained by using the proposed method were compared with previous studies under the same conditions, and it can be said that they were improved compared to those in [17].

In order to fairly compare the results of our proposed approach, the data and results of the previous study [17] were used, and the comparison results are given through figures and tables. In a previous study [17], electrical household appliances were divided into two groups—non-shiftable and shiftable—as shown in Table 2. As expected, when the shiftable home appliances are shifted to time slots when the electricity price is cheap, the total daily energy cost decreases along with the householder's comfort; however, this is not a desired case. Here, of course, the householder's comfort may be expected to be included in the optimization process, but this is beyond the scope of the study. The solar PV system consisting of monocrystalline panels generates a maximum power of 2 kW during the day, and its hourly power generation is shown in Figure 4. The energy storage system used for this study has the same physical characteristics with the one given in the previous investigation [17]. Unlike the previous study [17], the scheme we propose is flexible, and home appliances are first fed from the solar PV system in the daytime, and then from the energy storage system if it can provide sufficient power. The surplus energy is first evaluated in the energy storage system; the remaining energy is sold to the grid, thus reducing the cost.

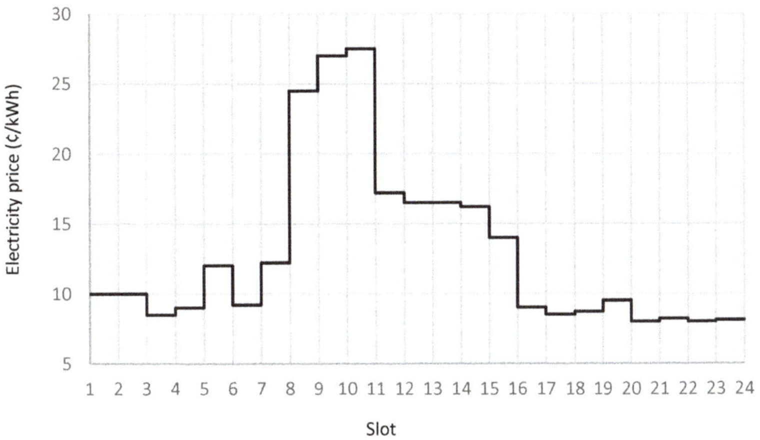

Figure 3. The hourly changing electricity prices, with the slots.

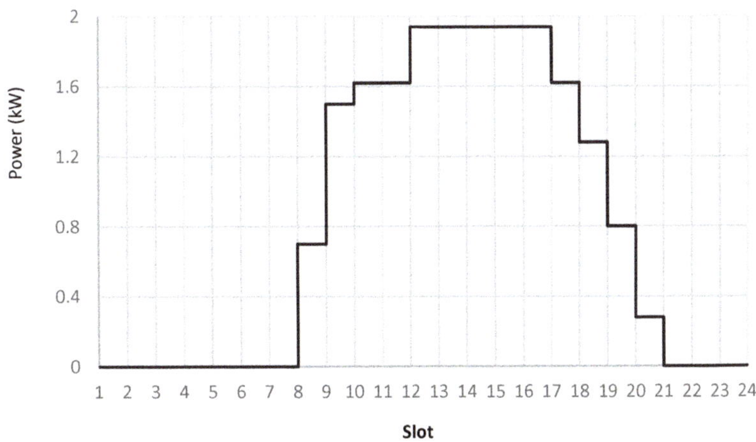

Figure 4. The hourly changing power generated by the solar PV power, with the slots.

3.1. Single Objective Optimization

In Equation (8), the weight numbers $\omega_1 = 1$ and $\omega_2 = 0$ are used as the objective function to minimize the total daily energy cost while calculating the optimal daily total cost, and in this calculation, the unit electricity price sold to the grid and the unit purchased from the grid are the same. The optimization process is detailed below.

Total Energy Cost

The comparison of the proposed approach with the results given in the previous study [17] for the total daily billing cost and PAR is shown in Figure 5. As seen in Figure 5a, the total cost of the daily electricity bill with the proposed method is 511.06 cents, which is better than those obtained with other methods [17]. There is an 8% reduction in the daily total electricity bill cost when the result is compared with the minimum result in previous studies [17]. As seen in Figure 5b, the PAR found is greater than only one of the results obtained from the algorithms used in the previous study [17]. However, in order to further improve the current results, it is necessary to solve a bi-objective optimization problem. In order to understand how this improvement was achieved, the graphs in Figure 6 should carefully be analysed. As seen in Figure 6a, there is a significant daily total cost difference between the optimized and non-optimized load dispatch. If attention is paid, in the scheduled home appliances, there is more power demand for them in the time intervals where the electricity price is lower, and less power demand in the intervals when the electricity price is higher. It can be said that this leads to a significant reduction in the daily total electricity bill. For example, while the electricity price is the lowest in the fourth time slot, the unscheduled home appliances of 1 kW are increased to 4 kW with an additional 3 kW. The scheduled average power demand between the 8 and 11 time slots, where the electricity price is the highest, is less than the unscheduled power demand, and this plays an important role in the reduction of the total daily billing cost.

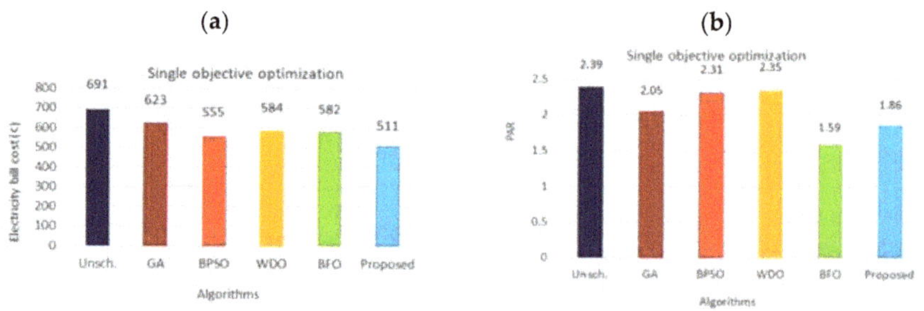

Figure 5. Comparison of the results with previous ones [17]: (**a**) daily total cost, and (**b**) PAR.

As seen in Figure 6b, while the energy received from the grid is less than others in the 8–11 time slots in the case of scheduled HA, the energy transferred from the energy storage system to home appliances is not available. Another important contribution to the reduction of the daily total electricity bill cost comes from the solar PV system; hence, this system feeds a limited number of the home appliances in times when electricity prices are high, leading to a decrease in the daily total electricity bill cost. In addition, it causes the daily total cost to decrease by transferring the surplus energy it produces to the ESS and the grid. However, some energy loss during the charging and discharging of the energy storage system from the solar PV system reduces the total efficiency of the entire system by 5%. In this regard, it is preferable to use the power obtained from the solar PV system to directly feed the household appliances. The operation times of the shiftable household appliances optimized in the performed simulation are given in Table 4.

Table 4. The optimized operation times of the appliances in SOO.

Type	Appliance	OT	SOO
	Washing machine	5	14
	Air conditioner	10	3
Shiftable	Dryer for clothes	4	19
	Water heater	8	3
	Dishwasher	3	3

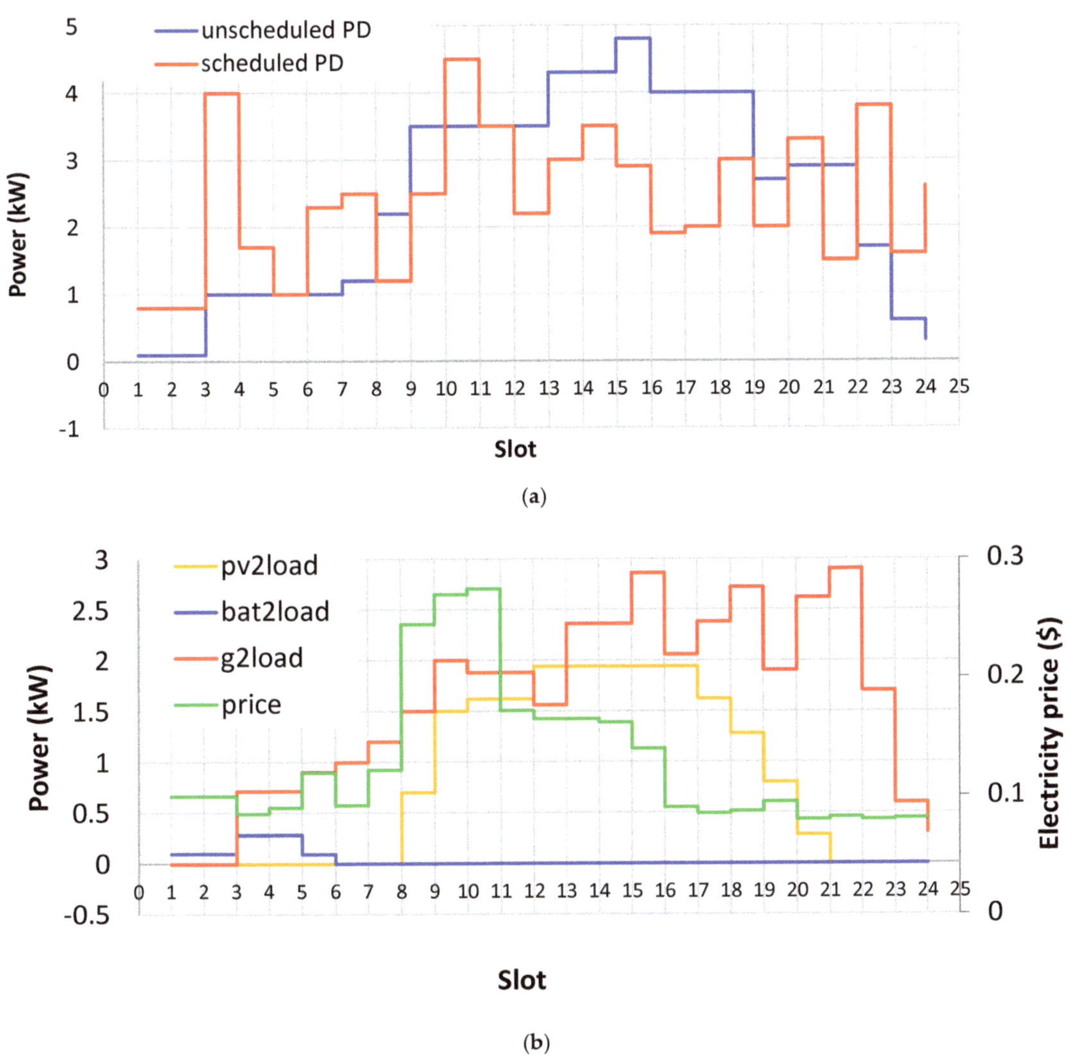

Figure 6. (**a**) The scheduled and unscheduled daily power demand, and (**b**) the powers delivered to the load. pv2load, PV to load; bat2load, battery to load; g2load, grid to load; price, electricity price in $.

3.2. Bi-Objective Optimization

The PAR is an important indicator of the behaviour of the householder's home appliances and significantly affects the performance of the main grid. The PAR rises as expected when we only focus on minimizing the daily total electricity bill cost. In order to overcome this problem, it is necessary to optimize both the daily total electricity bill cost and the PAR at the same time. It is obvious that the problem in question is a bi-objective optimization problem, and the two objective functions are converted to a single objective function with the help of constant weight coefficients. The selection of the coefficients is made according to the principle of the equalization of the maximum values; hence, the bi-objective functions take in a certain interval, with their sum being equal to unity. In the optimization process, the input values related to the method used for single optimization are taken as they are, except for the fitness function. The simulation time slightly increases because the problem is a little more complex than the single optimization problem.

While the simulation results illustrated in Figure 7 produce better results compared to the results obtained from the previously used methods [17], a better improvement was achieved in the result obtained in the single optimization. As can be seen from Figure 7, different weight coefficients produce slightly different results, and $w_1 = 1/3$ and $w_2 = 2/3$ yielded the best results among the three selected weight coefficients of $w_1 = 1/2$, $w_2 = 1/2$, $w_1 = 2/3$ and $w_2 = 1/3$. Contrary to what is expected here, in all three cases, the total daily cost has almost no change at all, and is very close to the result of the single optimization. The shiftable optimal operation hours obtained by bi-objective optimization by selecting appropriate weight coefficients are given in Table 5. As seen in Table 5, while the PAR is significantly reduced, the starting times of the shiftable household appliances are different from those obtained by single-objective optimization, and the best result was found when the weight numbers were $\omega_1 = 1/3$ and $\omega_2 = 2/3$.

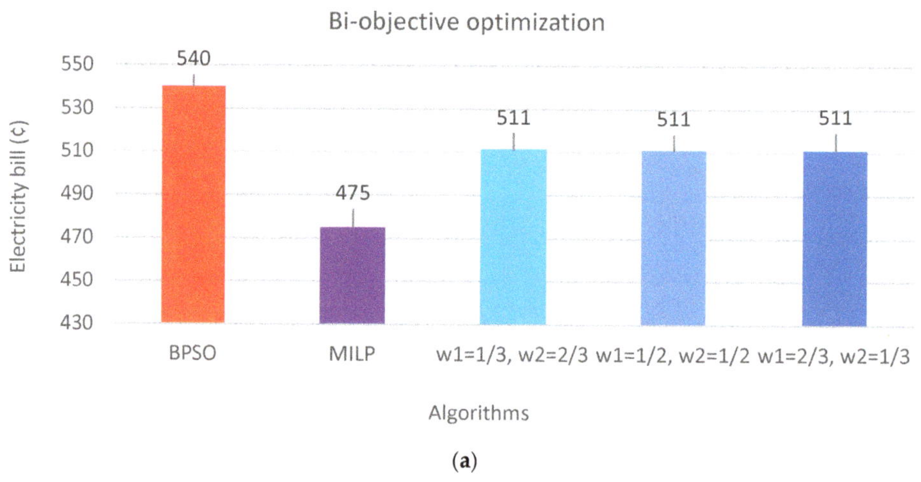

(**a**)

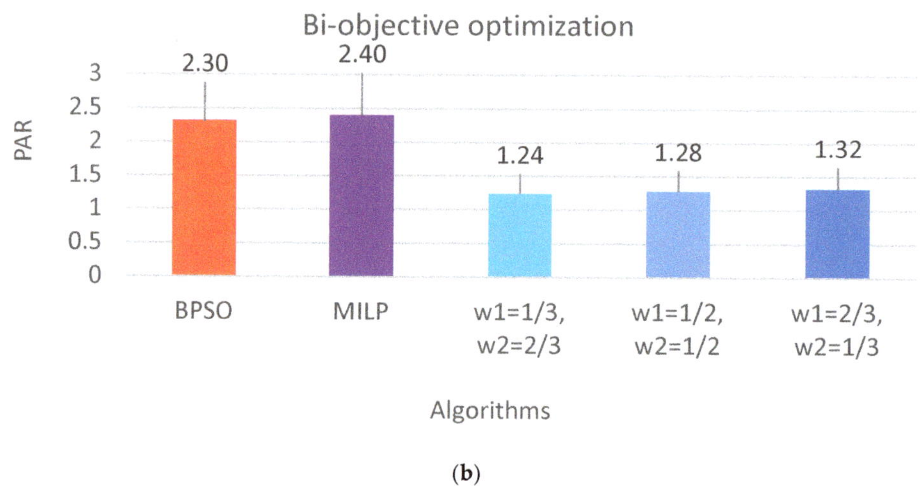

(**b**)

Figure 7. The comparison of the results with the previous ones [17]: (**a**) daily total cost, and (**b**) PAR.

Table 5. The optimized operation times of the appliances in BOO.

Type	Appliance	OT	BOO
	Washing machine	5	7
	Air conditioner	10	1
Shiftable	Dryer for clothes	4	12
	Water heater	8	1
	Dishwasher	3	1

In the designed bi-objective optimization, the PAR was reduced from 1.859 to 1.24; therefore, this is a very good improvement on the PAR drop. However, in this optimization process, the daily total electricity cost surprisingly remained constant at 511 US cents for three different weight coefficient sets. In the previous study [17], the cost of 475 cents obtained from the use of the mixed-integer linear programing (MILP) seemed to be a non-feasible solution, such that if all appliances are operated at the time slot when the electricity price is the lowest, their daily total cost is 464.80 US cents; this is really nonsense. Actually, this verifies that a PAR of 2.4 indicates that the household appliances are overlapped in a certain time slot. The variation of the hourly power demand of household appliances with the time slots and powers delivered to these appliances are shown in Figure 8. It can be seen from Figure 8a that the largest power demanded from the main grid is 4.5 kW in the 11th time period, and the large power demands are spread over low-priced time slots. Although this is the most influential factor that helps to significantly reduce the large PAR values, the additional power boost from the solar PV system at noon has a significant share in this decrease. As can be seen from Figure 8b, the reduction in the daily total cost is intended to meet the power needed by the appliances when the electricity price is the lowest. Note that here, the power transferred from the energy storage system to the appliances is not sufficient, as expected. The power boost from the solar PV generator appears to be effective at between 09 and 18 h, in contrast to the energy storage system in use.

In the light of all of these results, it can be seen that the proposed method, due to its nature, produces better results in terms of applicability when compared to other methods for the reduction of the daily total cost and PAR in the home energy management system. A one-to-one comparison with the results of a similar study was clearly demonstrated here.

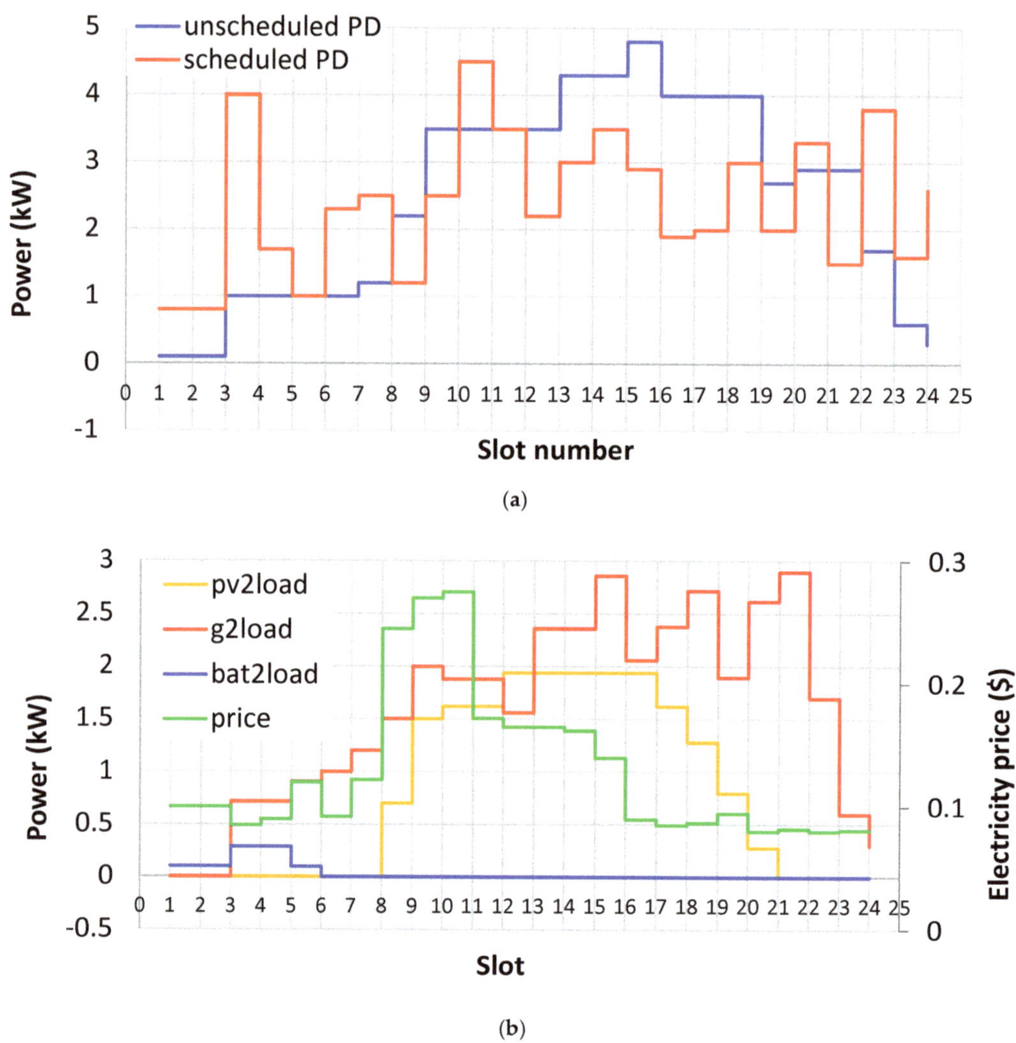

Figure 8. (**a**) The scheduled and unscheduled daily power demand, and (**b**) the powers delivered to the load.pv2load, PV to load; bat2load, battery to load; g2load, grid to load; price, electricity price in $.

4. Conclusions

Scheduling smart home appliances' operation times to simultaneously reduce both the electricity bill and the PAR based on demand response, as well as increasing user comfort, is an essential task. It is worth noting that in the scheduled operation, the cost is always lower than in the unscheduled operation. The cost value varies depending on the price of electricity, the rated power and number of household appliances, as well as the length of the operation range. Although it is desirable for the user to reduce the cost, this inevitably leads to a PAR increase in certain time intervals. This is not welcome for either the user or the utility, as it causes extra losses in power transmission lines. A gradual increase in the electricity price depending on the consumed power up to a certain level seemed to be a solution in the previous application, and it can be seen from the results that it is ineffective to decrease PAR. Instead, it was observed that the simultaneous optimization of both the cost and PAR offers better results to the householders. Taking user comfort into consideration, it was observed that as the user comfort increases, the PAR increases together with the cost when optimizing the cost, the PAR and the comfort simultaneously.

Author Contributions: Conceptualization, N.T.; methodology, N.T. and A.B.; software, N.T.; validation, M.J. and Z.L.; formal analysis, M.J. and E.J.; investigation, A.B.; writing—original draft preparation, N.T.; writing—review and editing, A.B.; visualization and supervision, Z.L.; project administration, N.T. and A.B.; funding acquisition, E.J. and Z.L. All authors have read and agreed to the published version of the manuscript.

Funding: This research was funded by Wroclaw University of Science and Technology, K38W05D02.

Institutional Review Board Statement: Not applicable.

Informed Consent Statement: Not applicable.

Data Availability Statement: The applied manuscript data are available on request from ntutkun@ticaret.edu.tr.

Conflicts of Interest: The authors declare no conflict of interests.

References

1. Setlhaolo, D.; Xia, X. Combined residential demand side management strategies with coordination and economic analysis. *Electr. Power Energy Syst.* **2016**, *79*, 150–160. [CrossRef]
2. Shakouri, H.; Kazemi, A. Multi-objective cost-load optimization for demand side management of a residential area in smart grids. *Sustain. Cities Soc.* **2017**, *32*, 171–180. [CrossRef]
3. Lokeshgupta, B.; Sivasubramani, S. Multi-objective home energy management with battery energy storage systems. *Sustain. Cities Soc.* **2019**, *47*, 101458. [CrossRef]
4. Zhao, Z.; Lee, W.C.; Shin, Y.; Song, K.B. An optimal power scheduling method for demand response in home energy management system. *IEEE Trans. Smart Grid* **2013**, *4*, 1391–1400. [CrossRef]
5. Hussain, I.; Ullah, M.; Ullah, I.; Bibi, A.; Naeem, M.; Singh, M.; Singh, D. Optimizing energy consumption in the home energy management system via a bio-inspired dragonfly algorithm and the genetic algorithm. *Electronics* **2020**, *9*, 406. [CrossRef]
6. Sharifi, A.H.; Maghouli, P. Energy management of smart homes equipped with energy storage systems considering the PAR index based on real-time pricing. *Sustain. Cities Soc.* **2019**, *45*, 579–587. [CrossRef]
7. Veras, J.M.; Silva, I.R.S.; Pinheiro, P.R.; Rabêlo, R.A.L.; Veloso, A.F.S.; Borges, F.A.S.; Rodrigues, J.J.P.C. A multi-objective demand response optimization model for scheduling loads in a home energy management system. *Sensors* **2018**, *18*, 3207. [CrossRef] [PubMed]
8. Hussain, H.M.; Javaid, N.; Iqbal, S.; Hasan, Q.U.; Aurangzeb, K.; Alhussein, M. An efficient demand side management system with a new optimized home energy management controller in smart grid. *Energies* **2018**, *11*, 190. [CrossRef]
9. Jordehi, A.R. Binary particle swarm optimisation with quadratic transfer function: A new binary optimisation algorithm for optimal scheduling of appliances in smart homes. *Appl. Soft Comput. J.* **2019**, *78*, 465–480. [CrossRef]
10. Zhu, J.; Lin, Y.; Lei, W.; Liu, Y.; Tao, M. Optimal household appliances scheduling of multiple smart homes using an improved cooperative algorithm. *Energy* **2019**, *171*, 944–955. [CrossRef]
11. Liu, Y.; Xiao, L.; Yao, G.; Bu, S. Pricing-based demand response for a smart home with various types of household appliances considering customer satisfaction. *IEEE Access* **2019**, *7*, 86463–86472. [CrossRef]
12. Waseem, M.; Lin, Z.; Liu, S.; Sajjad, I.A.; Aziz, T. Optimal GWCSO-based home appliances scheduling for demand response considering end-users comfort. *Electr. Power Syst. Res.* **2020**, *187*, 106477. [CrossRef]
13. Yahia, Z.; Pradhan, A. Multi-objective optimization of household appliance scheduling problem considering consumer preference and peak load reduction. *Sustain. Cities Soc.* **2020**, *55*, 102058. [CrossRef]
14. Bouakkaz, A.; Mena, A.J.G.; Haddad, S.; Ferrari, M.L. Efficient energy scheduling considering cost reduction and energy saving in hybrid energy system with energy storage. *J. Energy Storage* **2021**, *33*, 101887. [CrossRef]
15. Merdanoğlu, H.; Yakıcı, E.; Doğan, O.T.; Duran, S.; Karatas, M. Finding optimal schedules in a home energy management system. *Optim. Eng.* **2020**, *182*, 106229. [CrossRef]
16. Mouassa, S.; Bouktir, T.; Jurado, F. Scheduling of smart home appliances for optimal energy management in smart grid using Harris-hawks optimization algorithm. *Optim. Eng.* **2021**, *22*, 1625–1652. [CrossRef]
17. Ahmad, A.; Khan, A.; Javaid, N.; Hussain, H.M.; Abdul, W.; Almogren, A.; Alamri, A.; Niaz, I.A. An optimized home energy management system with integrated renewable energy and storage resources. *Energies* **2017**, *10*, 549. [CrossRef]

Article

Improving Detection of False Data Injection Attacks Using Machine Learning with Feature Selection and Oversampling

Ajit Kumar [1], Neetesh Saxena [2], Souhwan Jung [3] and Bong Jun Choi [1,*]

[1] School of Computer Science and Engineering, Soongsil University, Seoul 06978, Korea; ajitkumar.pu@gmail.com
[2] School of Computer Science and Informatics, Cardiff University, Cardiff CF10 3AT, UK; nsaxena@ieee.org
[3] School of Electronic Engineering, Soongsil University, Seoul 06978, Korea; souhwanj@ssu.ac.kr
* Correspondence: davidchoi@soongsil.ac.kr

Abstract: Critical infrastructures have recently been integrated with digital controls to support intelligent decision making. Although this integration provides various benefits and improvements, it also exposes the system to new cyberattacks. In particular, the injection of false data and commands into communication is one of the most common and fatal cyberattacks in critical infrastructures. Hence, in this paper, we investigate the effectiveness of machine-learning algorithms in detecting False Data Injection Attacks (FDIAs). In particular, we focus on two of the most widely used critical infrastructures, namely power systems and water treatment plants. This study focuses on tackling two key technical issues: (1) finding the set of best features under a different combination of techniques and (2) resolving the class imbalance problem using oversampling methods. We evaluate the performance of each algorithm in terms of time complexity and detection accuracy to meet the time-critical requirements of critical infrastructures. Moreover, we address the inherent skewed distribution problem and the data imbalance problem commonly found in many critical infrastructure datasets. Our results show that the considered minority oversampling techniques can improve the Area Under Curve (AUC) of GradientBoosting, AdaBoost, and kNN by 10–12%.

Keywords: Data Injection Attack; machine learning; critical infrastructure; smart grid; water treatment plant; power system

Citation: Kumar, A.; Saxena, N.; Jung, S.; Choi, B.J. Improving Detection of False Data Injection Attacks Using Machine Learning with Feature Selection and Oversampling. *Energies* **2022**, *15*, 212. https://doi.org/10.3390/en15010212

Communicated by: Silvio Simani and Zbigniew Leonowicz

Received: 9 November 2021
Accepted: 27 December 2021
Published: 29 December 2021

Publisher's Note: MDPI stays neutral with regard to jurisdictional claims in published maps and institutional affiliations.

1. Introduction

Today, the umbrella term 'Industry 4.0' represents the integration of digital control, Information and Communications Technology (ICT), and intelligent decision-making into critical infrastructures. This upgrade is possible due to the amalgamation of information and industrial technologies into standard components and processes [1,2]. This shift, from the traditional system to Industry 4.0, has helped improve the overall performance and productivity of critical infrastructures that have become the fundamental building blocks of modern society. For instance, electricity distribution and usage can be optimized in smart grids. In water systems, in-time data about usage and plant treatment capacity can reduce water wastage. Along with various benefits and improvements, the addition of new components into critical infrastructures presents new vulnerabilities [3–5]. This critical infrastructure is especially sensitive to cyberattacks. Even a low-scale attack that causes a few critical infrastructure components to malfunction can impact the whole system. For example, even a short disruption in the power grid can halt the functioning of many industries and infrastructures, from food processing plants to hospitals. The attack on the Ukraine grid infrastructure and a recent ransomware attack on a colonial pipeline are some of the many alarming examples that call for improvements to be made to the defense techniques which protect critical infrastructures [6,7].

The injection of data or commands at the source or during communication is collectively called False Data Injection Attack (FDIA). Data injection refers to the manipulation of the value generated from sensors, actuators, and other devices, while the command 'injection' refers to changing server-issued instructions. The FDIA is one of the most common attacks and can be launched on any critical infrastructure by penetrating the communication sessions between different devices. FDIAs can damage physical components, induce huge economic losses, and even create life-threatening scenarios [8,9]. Therefore, it is essential to prevent and detect FDIAs in any critical infrastructure. However, most of the existing solutions are only theoretical or adopt techniques from the cyber environment, such as Intrusion Detection Systems (IDS) and Intrusion Prevention Systems (IPS), which are used to protect conventional computer networks. These existing approaches are missing the specific security requirements and properties of critical infrastructure, such as a higher rate of event, the need for real-time interaction and detection, the need for proactive defense, and complex physical and cyber interfaces. Therefore, this study addresses these limitations by applying machine learning techniques to detect injection attacks.

This study work addresses two key issues to fulfill the requirements of proactive detection, low response time, and detection of minority classes of critical infrastructures: (1) finding the set of best features under different selection techniques and (2) resolving the class imbalance problem using oversampling. Our work uses a smart power system [10] and a water treatment plant [11] as case studies. We found the best features using different filter and wrapper selection methods (top- and bottom-ranked features). We also tested the performance of various machine-learning algorithms with different sets (with varying sizes) of the best features. Further, to improve the detection performance, especially for the minority class, i.e., attack class, we applied different oversampling methods to increase the sample for the minority class. We also solved common issues in the datasets, such as missing and corrupted values.

The contribution of this study is summarized below.

1. We provide a comprehensive analysis of machine-learning algorithms for FDIA detection using the two representative datasets, namely power system and water treatment datasets.
2. We determine the subset of features which can be used to achieve the best performance using different filter and wrapper approaches.
3. We mitigate performance bias in imbalanced datasets using four different oversampling methods.

The remainder of the paper is organized as follows. The related works are presented in Section 2. A detailed explanation of both critical infrastructures (power system and water treatment plant) from which data about events have been recorded for datasets creation is presented in Section 3. Section 4 presents the three feature selection approaches and provides a ranking of the features. The class imbalance issue and oversampling methods are discussed in Section 5. The results of training and testing and the outcome of oversampling are presented in Section 6. Lastly, Section 7 provides the conclusion and future research directions.

2. Related Work

In the section, we present the details of FDIA and other attacks targeted to the Cyber-Physical System (CPS) and provide a summary of existing FDIA detection methods based on machine learning. Further, the limitations and research gaps in the existing literature, which motivated the current study, are discussed.

With the fast transition of the traditional grid to the smart grid, the effective implementation of FDIA is critical to the success of the smart grid. There have been many FDIA attacks demonstrated in the literature. In the last five years (2015–2020), some surveys provided discussion and a comprehensive summary of challenges and countermeasures regarding FDIA. The role and importance of Artificial Intelligence (AI) and big data technologies for FDIA detection were also highlighted [9,12–14]. The financial impact of FDIAs was demonstrated in [8]. The authors assumed an insider attack and simulated an injection attack by changing the value of the memory location of the Programmable Logic Controller

(PLC). Experimental results showed that the injection attack could directly impact the electric usage billing system, generating a manipulated final bill.

Traditionally, state estimation and time-series analysis are the main methods used for FDIA detection. Recently, many AI-based approaches have been adapted to improve detection performance [15]. Class labeling and class-balanced datasets are two critical challenges for developing a machine learning-based FDIA detection system for the smart grid because of the small sample size for FDIA class and complex labeling. Maglaras et al. [16] used a One-Class Support Vector Machine (OCSVM) with *normal events* to resolve these two challenges for a Supervisory Control and Data Acquisition (SCADA)-based critical infrastructure. Due to the challenges involved in dataset preparation, FDIA detection with minimum training and prediction time is required to handle the high rate of data generation in the smart grid. Reducing the vectors of features using Principal Component Analysis (PCA) and speeding up the training time using Distributed SVM are used to achieve low computation requirements of the smart grid [17]. Further, FDIAs in the smart grid are grouped into 'direct' and 'stealth', where 'stealth FDIAs' are more challenging to detect than 'direct FDIAs'. Yan et al. [18] used supervised machine learning to build FDIA detection systems by formulating the detection as binary classification (direct and stealth). The authors also tested detection performance for balanced vs. imbalanced class distribution using the IEEE 30-bus simulation dataset. More recently, the Artificial Neural Network (ANN) has been applied for FDIA detection. Khanna et al. [4] used ANN and Extreme Learning Machine (ELM) to detect Data Injection Attacks on the consumer side of the smart grid and classified electric meters as either benign malicious. The NYISO load data was mapped to an IEEE 14-bus system for performing simulation, experiments, and validation of results. Data generation sources in the smart grid can be grouped into cyber or physical space. Wang et al. [19] have collected simulated and real-world measurements of synchronized PMUs and applied the Margin Setting Algorithm (MSA) for detection. The ensemble of Machine Learning (ML) algorithms was shown to improve the detection performance in [15]. In this direction, the performance of ensemble learning for multi-class classification was tested for a total of 37 classes, including FDIA in [20]. The experiments were executed using a dataset containing measurements of four Phasor Measurement Units (PMUs) and network communication data to and from the firewall and IDS of the experimental power system [10]. FDIA detection is also formulated as a three-class problem, rather than a binary classification, in the literature. Panthi et al. [21] used machine-learning algorithms and the publicly available power dataset [10] to build a classifier to group events into natural, no-event, or attack classes.

A fingerprinting-based detection of stealthy cyber-attacks in water treatment plants was proposed in [22]. An IDS using a semi-supervised system for attack localization and deep neural network learning for anomaly detection was proposed in [23]. More recently, a two-level attack-detection framework using a decision tree for detection and deep learning for attribution was proposed in [24].

Based on the summary of existing literature, we observe that machine learning-based FDIA detection approaches can improve detection performance and address some of the key requirements, such as real-time large-scale data generation in the smart grid. Such improvement will promote machine learning models for FDIA detection in smart grids and other critical infrastructure. Our literature also indicates that most existing research works have used the power system dataset and formulated FDIA classification as a multi-class problem. So, to explore a novel dimension, we consider FDIA classification as a binary problem and made pre-processing necessary to the dataset to experiment under various environments. The power system dataset [10] used also contains a binary version that was formulated the classification as 'Attack' and 'Normal'. In contrast, the classification problem was formulated as 'FDIA' and 'non-FDIA' in this study. Feature selection is useful and required to reduce time complexity but is seldom used with critical infrastructure datasets. Therefore, we experimented with feature selection methods and machine-learning algorithms. We aim to find the best performance of classifiers given the selected features.

As shown in Table 1, the data imbalance issue is rarely addressed. Therefore, we also performed minority class oversampling to balance the class distribution beyond identifying the imbalanced dataset's effect on detection accuracy.

Table 1. Summary of related work on FDIA detection using machine learning

Ref.	Method	Dataset	Samples Ratio (Normal, Attack)	Feature Selection
[18]	Supervised Learning (SVM, kNN, and ENN)	Simulations IEEE 30-bus system	0.1	No
[4]	ANN and ELM	NYISO load data IEEE 14-bus system	NA	NA
[19]	Data Centric (Big Data and MSA)	Simulated (6 bus power system) and real-world (Texas synchrophasor network)	100 K, 0.334, 0.196, 0.086	No
[20]	Voting on ML-classifier, dataset divided per PMUs	4 PMUs events and firewall log [10]	Balance	Yes
[16]	OCSVM	Network traces	1570, NA	NA
[15]	Ensemble Learning	Measurement data and power system audit logs	Balance	Yes
[17]	Distributed SVM and PCA	IEEE standard test systems	NA	Yes
[21]	Machine Learning (One R, J-Ripper, NB, RF)	Power system [10]	NA	No
[22]	Fingerprinting and OCSVM	Water treatment (SWaT)	NA	NA
[23]	AutoEncoder	SWaT	NA	Yes
[24]	DT and Deep learning	SWaT and gas pipeline	214 K, 21.86%	Yes

3. Critical Infrastructure Experimental Framework

The critical infrastructure is any physical infrastructure, such as a power system, healthcare [25], or gas pipeline, that is essential to support our daily lives [26]. Hence, disturbance to these systems has a huge impact on society, the economy, and the environment [27]. Recently, the development in computer and network technologies has enabled the fast adoption of ICT in such critical infrastructures. For example, the traditional power grid is now controlled, operated, and monitored using ICT, migrating the century-old power grid into a smart grid. Such integration is also evident in almost all critical infrastructures. However, cyber involvement in the physical system makes it vulnerable to various cyber attacks such as FDIA, unauthorized access, etc. In our study, we considered FDIA detection in power systems and water treatment plants using two very popular open datasets.

3.1. The Experimental Framework for Power System Data

Power generation, storage, and distribution tasks are continuously performed in the smart grid. Moreover, the complexity and large scale make it infeasible to experiment with the real infrastructure. Moreover, data access in the smart grid environment is highly restricted due to privacy and security concerns. Research is often performed on a reduced scale or using simulated datasets, such as IEEE 14/30 buses, to accommodate the abovementioned limitations. In 2015, the Mississippi State University and Oak Ridge National Laboratory dataset (https://www.sites.google.com/a/uah.edu/tommy-morris-uah/ics-data-sets (accessed on 8 November 2021)) produced a scaled-down version of the power system and recorded a dataset with various simulated attacks in addition to normal events [10]. The experimental power system has two power generators (*G*1 and *G*2), four Intelligent Electronic Devices (IEDs) (R1 to R4), and four breakers (BR1 to BR4). Two lines were created in the power system using the pairs of breakers (BR1 and BR2; BR3 and BR4). The four IEDs, R1 to R4, were configured to open or close the four breakers, BR1 to BR4, respectively. A server controls the physical part of the power framework in the control room, and these cyber and physical parts are connected using a switch and a Power Distribution Center (PDC).

3.1.1. Dataset Pre-Processing

The power system environment discussed in Section 3.1 helped to create a suitable dataset for conducting machine learning-based detection experiments [10]. The complete dataset is available and being distributed as 15 sets. The dataset comprises 37 power system events that can be grouped into three main scenarios: (1) *Natural*, (2) *No Events*, and (3) *Attack Events (data injection and command injection)* containing 8, 1, and 28 events, respectively. Regrouping and resampling are performed using these three types of events, and three datasets are created for binary, three-class, and multi-class classifications. For the multi-class classification, each event type is considered a class therefore, it has 37 classes. The binary and three-class datasets are distributed in CSV file format. However, the multi-class dataset is available only as an ARFF (an Attribute-Relation File Format (ARFF) is a file format created to be used by the Waikato Environment for Knowledge Analysis (WEKA) tool. It is a Graphical User Interface (GUI) tool for performing machine learning tasks such as per-processing, training, exporting models, and creating an ML pipeline.).

Building a multi-class machine learning classifier is complex and resource-consuming. It also creates a dataset's class imbalance problem. Considering this, we reformulated the FDIA detection as a *binary classification* with 'FDIA' and 'Non-FDIA' classes. However, the existing dataset was unsuitable for this study, so we grouped samples based on the type of events. Before resampling, we converted the multi-class ARFF to CSV format to simplify further pre-processing, training, and testing. Filtering and merging were performed on all 15 sets to group all scenarios into two predefined classes; Normal/non-FDIA classes were 1–6, 13, 14, and 41, while "FDIA" classes were 7–12. Further, the Normal/non-FDIA sample was labeled as 0, and the FDIA sample was labeled as 1. The total number of samples that the final pre-processed and resampled dataset contained was 32,296. Figure 1 shows the class distribution. There are 22,714 samples in the 'Normal/non-FDIA' class and 9582 in the 'FDIA' class. As shown in Figure 1 and listed in Table 2, it is clear that the power system dataset is an imbalanced dataset where 'Normal/non-FDIA' is the majority class. All four features related to impedance for IEDs relays such as 'R1-PA:Z' had infinite value, and so, as a pre-processing step, they were replaced with 0.

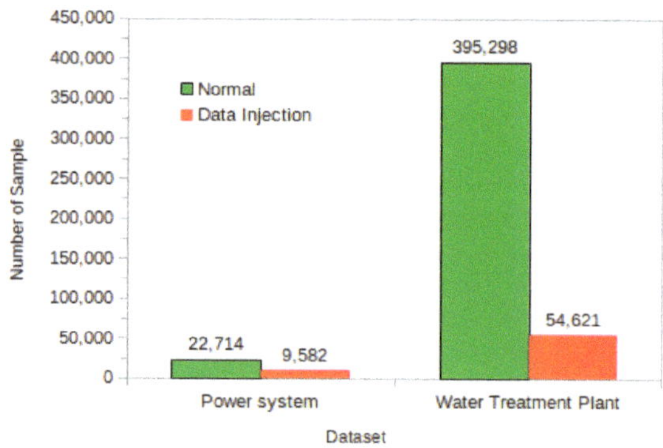

Figure 1. Number of samples per class in the dataset.

Table 2. Distribution of the classes (normal and FDIA) in the dataset.

Class Type	Number of Samples	
	Power System	Water Treatment
Normal	22,714	395,298
FDIA	9582	54,621
Total	32,296	449,919

3.1.2. Description of Features

There is a total of 128 features in the dataset, consisting of PMUs measurement and software logs. A total of 29 measurements were recorded for each PMU, so a total of 116 features were collected from 4 PMUs. The logs were recorded from three sources: snort, relay, and control panel. Each had 4 values, so logs contributed a total of 12 features. Each feature was given a name based on the combination of a source of data and type of value. For example, PMUs features begin with *R#−Signal Reference* and log features start with the source of logs such as snort, control_panel, and relay. The # for PMU features was a number between 1–4, indicating the PMUs number, while postfix *Signal Reference* was the type of measurement. These measurements fall into two groups: first, phase angle and magnitude for voltage and current, and, second, frequency, frequency delta, appearance impedance, and appearance impedance angle for relays. Details of these features are presented and explained in the original dataset description document (http://www.ece.uah.edu/~thm0 009/icsdatasets/PowerSystem_Dataset_README.pdf (accessed on 8 November 2021)).

It is important to understand the impact of *false data injection* on individual features. We used a data distribution approach and created an overlapped histogram for individual features. For example, a histogram for *R1-PA1:VH* features is plotted for all normal and FDI samples. We can observe from Figure 2 that the count for the specific value range is higher in the FDI sample, which indicates data injection.

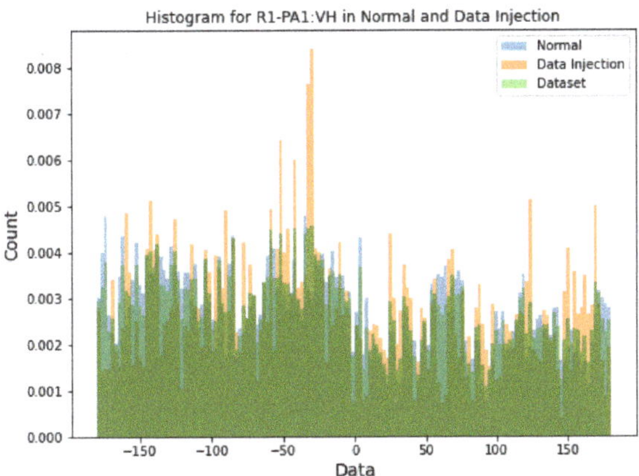

Figure 2. Histogram comparison of normal, injection, and all dataset events for *R1-PA1:VH*.

3.2. Water Treatment Plant

For the same reason as for a power system, a testbed, i.e., a scaled-down version of a real water treatment plant or pipeline, is normally created to experiment and collect data. In this study, we used a similar dataset, the Secure Water Treatment (SWaT) testbed [11] (a fully operational scaled-down water treatment plant), for FDIA detection. The configuration and framework of the experimental water treatment plant are depicted in Figure 3. It has six processing stages for water treatment labeled P1–P6. In total, the testbed has 24 sensors, 27 actuators, and 6 PLCs (one for each stage). The count for each type of sensor and actuator is listed in Table 3.

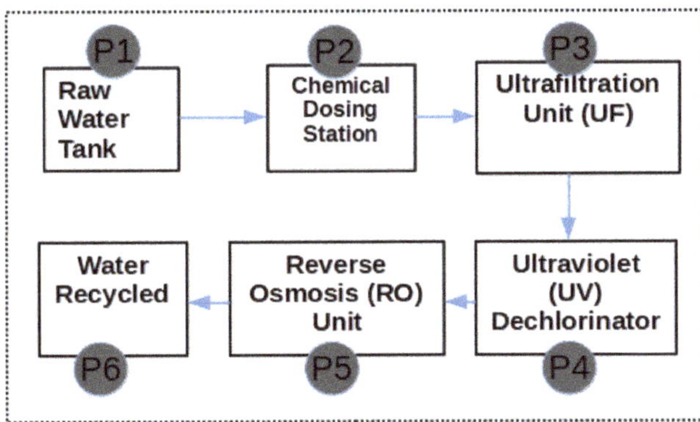

Figure 3. The water treatment plant in [11], where P1–P6 denotes total six stages of processing in the plant.

3.2.1. Dataset Pre-Processing

In the SWaT testbed, there are PLCs, Human Machine Interfaces (HMIs), SCADA, and a Historian in a layered communication network. Data from field devices are available to SCADA via PLCs and transferred to the Historian for analysis. The dataset contains events from *physical* and *network activities* against the 36 predefined attacks. The complete dataset was collected during a period of 11 days, during which the plant was running continuously for 24 h each day.

Table 3. Description of different sensors and actuators used for data generation. Features are named as combination of type (i.e., MV, P, FIT etc.) and suffix (process number and device number). For example, FIT-101 can be read as being a first flow meter sensor of process stage 1.

Field Device	Type	Description	Total (51)
Actuators (27)	MV	Motorized Valve	6
	P	Pump	19
	LIT	Level Transmitter	1
	UV	Dechlorinator	1
Sensors (24)	FIT	Flow Meter	9
	LIT	Level Transmitter	2
	AIT	Analyzer	9
	DPIT	Differential Pressure Indicating transmitter	1
	PIT	Pressure meter	3

3.2.2. Feature Description

The collected dataset contains 449,919 physical events and 51 features mainly generated from 24 sensors and 27 actuators. Table 3 provides the details of the sensors and actuators used in the water treatment process. The network data are packets communication between PLCs and SCADA. They have 18 features based on network attributes such as date, time, IPs, etc. This sub-part of the dataset is not used in this study. The dataset was collected, stored, and distributed on CSV files. The attacks on both physical and network were *injection* type attacks, i.e., on either the value of sensor or actuators.

The power and water system datasets were obtained using different field devices and operational environments. In the power system dataset, the majority of features are measurements of PMUs, whereas, in the water treatment dataset the events were collected by sensors and actuators. Differences in the data source provide varying data types: PMUs provide voltage, current phase angles, and magnitude, while sensors and actuators provide numerical or Boolean values.

4. Feature Selection

A feature represents a characteristic of any object. In ML, a sample is decomposed into a set of features before training and testing for tasks, such as classification, prediction, or clustering. The dimension of a feature vector can be small to large, and each feature has unequal discriminative potential. So, there is a need to select the best possible set of features without significantly impacting the model's performance. The different feature-selection approaches provide various ways to rank and select a set of features. The selection of features is performed in relation to the output variable that can be a class for classification or the predictive variable. Feature selection provides two key benefits. First, it helps to improve the model's performance in terms of accuracy, precision, and recall. Second, it reduces the computation cost (time and space) for the training, testing, and deployment of ML models. As a result of these two benefits, feature selection (as a part of feature engineering) is critical in the ML model. Based on the technique of feature ranking and selection, various feature selection methods are grouped into three main classes: the filter, wrapper, and embedded methods [28].

4.1. Filter Method

The filter method examines the dependency relationship of features X and class labels Y to select features based on their strength level with Y. The dependency strength level of the variables is calculated using traditional statistical tests, such as ANOVA, Z-test, T-test, chi-square, and Pearson Correlation Coefficient. Due to the individual evaluation of each feature, the filter method is also called univariate selection; it also speedily calculates and easily interprets results [28]. In this study, under the filter method, the ANOVA F-value was used as the statistical test; the dataset features were ranked, and the best set of features was selected.

Figure 4 and the left four columns of Table 4 show the results of feature selection using a filter-for-power system and water treatment plant (SWaT dataset) respectively. The results are represented as feature name and score from the top (ten features) and bottom (ten features) of the feature rank list. In the case of the power system dataset, from Figure 4, we can observe that *magnitude*-related measurement of PMUs achieves larger scores and is top-ranked using the filter approach. Based on the value of the *magnitude* features in the dataset, we can observe that the larger values influence the statistical test. In contrast, *angle*-based features fall into a smaller value range (negative to a small positive value), and the statistical test was given a low score and was lower-ranked.

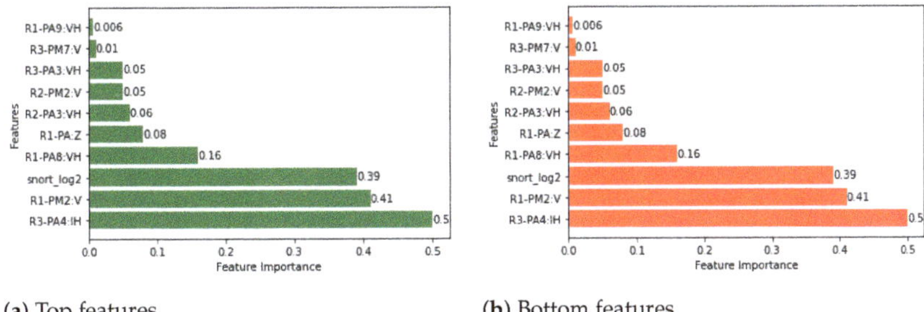

(**a**) Top features (**b**) Bottom features

Figure 4. Ten top and bottom features with individual score using filter method.

Table 4. Water Treatment Plant: Ten top and bottom features with their importance values using the filter and wrapper methods.

Filter Method				Wrapper Method			
Top Features		Bottom Features		Top Features		Bottom Features	
Feature	Value	Feature	Value	Feature	Value	Feature	Value
FIT401	6.281	FIT601	0.00066	FIT504	0.223181	P204	0.000403
FIT504	6.218	P602	0.00058	FIT401	0.125924	P206	0.000313
FIT503	6.105	P403	0.00008	P501	0.105114	P402	0.000114
UV401	6.076	P202	0.0	PIT502	0.070205	P403	0.000035
P501	6.075	P301	0.0	FIT503	0.063296	P202	0.000000
PIT501	5.972	P401	0.0	P102	0.040979	P301	0.000000
FIT501	5.906	P404	0.0	LIT301	0.040890	P401	0.000000
PIT503	5.899	P502	0.0	LIT101	0.030181	P404	0.000000
FIT502	5.860	P601	0.0	LIT401	0.027320	P502	0.000000
P402	5.550	P603	0.0	DPIT301	0.022423	P601	0.000000

In the SWaT dataset, the top three features are FIT401, FIT504, and FIT503, and all these are flow control sensors placed in the crucial stages, i.e., the 4-th and 5-th stages of a 6-stage process. Similarly, other top features also have critical roles and are found in later stages of the plant process. From Table 4, we can observe that the two bottom features are P601 and P60. These are two actuators placed in the last stage. Interestingly, these two were not implemented in SWaT, and the features selection correctly placed these at last. Other bottom features, P401, P404, and P502, are actuators. These were implemented as backups, and so, for this reason, they are not considered during attack events.

4.2. Wrapper Method

Compared to the filter method, features ranking is performed concerning a particular algorithm in the wrapper method. So, the best-selected feature set works well with the machine-learning algorithm, and the feature set differs when the selection is made using another algorithm. Unlike the filter method, the wrapper feature selection process is costly in terms of time and space. Most wrapper methods use greedy search, which is not optimal, and suffer from false starts (wrongly choosing the first best feature) [28]. Figure 5 and the right four columns of Table 4 show the best-selected feature sets and their importance for the wrapper method for the power system dataset and the water treatment plant dataset (SWaT dataset), respectively.

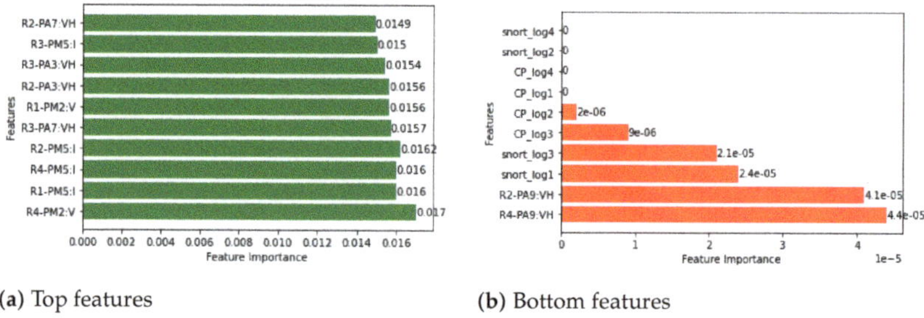

(**a**) Top features   (**b**) Bottom features

Figure 5. Ten top and bottom features with individual score using wrapper method.

In the case of the power system dataset, as shown in Figure 5, the features based on *magnitude* and *angle* are top-ranked in a nearly equal proportion, i.e., 6 and 4, respectively. So, the feature rank list differs from the filter method in which *magnitude*-related features were dominating. This study used a tree-based classifier as a wrapper method. In this approach, features are selected based on their impact on classification accuracy, rather than *number of features*. Similar behavior can be observed in the control log features listed at the bottom of the feature list. These features are Boolean and sparse, and their contribution to the classification is negligible, i.e., with an importance value of zero.

In the SWaT dataset, FIT401 and FIT504 are ranked as top features. The other top features, i.e., P501 and PIT502, are the actuator and the sensor for a pump and a pressure meter, respectively. The bottom features are similar to those from the filter methods, which verify the importance of ranking these features.

4.3. Embedded Method

The embedded method combines the techniques of the filter and wrapper approaches. The purpose of the combination is to take advantage of both approaches in terms of the speed and performance of the filter and wrapper methods, respectively. From the implementation perspective, feature selection becomes part of training in the embedded method. The algorithm starts training with the seed feature set (i.e., all features) and recursively selects a set of best features for the next round of training based on the importance of the features in the trained model [28]. The retraining continues until the predefined termination condition, e.g., based on the algorithm's convergence criteria or expected performance. The commonly used embedded methods are LASSO and RIDGE regression.

5. Imbalance Dataset: Issue and Solution

In a supervised ML case, if the training sample for each classification class is approximately equal, then the given dataset is considered imbalanced [29–31]. If the dataset is imbalanced, the training is highly influenced by the majority class sample (i.e., the class with the largest samples). Hence, the trained model lacks generalization in the real world and misclassifies the minor class.

The imbalance issue is more prominent in the cases such as this study, where the task is to detect a rare event, i.e., an anomaly, maliciousness, an attack, etc. In contrast, the normal events contribute the majority of the dataset. Both of our datasets are of an imbalanced nature, as can be verified from Table 2 and the bar plots in Figure 1. The power system dataset has an imbalance ratio of 1:2.5. In contrast, the water treatment plant dataset has an imbalance ratio of 1:7.23, meaning that the samples for normal events are 70% and 88% of the total sample, respectively.

Imbalanced datasets are a major issue and create a bottleneck in machine learning, so there have been many methods to address and resolve the problem of training with imbalanced datasets. These techniques mainly work on two principles: oversampling and under-sampling. Oversampling suggests increasing the sample in the minority class, while under-sampling is the process of reducing the sample in the majority class. The under-sampling method goes against the basic principle of machine learning, which mainly aims to obtain more samples to achieve better performance. So, under-sampling is suitable only when the dataset has a very large sample for the majority class, and removing the sample will have a very low or no impact on training. In this study, we adopted the oversampling technique, given the limited number of samples, and focused on increasing the sample of the minority class.

5.1. Synthetic Minority Oversampling Technique (SMOTE)

SMOTE is a minority-class oversampling method that creates synthetic examples. The synthetic examples are created by performing operations in data space and are nearly free from any particular application domain. The synthetic examples are plotted against the minority class samples, and the required samples, denoted as k, are randomly selected as k nearest neighbors [30].

5.2. Borderline-SMOTE

Borderline-SMOTE is also a minority oversampling method. There are two variants [32]: borderline-SMOTE1 and borderline-SMOTE2. Both the methods only oversample those minority samples on the borderline of class separation. The algorithm first finds the borderline samples from minority groups, then generates synthetic examples. It is assumed that borderline samples of minority classes are more prone to misclassification than samples far from the classification line.

5.3. Borderline Oversampling

Borderline Oversampling is similar to other oversampling methods that try to create synthetic samples around the classification line. Support Vector Machine (SVM) can be used to create a classification line and select boundary samples for oversampling [33]. First, the SVM model is trained on the complete dataset. Later, the trained model is used to identify the borderline, and synthetic/new samples for minority class are generated around the borderline. The number of samples, i.e., nearest neighbors, are generated either using interpolation or extrapolation depending upon the density of majority class instances around the borderline. This method differs from SMOTE, mentioned above, by choosing a new sample (i.e., nearest neighbor). SMOTE chooses randomly, while this method chooses the first k nearest neighbors.

5.4. Adaptive Synthetic (ADASYN) Sampling

Adaptive Synthetic (ADASYN) sampling incorporates *weighting* for oversampling of the minority class as per the difficulty level in learning [34]. The method claims to improve learning in two aspects, first, by reducing the bias induced by class imbalance, and second, by adapting to the classification decision boundary as per the difficult examples. The difficult examples are those samples from minority classes close to the decision boundary. They often result in false classifications. The algorithm starts by calculating the degree of class imbalance (d), i.e., $d = m_s / m_l$, where $d \in (0, 1)$ and m_s and m_l are the number of samples in the minority and majority classes respectively. The values of d in the used power system and water treatment plant datasets, respectively, are $0.4, 219$ and $0.1, 382$. The value of d is compared with the preset threshold value d-th for the maximum tolerated degree of class imbalance. Further, the total required synthetic samples (*G*) is calculated, and then the algorithm generates those samples.

Apart from the aforementioned oversampling methods, new methods have been proposed in the recent literature for improving model performance with an imbalanced dataset. Elyan et al. [35] have proposed class decomposition-based SMOTE (CDSMOTE). The proposed method improves performance by taking two actions: first, to reduce the dominance of the majority class by applying class decomposition, and second, to increase the representation of the minority class by oversampling. Moreover, a two-step hybridization of minority oversampling (SMOTE) and a novel data cleaning method (Weighted Edited Nearest Neighbor rule, or WENN) was proposed in [36]. Fajardo et al. [37] have applied deep conditional generative models for learning to the distribution of minority classes and then generated synthetic samples for solving the class imbalance in the dataset to improve the model's performance. Similarly, Bellinger et al. [38] have proposed a new training approach of a deep learning model (CNN) which mixes three techniques (batch resampling, instance mixing, and soft labels) to create a robust model from a long-tailed or imbalanced dataset. Krawczyk et al. [39] have studied the issues of the imbalanced dataset for multi-class classification. The authors have proposed a two-step under-sampling approach; in the first step, a one-class SVM is trained for all classes. An evolutionary under-sampling approach is applied to each learned classifier in the second step. Using under-sampling on the set of support vectors instead of on the original dataset, the authors claimed significant computational and performance improvements.

All the methods mentioned above for handling class imbalances in learning are suitable for single-model-based learning algorithms. They can be extended to suit ensemble-based learning algorithms [39]. SVM is a good choice for dealing with imbalanced datasets [33]. All these oversampling methods were tested on the power system dataset, and the results are presented in Section 6.4 along with explanations.

6. Experiments and Results

This section provides details of various experiments conducted to analyze the performance of ML algorithms for feature selection and improvement of minority class detection for imbalance datasets. Figure 6 illustrates the steps, structures, and components of the conducted experiments.

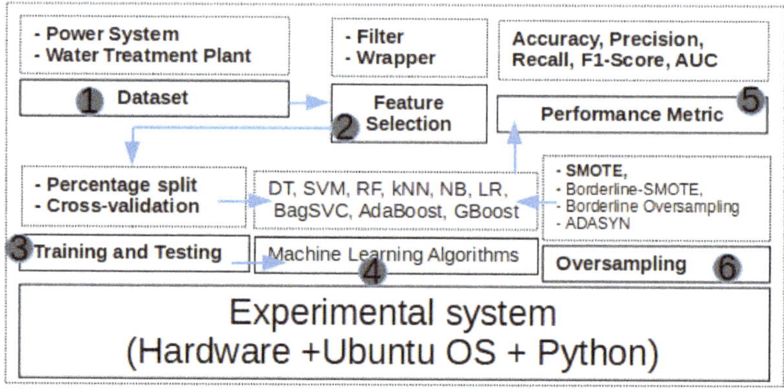

Figure 6. A summary of the steps, structures, and components of the conducted experiments.

These experiments were designed to test and validate the different hypotheses. For example, performance comparison for the train–test split vs. cross-validation, identifying the impact of the imbalanced dataset on performance, testing oversampling techniques to improve the performance, and finding the classifiers' performance on a different subset of features (ALL, Top10, Top20, Top30, Top40, and Top50). All the experiments were tested against standard performance metrics such as accuracy, precision, recall, and F1-score. The Receiver Operating Characteristic(ROC) and Area Under ROC(AUC) are additionally used to show performance.

6.1. Experimental System

The experimental system was prepared with Ubuntu OS in the Python development environment. The python environment was prepared with required machine learning modules and frameworks such as Pandas, NumPy, matplotlib, CSV, and Scikit-learn [40].

6.2. Machine Learning Algorithms

In our work, ML algorithms were chosen based on their working principles. We tried to keep a diverse set of algorithms for a better understanding and performance comparison. For example, Naive Bayes (NB) works on conditional probability, while kNN applies a distance function to associate a node to a group or cluster [41]. Based on this, we initially selected nine algorithms and later, due to relatively much large training time, dropped the Bagging (SVC) and XBoost algorithms from further experiments. The selected algorithms were Decision Tree (DT), Support Vector Machine (SVM), Random Forest (RF), k-Nearest Neighbour (kNN), Naive Bayes (NB), Regression, Bagging, and Boosting. Training and testing with such a diverse set of algorithms helped the authors to understand and find suitable features and algorithms. All the algorithms were tested with default parameters available in the scikit-learn framework. However, parameter setting is explicitly mentioned wherever the default value changes. Some critical parameters for the best performing model, i.e., random forest, are the number of trees: 100; split method: Gini; and the minimum number of samples required to split: 2. Hyper-parameter-tuning finds the best value for the algorithm's parameters from the search space. This study did not perform hyper-parameter-tuning; however, this is a possible area of research for future work.

6.3. Training and Testing

The power system and SWaT datasets were divided into training and testing sets. Each algorithm was trained on the training set, while the performance evaluation of the model was completed on the testing set. Percentage split and cross-validation are two main methods for splitting the dataset into training and testing sets. The percentage split simply divides the original samples into two sets based on the given *percentage of the sample* to be considered for training and testing. However, the cross-validation divides the original samples into N folds containing equal numbers of samples. We used a 70/30 ratio for percentage split, while 10 folds (i.e., $N = 10$) were used with cross-validation. As cross-validation is an iterative process, the algorithm's performance was taken as the mean of N rounds of training and testing. In each round, $N - 1$ folds were used for training, and the remaining fold was used for testing. Training and testing in multiple folds provide diversity to the dataset, cross-validation provides robust training, and the trained model was generalized well on an unseen sample [42].

6.3.1. Percentage Split (70–30)

As mentioned earlier, based on the required percentage for training and testing samples, the percentage split method divides the samples into two sets. We used 70% for training and the remaining 30% for testing. This split method randomly selects the sample from the dataset for training. Training on a train–test dataset (from the split method) provides an approximate model, because randomly selected training samples do not represent actual data distribution. As such, the trained model suffers from over-fitting, i.e., it performs poorly on unseen samples. We trained and tested all nine algorithms on training and testing sets obtained from a percentage split (70–30%) to measure training time and approximate performance of FDI classification. Tables 5 and 6 show the precision, recall, F1-score, and accuracy of all algorithms for the power system and SWaT dataset, respectively. Except for accuracy, the other three metrics provide results for both classes

(Normal and FDI). From Table 5, we can observe that, with an accuracy score 92%, Random Forest performance is the best performed as an ensemble algorithm, while Decision Tree has an accuracy score of 85% and is the best performer as a single model. The accuracy value is biased towards the majority class, and the model suffers for the minority class. This is evident from the precision, recall, and F1-score value from Table 5 for both classes. The performance reduced by about 6–10% for best-performing classifiers. In the case of the SWaT dataset, all the classifiers have an accuracy above 95%. The precision, recall, and F1-score for the normal class is as per the accuracy but reduced for the FDI class for many classifiers. From Table 6, we can observe that kNN, DT, and RF show a perfect 100% score for all metrics; this is indicating over-fitting. Over-fitting can be attributed to the water treatment dataset having fewer features and many samples. Therefore, kNN, DT, and RF can memorize the class distribution for training data. We have investigated these three algorithms further with cross-validation, and the results are presented in Section 6.3.2.

Table 5. Classifiers' performance in the power dataset.

ML Model	Precision		Recall		F1-Score		Accuracy
	Normal	FDI	Normal	FDI	Normal	FDI	–
NB	0.71	0.33	0.98	0.02	0.82	0.04	0.70
SVM	**0.97**	0.29	0.01	**1.0**	0.01	0.46	0.30
kNN	0.86	0.70	0.88	0.66	0.87	0.68	0.82
DT	0.90	0.75	0.90	0.75	0.90	0.75	0.85
RF	0.91	**0.93**	0.98	0.78	**0.94**	**0.85**	**0.92**
Ada	0.72	0.53	0.96	0.10	0.82	0.16	0.71
LR	0.71	0.49	**1.0**	0.01	0.83	0.02	0.71

Table 6. Classifiers' performance in the water treatment dataset.

ML Model	Precision		Recall		F1-Score		Accuracy
	Normal	FDI	Normal	FDI	Normal	FDI	–
NB	0.96	0.98	1.0	0.70	0.98	0.82	0.96
SVM	0.96	0.99	1.0	0.71	0.98	0.83	0.96
kNN	1	1	1	1	1	1	1
DT	1	1	1	1	1	1	1
RF	1	1	1	1	1	1	1
BagSVC	0.96	0.99	1	0.66	0.98	0.80	0.96
LR	0.95	0.99	1	0.62	0.97	0.76	0.95

In critical infrastructure, decision making needs to be quick. So, a low prediction time is required from the machine learning model. In this study, time is one of the key performance metrics. So, the time taken for training and testing by each algorithm is measured. The training time will help find a suitable algorithm for the power system or other critical infrastructure. Training time is also essential because data generation is fast, and the models often require retraining. A model with a lower training time will be more suitable. Table 7 presents the training and testing times (in seconds) of all nine algorithms. As mentioned earlier, Naive Bayes (NB) is a fast and probabilistic algorithm, because it uses prior probability values to calculate the posterior. Probability values can be calculated in advance, so NB training is faster than others. However, *conditional independence* is one of the critical assumptions that attributes need in order to be satisfied. *Bagging* is an ensemble method that creates multiple base models on the subset of the dataset. These subsets are created using random sampling. Table 7 shows that NB has the lowest training time of all algorithms, while Bagging (with SVC) has the highest training time. Hence, the highest training time is the accumulation of time taken for dataset generation, multiple model training, and testing.

Table 7. Training and testing time of machine-learning algorithms.

Classifiers	NB	SVC	kNN	DT	RF	Ada	BSVC	LR	XGB
Time (S)	0.132	9.770	7.280	3.580	17.900	11.800	533.00	1.140	62.0

6.3.2. Testing with Cross-Validation

The previous section shows the detection performance and training time of algorithms for the training and testing set created using a percentage split. With this initial estimation, the algorithms were further trained and tested with 10-folds cross-validation to know the trained model's generalization capacities. We compared the detection accuracy of all algorithms for the percentage split and cross-validation of the power-system dataset. We have two key observations; first, five algorithms (DT, SVC, kNN, GB, and RF) achieved lower accuracy in the 10-fold cross-validation than the percentage split. Second, four other algorithms (Adaboost, Bagging (ensemble), LR, and NB) had minimum impact, i.e., accuracy either reduced with smaller margin or remained constant. Based on these two observations, we can conclude that the former five algorithms over-fit inherently, while the latter four algorithms have the inbuilt function to overcome over-fitting during training. Considering this outcome, for the SWaT dataset, we performed training with 10-fold cross-validation for the three most over-fitted classifiers, i.e., kNN, DT, and RF. Figure 7 shows the AUC of all classifiers for 10 folds training. For RF, although 4 passes have an AUC value of 1.0, the mean performance of all three classifiers reduced to 0.79, 0.75, and 0.81 from 1.0 for kNN, DT, and RF, respectively.

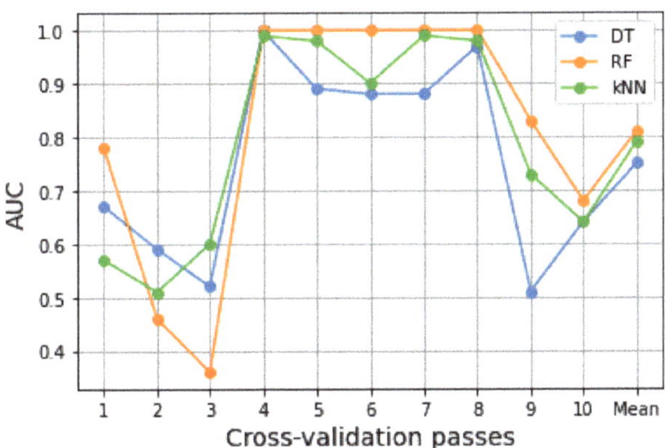

Figure 7. AUC with 10-fold cross-validation for the water treatment plant dataset.

The previous section presented the outcome of the filter and wrapper methods. One further key objective of this study is to test the performance of all algorithms on the selected set of features. For this, we experimented only on the power-system dataset. A total of ten datasets were created to train and test different machine-learning algorithms using sets of the selected top features, five sets each from the filter and wrapper method. The experimental results are shown as Top10, Top20, Top30, Top40, and Top50 for feature selection sets, and ALL represents all features. These 11 sets of the dataset were used to train and test all algorithms in 10-fold cross-validation. Figure 8a shows the results for the filter method sets. The detection accuracy in Adaboost, SVC, and GB decreased, while the detection accuracy in DT, kNN, and LR increased. The performance of RF and NB was unaffected (insignificant change) by the change in the number of features.

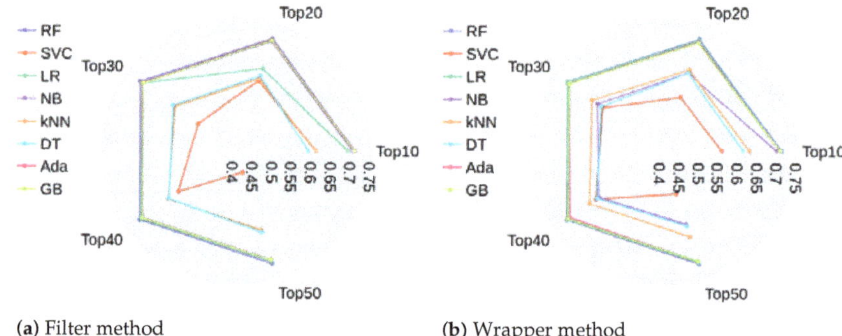

(**a**) Filter method (**b**) Wrapper method

Figure 8. Accuracy of classifiers with selected features for power system dataset.

Similarly, five datasets were created using the feature ranking of the wrapper method. Further, using the 10-fold cross-validation approach, all algorithms were trained and tested on all five sets. Figure 8b shows the performance of all algorithms. Unlike the filter method case, there was no change in the performance of Logistic Regression; the performance of Naive Bayes decreased significantly. In either selection method, kNN had a similar pattern, i.e., accuracy increased with the number of selected features. However, there was no clear pattern in the performance change of Adaboost, RF, GB, DT, and SVC. The performance of RF and DT decreased and had the lowest accuracy with thirty top features, while there were no significant changes in accuracy with other sets of features (top10, top20, top40, and top50). In other groups, the accuracy of SVC and Adaboost did not seem related to the number of features.

6.4. Imbalance Dataset and Impact

As discussed in Section 5, with an imbalanced dataset, ML models suffer performance degradation while making predictions about minority classes. This is because a model learns mainly from the majority class or is over-fitted to the majority class. Accuracy is the most-used metric for measuring the performance of machine-learning algorithms, but it is not suitable for imbalanced datasets [30]. Values from Tables 5 and 6 verify the performance degradation of the model with more robust metrics such as precision, recall, and F1-score.

The two main approaches for handling the imbalanced dataset are oversampling the minority class and under-sampling the majority class. While oversampling is suitable for maximum use-cases, under-sampling only suits when the majority class has many samples and the minority class also has enough samples to represent the nature of the distribution. In this study, we have adopted the oversampling approach, and the minority class is oversampled using four different sampling techniques. SMOTE is the main technique for oversampling, and the other three, i.e., Borderline-SMOTE, Borderline-SMOTE with SVM, and Adaptive Synthetic Sampling, are variants of SMOTE. As explained in Section 5, in borderline-SMOTE, only the borderline sample from the minority class is used for oversampling. In the original borderline-SMOTE algorithm [32], kNN is used for sample selection, while in the modified version (borderline-SMOTE with SVM [33]), SVM is used for sample selection. To understand and highlight the impact of the imbalanced dataset, we have used AUC as a performance metric. After applying each oversampling technique, we trained all ML algorithms on the imbalanced dataset and again trained the algorithms. Figure 9 depicts the algorithms' AUC values with the imbalanced dataset, i.e., Figure 9a and after balancing the dataset (making an equal sample for both classes by oversampling minority class, i.e., FDI class) using SMOTE, shown in Figure 9b. From Figure 9, it is evident that imbalance datasets have varying impacts on the different types of ML algorithms. These are obvious, because each performs training differently. Further, this observation can be broken down into two: first, some algorithms are not impacted (they can have high performance, i.e., DT and RF, or low-performance, i.e., NB and LR) by the ratio of samples for each class, so oversampling also fails to impact performance. Second, some algorithms (GradientBoost, AdaBoost, and kNN) have a high impact on the imbalanced dataset, and so the performance of these algorithms improves after oversampling. Table 8 shows the AUC values of different classifiers for the imbalanced dataset and after applying four selected oversampling methods. In Table 8, borderline-SMOTE and SVM-based borderline-SMOTE is coded as BSMOTE and BSMOTE-SVM. From Table 8 we can observe that all oversampling techniques improve the AUC values of almost all classifiers. The *magnitude* of improvement depends upon the type of algorithm used. As mentioned previously, the best improvement, i.e., 10–12%, was observed for GradientBoost, and AdaBoost algorithms, while kNN had a 6–8% improvement with different oversampling techniques. DT and RF are considered robust against imbalanced datasets, but these algorithms achieved a 2–3% performance improvement after applying oversampling.

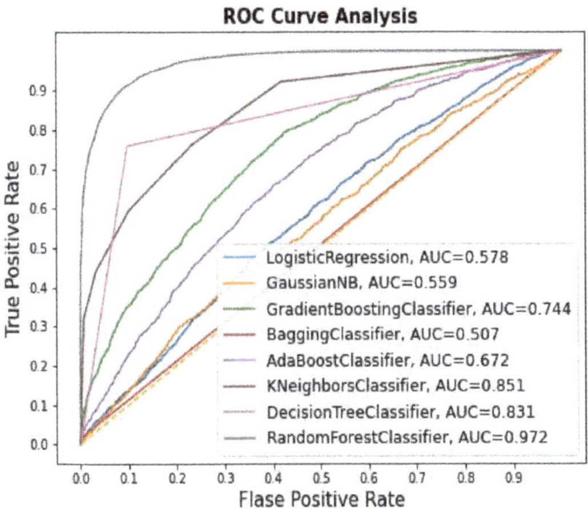

(**a**) ROC with imbalance dataset

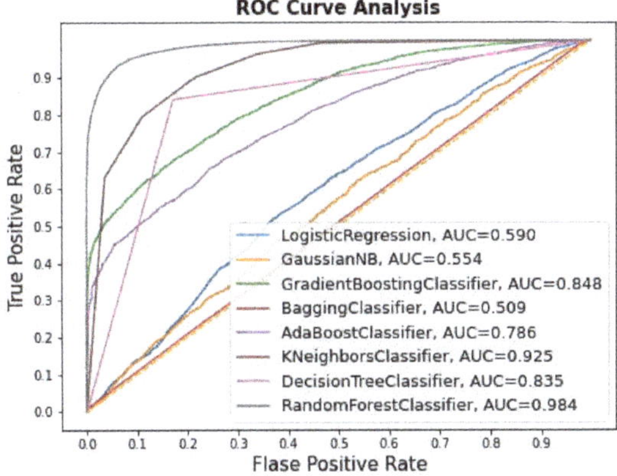

(**b**) ROC after SMOTE

Figure 9. ROC for classifiers with imbalance and SMOTE balance power system dataset.

Table 8. Comparison of classifiers (on AUC %) for imbalanced datasets and different oversampling methods for the power system dataset.

Classifiers	Imbalance	Smote	Bsmote	Bsmote-Svm	Adasyn
LR	0.592	0.590	0.550	0.595	0.565
GNB	0.559	0.548	0.578	0.544	0.596
GBoost	0.747	**0.847**	**0.839**	**0.839**	**0.845**
BagSVC	0.504	0.505	0.512	0.515	0.513
AdaBoost	0.672	**0.783**	0.777	0.747	0.785
kNN	0.855	**0.924**	**0.917**	**0.926**	**0.910**
DT	0.826	0.830	0.835	0.847	0.831
RF	**0.974**	**0.984**	0.982	0.988	0.983

In this study, we used robust performance metrics such as precision, recall, and F1-score calculated using a confusion matrix. In addition to this, to visualize the model's performance, we selected the best performing model in both the imbalanced and balanced dataset and plotted the confusion matrix. Figure 10 shows the results for both cases. As shown in Figure 10a,b, the detection performance of the attack class improved with a balanced dataset but decreased for normal class. However, improved attack detection is critical and required for critical infrastructures.

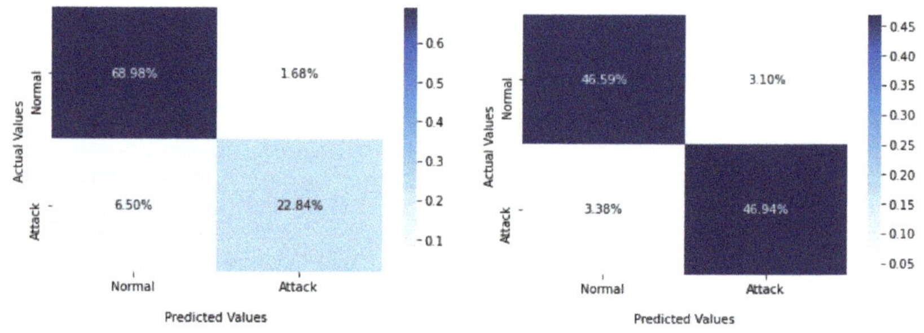

(**a**) Imbalance dataset   (**b**) Balance dataset (SMOTE)

Figure 10. Confusion matrix for best model (RF) on imbalance and SMOTE balanced power system dataset.

6.5. Comparison with Previous Works

In this section, we compare the performance of our best model with those described in the existing literature. Jingyu wang et al. [43] used deep autoencoder to detect data manipulation attacks in power systems. Adhikari et al. [44] combined Non-Nested Generalized Exemplars (NNGEs) and the STate Extraction Method (STEM) for cyber-attack event detection. Defu Want et al. [20] divided the features as per each PMU and then used an ensemble approach to combine the results of five classifiers (the authors trained four classifiers on four PMUs data and the fifth with combined features). Table 9 shows the performance comparison of the best model of earlier studies and this study. The performance of this study is shown as an AUC value. AUC is a robust metric that represents the total area under ROC (trade-off between True Positive Rate (TPR) and False Positive rate (FPR)). A higher AUC value indicates better classification performance of the model. This study has achieved an AUC value of 0.984, which is better than the existing literature.

Table 9. Comparison with previous works on FDIA detection for the power system dataset.

Work	Algo.	Accuracy	Precision	Recall	F1-Measure	AUC
[43]	DAE	0.941	0.996	0.886	0.938	-
	XGBoost	0.848	0.990	0.703	0.82	-
[44]	–	-	0.96	-	0.95	-
[20]	Ensemble	0.9391	-	-	-	-
Current Study	RF	-	-	-	-	**0.984**

7. Conclusions and Future Scope

We examined and improved the performance of machine-learning algorithms for detecting FDIAs in critical infrastructure by determining the best features and mitigating imbalanced dataset problems. Performance improvement was tested and validated through various experimental results. These experiments included feature selection methods, oversampling techniques, and training and testing ML algorithms on two popular datasets related to power systems and water treatment plants. Our results show that the performance of algorithms varies significantly depending on the feature selection and the number of features. For example, the performance of NB is unaffected by increasing the number of features in the filter method while decreasing the number wrapper features. We also found that selection methods rank features differently. We found that RF is generally suitable for building an FDIA detector based on detection performance and training time trade-offs. Additionally, model training with 10-fold cross-validation is suitable because it highlights the over-fitting issues. Moreover, we analyzed the impact of the imbalanced dataset and applied minority oversampling techniques to improve detection performance.

New sampling techniques based on deep learning and hybrid sampling approaches are proposed in the literature [35–39]. Future studies can explore these recent techniques with the power system and other critical infrastructure datasets. The binary classification formulation in this study can be further divided and reformulated as a multi-class classification for training machine-learning algorithms. Moreover, consideration of the space and computation requirements of critical infrastructures can motivate new research objectives.

Author Contributions: Conceptualization: A.K. and B.J.C.; data curation: A.K.; formal analysis: A.K.; funding acquisition: S.J.; methodology: A.K. and B.J.C.; project administration: S.J. and B.J.C.; resources: A.K.; software: A.K.; supervision: B.J.C.; validation: A.K., N.S. and B.J.C.; visualization: A.K.; writing—original draft: A.K.; writing—review and editing: B.J.C. All authors have read and agreed to the published version of the manuscript.

Funding: This research was supported by the MSIT (Ministry of Science and ICT), Korea, under the National Research Foundation (NRF), Korea (2019R1C1C1007277) and the ITRC (Information Technology Research Center) support program (IITP-2020-2020-0-01602) supervised by the IITP (Institute for Information and Communications Technology Planning and Evaluation). This research was also supported by the Cardiff University HEFCW GCRF Small Project: Secure, Low-Cost, and Efficient Energy Solution (SP113).

Sample Availability: The modified version of the Power system dataset is available on Github for comparison, but citing the original authors' work while using the dataset is also suggested. The SWaT dataset is only available on request to the original author.

Abbreviations

The following abbreviations are used in this manuscript:

ANN	Artificial Neural Network
AI	Artificial Intelligence
ARFF	Attribute-Relation File Format
AUC	Area Under Curve
CPPS	Cyber-Physical Power System
CPU	Central Processing Unit
CSV	Comma-Separated Value
DT	Decision Tree
FDIA	False Data Injection Attack
GB	Gradient Boost
GUI	Graphical User Interface
HMI	Human–Machine Interfaces
ICT	Information and Communication Technology
IDS	Intrusion Detection System
IPS	Intrusion Prevention System
IED	Intelligent Electronic Device
kNN	k Nearest Negibour
LR	Linear Regression
LTS	Long Term Support
ML	Machine Learning
NB	Naive Bayes
OCSVM	One-Class Support Vector Machine
PCA	Principal Component Analysis
PDC	Power Distribution Center
PLC	Programmable Logic Controllers
PMU	Phasor Measurement Unit
RF	Random Forest
ROC	Receiver Operating Characteristic
SCADA	Supervisory Control and Data Acquisition
SVC	Support Vector Classifier
SVM	Support Vector Machine

References

1. Corallo, A.; Lazoi, M.; Lezzi, M. Cybersecurity in the context of industry 4.0: A structured classification of critical assets and business impacts. *Comput. Ind.* **2020**, *114*, 103165. [CrossRef]
2. Griffor, E.R.; Greer, C.; Wollman, D.A.; Burns, M.J. Framework for cyber-physical systems: Volume 1, overview. *NIST SP* **2017**. [CrossRef]
3. Rodofile, N.R.; Radke, K.; Foo, E. Extending the cyber-attack landscape for SCADA-based critical infrastructure. *Int. J. Crit. Infrastruct. Prot.* **2019**, *25*, 14–35. [CrossRef]
4. Khanna, K.; Panigrahi, B.K.; Joshi, A. AI-based approach to identify compromised meters in data integrity attacks on smart grid. *IET Gener. Transm. Distrib.* **2017**, *12*, 1052–1066. [CrossRef]
5. Maleh, Y.; Shojafar, M.; Darwish, A.; Haqiq, A. *Cybersecurity and Privacy in Cyber Physical Systems*; CRC Press: Boca Raton, FL, USA, 2019.
6. Liang, G.; Weller, S.R.; Zhao, J.; Luo, F.; Dong, Z.Y. The 2015 ukraine blackout: Implications for false data injection attacks. *IEEE Trans. Power Syst.* **2016**, *32*, 3317–3318. [CrossRef]
7. Reeder, J.R.; Hall, C.T. *Cybersecurity's Pearl Harbor Moment: Lessons Learned from the Colonial Pipeline Ransomware Attack*; Government Contractor Cybersecurity: Washington, DC, USA, 2021.
8. Gönen, S.; Sayan, H.H.; Yılmaz, E.N.; Üstünsoy, F.; Karacayılmaz, G. False Data Injection Attacks and the Insider Threat in Smart Systems. *Comput. Secur.* **2020**, *97*, 101955. [CrossRef]
9. Aoufi, S.; Derhab, A.; Guerroumi, M. Survey of false data injection in smart power grid: Attacks, countermeasures and challenges. *J. Inf. Secur. Appl.* **2020**, *54*, 102518. [CrossRef]
10. Pan, S.; Morris, T.; Adhikari, U. Developing a hybrid intrusion detection system using data mining for power systems. *IEEE Trans. Smart Grid* **2015**, *6*, 3104–3113. [CrossRef]
11. Goh, J.; Adepu, S.; Junejo, K.N.; Mathur, A. A dataset to support research in the design of secure water treatment systems. In *International Conference on Critical Information Infrastructures Security*; Springer: Berlin/Heidelberg, Germany, 2016; pp. 88–99.

12. Guan, Z.; Sun, N.; Xu, Y.; Yang, T. A comprehensive survey of false data injection in smart grid. *Int. J. Wirel. Mob. Comput.* **2015**, *8*, 27–33. [CrossRef]
13. Liang, G.; Zhao, J.; Luo, F.; Weller, S.R.; Dong, Z.Y. A review of false data injection attacks against modern power systems. *IEEE Trans. Smart Grid* **2016**, *8*, 1630–1638. [CrossRef]
14. Musleh, A.S.; Chen, G.; Dong, Z.Y. A survey on the detection algorithms for false data injection attacks in smart grids. *IEEE Trans. Smart Grid* **2019**, *11*, 2218–2234. [CrossRef]
15. Cao, J.; Wang, D.; Qu, Z.; Cui, M.; Xu, P.; Xue, K.; Hu, K. A Novel False Data Injection Attack Detection Model of the Cyber-Physical Power System. *IEEE Access* **2020**, *8*, 95109–95125. [CrossRef]
16. Maglaras, L.A.; Jiang, J. Intrusion detection in SCADA systems using machine learning techniques. In Proceedings of the 2014 Science and Information Conference, Las Vegas, NV, USA, 25–26 April 2014; pp. 626–631.
17. Esmalifalak, M.; Liu, L.; Nguyen, N.; Zheng, R.; Han, Z. Detecting stealthy false data injection using machine learning in smart grid. *IEEE Syst. J.* **2014**, *11*, 1644–1652. [CrossRef]
18. Yan, J.; Tang, B.; He, H. Detection of false data attacks in smart grid with supervised learning. In Proceedings of the 2016 International Joint Conference on Neural Networks (IJCNN), Vancouver, BC, Canada, 24–29 July 2016; pp. 1395–1402.
19. Wang, Y.; Amin, M.M.; Fu, J.; Moussa, H.B. A novel data analytical approach for false data injection cyber-physical attack mitigation in smart grids. *IEEE Access* **2017**, *5*, 26022–26033. [CrossRef]
20. Wang, D.; Wang, X.; Zhang, Y.; Jin, L. Detection of power grid disturbances and cyber-attacks based on machine learning. *J. Inf. Secur. Appl.* **2019**, *46*, 42–52. [CrossRef]
21. Panthi, M. Anomaly Detection in Smart Grids using Machine Learning Techniques. In Proceedings of the 2020 First International Conference on Power, Control and Computing Technologies (ICPC2T), Raipur, India, 3–5 January 2020; pp. 220–222.
22. Ahmed, C.M.; Zhou, J.; Mathur, A.P. Noise matters: Using sensor and process noise fingerprint to detect stealthy cyber attacks and authenticate sensors in cps. In Proceedings of the 34th Annual Computer Security Applications Conference, San Juan, PR, USA, 3–7 December 2018; pp. 566–581.
23. Dutta, A.K.; Negi, R.; Shukla, S.K. *Robust Multivariate Anomaly-Based Intrusion Detection System for Cyber-Physical Systems. International Symposium on Cyber Security Cryptography and Machine Learning*; Springer: Berlin/Heidelberg, Germany, 2021; pp. 86–93.
24. Jahromi, A.N.; Karimipour, H.; Dehghantanha, A.; Choo, K.K.R. Toward Detection and Attribution of Cyber-Attacks in IoT-enabled Cyber-physical Systems. *IEEE Internet Things J.* **2021**. [CrossRef]
25. Begli, M.; Derakhshan, F.; Karimipour, H. A layered intrusion detection system for critical infrastructure using machine learning. In Proceedings of the 2019 IEEE 7th International Conference on Smart Energy Grid Engineering (SEGE), UOIT, ON, Canada, 12–14 August 2019; pp. 120–124.
26. Dick, K.; Russell, L.; Souley Dosso, Y.; Kwamena, F.; Green, J.R. Deep learning for critical infrastructure resilience. *J. Infrastruct. Syst.* **2019**, *25*, 05019003. [CrossRef]
27. Rodofile, N.R.; Schmidt, T.; Sherry, S.T.; Djamaludin, C.; Radke, K.; Foo, E. Process control cyber-attacks and labelled datasets on S7Comm critical infrastructure. In *Australasian Conference on Information Security and Privacy*; Springer: Berlin/Heidelberg, Germany, 2017; pp. 452–459.
28. Kotsiantis, S. Feature selection for machine learning classification problems: a recent overview. *Artif. Intell. Rev.* **2011**, *42*, 157–176. [CrossRef]
29. He, H.; Ma, Y. *Imbalanced Learning: Foundations, Algorithms, and Applications*; Wiley-IEEE Press: Hoboken, NJ, USA , 2013.
30. Chawla, N.V.; Bowyer, K.W.; Hall, L.O.; Kegelmeyer, W.P. SMOTE: synthetic minority over-sampling technique. *J. Artif. Intell. Res.* **2002**, *16*, 321–357. [CrossRef]
31. Chawla, N.V. Data mining for imbalanced datasets: An overview. In *Data Mining and Knowledge Discovery Handbook*; Springer: Berlin/Heidelberg, Germany, **2009**; pp. 875–886.
32. Han, H.; Wang, W.Y.; Mao, B.H. Borderline-SMOTE: A new over-sampling method in imbalanced data sets learning. In *International Conference on Intelligent Computing*; Springer: Berlin/Heidelberg, Germany, **2005**; pp. 878–887.
33. Nguyen, H.M.; Cooper, E.W.; Kamei, K. Borderline over-sampling for imbalanced data classification. *Int. J. Knowl. Eng. Soft Data Paradig.* **2011**, *3*, 4–21. [CrossRef]
34. He, H.; Bai, Y.; Garcia, E.A.; Li, S. ADASYN: Adaptive synthetic sampling approach for imbalanced learning. In Proceedings of the 2008 IEEE International Joint Conference on Neural Networks (IEEE World Congress on Computational Intelligence), Hong Kong, China, 1–6 June 2008; pp. 1322–1328.
35. Elyan, E.; Moreno-Garcia, C.F.; Jayne, C. CDSMOTE: class decomposition and synthetic minority class oversampling technique for imbalanced-data classification. *Neural Comput. Appl.* **2021**, *33*, 2839–2851. [CrossRef]
36. Guan, H.; Zhang, Y.; Xian, M.; Cheng, H.D.; Tang, X. SMOTE-WENN: Solving class imbalance and small sample problems by oversampling and distance scaling. *Appl. Intell.* **2021**, *51*, 1394–1409. [CrossRef]
37. Fajardo, V.A.; Findlay, D.; Jaiswal, C.; Yin, X.; Houmanfar, R.; Xie, H.; Liang, J.; She, X.; Emerson, D. On oversampling imbalanced data with deep conditional generative models. *Expert Syst. Appl.* **2021**, *169*, 114463. [CrossRef]
38. Bellinger, C.; Corizzo, R.; Japkowicz, N. Calibrated Resampling for Imbalanced and Long-Tails in Deep Learning. In *International Conference on Discovery Science*; Springer: Berlin/Heidelberg, Germany, 2021; pp. 242–252.

39. Krawczyk, B.; Bellinger, C.; Corizzo, R.; Japkowicz, N. Undersampling with support vectors for multi-class imbalanced data classification. In Proceedings of the 2021 International Joint Conference on Neural Networks (IJCNN), Shenzhen, China, 18–22 July 2021; pp. 1–7.
40. Pedregosa, F.; Varoquaux, G.; Gramfort, A.; Michel, V.; Thirion, B.; Grisel, O.; Blondel, M.; Prettenhofer, P.; Weiss, R.; Dubourg, V.; et al. Scikit-learn: Machine Learning in Python. *J. Mach. Learn. Res.* **2011**, *12*, 2825–2830.
41. Shalev-Shwartz, S.; Ben-David, S. *Understanding Machine Learning: From theory to Algorithms*; Cambridge University Press: Cambridge, UK, 2014.
42. Shao, J. Linear model selection by cross-validation. *J. Am. Stat. Assoc.* **1993**, *88*, 486–494. [CrossRef]
43. Wang, J.; Shi, D.; Li, Y.; Chen, J.; Ding, H.; Duan, X. Distributed framework for detecting PMU data manipulation attacks with deep autoencoders. *IEEE Trans. Smart Grid* **2018**, *10*, 4401–4410. [CrossRef]
44. Adhikari, U.; Morris, T.H.; Pan, S. Applying non-nested generalized exemplars classification for cyber-power event and intrusion detection. *IEEE Trans. Smart Grid* **2016**, *9*, 3928–3941. [CrossRef]

MDPI

St. Alban-Anlage 66

4052 Basel

Switzerland

Tel. +41 61 683 77 34

Fax +41 61 302 89 18

www.mdpi.com

Energies Editorial Office

E-mail: energies@mdpi.com

www.mdpi.com/journal/energies